AF361878

Vehicular Sensor Networks

Vehicular Sensor Networks

Applications, Advances and Challenges

Editors

Fatih Kurugollu
Syed Hassan Ahmed
Rasheed Hussain
Farhan Ahmad
Chaker Abdelaziz Kerrache

MDPI • Basel • Beijing • Wuhan • Barcelona • Belgrade • Manchester • Tokyo • Cluj • Tianjin

Editors
Fatih Kurugollu
University of Derby
UK

Syed Hassan Ahmed
MA Wireless
USA

Rasheed Hussain
Innopolis University
Russia

Farhan Ahmad
University of Derby
UK

Chaker Abdelaziz Kerrache
University of Ghardaia
Algeria

Editorial Office
MDPI
St. Alban-Anlage 66
4052 Basel, Switzerland

This is a reprint of articles from the Special Issue published online in the open access journal *Sensors* (ISSN 1424-8220) (available at: https://www.mdpi.com/journal/sensors/special_issues/Vehicular_Sensor_Networks).

For citation purposes, cite each article independently as indicated on the article page online and as indicated below:

LastName, A.A.; LastName, B.B.; LastName, C.C. Article Title. *Journal Name* **Year**, *Article Number*, Page Range.

ISBN 978-3-03936-762-7 (Hbk)
ISBN 978-3-03936-763-4 (PDF)

Contents

About the Editors

Fatih Kurugollu obtained his BSc and MSc in Computer and Control Engineering degree from Istanbul Technical University, Turkey, in 1989 and 1994, respectively. He was awarded with a PhD degree in Computer Engineering from the same university in 2000. He was employed as a research fellow by the Marmara Research Centre, which is the main governmental research unit of the Turkish Scientific Research Council (TUBITAK) in 1991. He joined the School of Electronics, Electrical Engineering and Computer Science at Queen's University, Belfast, in 2000, initially as a Postdoctoral Research Fellow. In 2003, he was appointed to a lectureship at the same department and, later on, was promoted to Senior Lecturer in Computer Science. He is now a full Professor of Cyber Security at University of Derby. His current research interests are centered around security and privacy in Internet of Things, cloud security, imaging for forensics and security, security-related multimedia content analysis, big data in cyber security, homeland security, security issues in healthcare systems, biometrics, and image and video analysis. He has been principal investigator and co-investigator of several projects funded by EPSRC, Royal Academy Engineering (RAEng), Leverhulme Trust, Action Medical Research as well as principal supervisor of KTP projects. He has supervised 11 PhD projects and has authored more than 130 publications. He is a Senior Member of IEEE, Member of Associate College of Engineering and Physical Sciences Research Council (EPSRC), Fellow of the Higher Education Academy (HEA), Voting Member of IEEE Communication Society Multimedia Communications Technical Committee, and Affiliate Member of IEEE Signal Processing Society Information Forensics and Security Technical Committee.

Syed Hassan Ahmed (SM'18) is currently working at JMA Wireless as a Product Specialist for distributed antenna system (DAS), CBRS, small cell, and the virtualized RAN product line. Previously, he was Assistant Professor at the Department of Computer Science, Georgia Southern University, USA. He also founded the Wireless Internet and Networking Systems (WINS) lab. Prior to this, he was a Postdoctoral Fellow at the Department of Electrical and Computer Engineering, University of Central Florida, Orlando, USA. Before moving to the United States, he completed his BS with honors in CS from Kohat University of Science & Technology (KUST), Pakistan, and master's as well as PhD degree from School of Computer Science and Engineering (SCSE), Kyungpook National University (KNU), Republic of Korea (South Korea). In summer 2015, he was also a Visiting Researcher at Georgia Tech, Atlanta, USA. Overall, he has authored/co-authored over 200 international publications including journal articles, conference proceedings, and book chapters in addition to 3 books. In 2016, his work on robust content retrieval in future vehicular networks led to him winning the Qualcomm Innovation Award at KNU, Korea. Dr. Hassan's research interests include sensor and ad hoc networks, cyberphysical systems, vehicular communications, and future internet. He has been an appointee of the Board of Governors of the IEEE Vehicular Technology Society as liaison to the IEEE Young Professionals society for 2018–2019. Since 2018, he is also an ACM Distinguished Speaker.

Rasheed Hussain received his BS Engineering degree in Computer Software Engineering from University of Engineering and Technology, Peshawar, Pakistan, in 2007 and MS and PhD degrees in Computer Science and Engineering from Hanyang University, South Korea in 2010 and 2015, respectively. He served as a Postdoctoral Fellow at Hanyang University, South Korea, from March 2015 to August 2015. He was also Guest Researcher and Consultant at University of Amsterdam (UvA), The Netherlands, from September 2015 until May 2016 and Assistant Professor at Innopolis University, Innopolis, Russia, from June 2016 until December 2018. Currently, he is Associate Professor and Director of the Institute of Information Security and Cyber-Physical Systems at Innopolis University, Innopolis, Russia. He is also the Director of Networks and Blockchain Lab at Innopolis University and serves as an ACM Distinguished Speaker. He is a senior member of IEEE, member of ACM, and serves as an editorial board member for various journals including IEEE Access, IEEE Internet Initiative, Internet Technology Letters, Wiley, Cluster Computing, and Springer, and serves as a reviewer for most of the IEEE Transactions, Springer, and Elsevier journals. He also serves as technical program committee member of various conferences including IEEE VTC, IEEE VNC, IEEE Globecom, IEEE ICCVE, IEEE ICC, and ICCCN. He is a certified trainer for the Instructional Skills Workshop (ISW) and a recipient of Netherland's University Teaching Qualification (Basis Kwalificatie Onderwijs, BKO). His research interests include information security and privacy and particularly security and privacy issues in vehicular ad hoc networks (VANETs), vehicular clouds, vehicular social networking, applied cryptography, Internet of Things, content-centric networking (CCN), cloud computing, API security, and blockchain. Currently, he is working on machine and deep learning for IoT security and API security.

Farhan Ahmad received his MSc in Communication and Information Technology from the University of Bremen, Germany, and PhD in Computer Science from the College of Engineering and Technology, University of Derby, UK, in 2014 and 2019, respectively. He is currently working as a Postdoctoral Research Fellow within the Cyber Security Research Group of University of Derby, UK. His research focuses on cyber security and trust management issues in vehicular ad-hoc networks, vehicular cloud networks, M2M communications, smart cities, and Internet of Things (IoT), where he has authored/co-authored over 30 international research publications. Currently, he is working on the application of security and trust management in Industrial IoT (IIoT), e-healthcare, and Internet of Vehicles.

Chaker Abdelaziz Kerrache is an Associate Professor at the department of Mathematics and Computer Science, University of Ghardaia, Algeria. He received his MSc degree in Computer Science at the University of Laghouat, Algeria, in 2012, and his PhD degree in Computer Science degree at the University of Laghouat, Algeria, in 2017. In 2013, he joined the Informatics and Mathematics Laboratory (LIM) as Research Assistant and the Computer Networks Group (GRC) in 2015 as a visiting PhD student. His research activity is related to trust and risk management, secure multi-hop communications, vehicular networks, named data networking (NDN), and UAVs. He also serves as Associate Editor of Elsevier Computer and Electrical Engineering and Frontiers in Space Technologies, and a reviewer and TPC member for several international journals and conferences including IEEE TVT, IEEE TITS, IEEE IoT, Elsevier VehCom, Elsevier CEE, Elsevier COSE, IEEE Access Wiley ETT, IEEE Future Directions, Ad Hoc Sensors and Wireless Networks, and Internet Technology Letters, CCNC, ICC, and ICCCN. He has also served as a Guest Editor for Special Issues in Elsevier Computers & Electrical Engineering, Elsevier Computer Communications, Wiley Transactions on Emerging Telecommunications Technologies, MDPI Electronics, and MDPI Sensors..

Editorial

Vehicular Sensor Networks: Applications, Advances and Challenges

Fatih Kurugollu [1,*], **Syed Hassan Ahmed** [2], **Rasheed Hussain** [3], **Farhan Ahmad** [1] **and Chaker Abdelaziz Kerrache** [4]

1   Cyber Security Research Group, College of Engineering and Technology, University of Derby, Derby DE22 3AW, UK; f.ahmad@derby.ac.uk
2   JMA Wireless, Liverpool, NY 13088, USA; sh.ahmed@ieee.org
3   Institute of Information Systems, Innopolis University, 420500 Innopolis, Russia; r.hussain@innopolis.ru
4   Department of Mathematics and Computer Science, University of Ghardaia, Ghardaia 4700, Algeria; ch.kerrache@univ-ghardaia.dz
*   Correspondence: f.kurugollu@derby.ac.uk

Received: 22 June 2020; Accepted: 29 June 2020; Published: 1 July 2020

Abstract: Vehicular sensor networks (VSN) provide a new paradigm for transportation technology and demonstrate massive potential to improve the transportation environment due to the unlimited power supply of the vehicles and resulting minimum energy constraints. This special issue is focused on the recent developments within the vehicular networks and vehicular sensor networks domain. The papers included in this Special Issue (SI) provide useful insights to the implementation, modelling, and integration of novel technologies, including blockchain, named data networking, and 5G, to name a few, within vehicular networks and VSN.

Keywords: vehicular sensor networks (VSN); vehicular ad-hoc networks (VANET); security; privacy and trust; cyber security; multimedia and cellular communication; emerging IoT applications in VANET and VSN; blockchain within VANET and VSN

---

1. Introduction

Recent years have witnessed tremendous growth in connected vehicles due to the major interest in vehicular ad-hoc networks (VANET) technology from both the research and industrial communities. VANET involves the generation of data from on-board sensors and its dissemination in other vehicles via vehicle-to-everything (V2X) communication, thus resulting in numerous applications such as steep-curve warnings. However, to increase the scope of applications, VANET has to integrate various technologies including sensor networks, which results in a new paradigm, commonly known as vehicular sensor networks (VSN).

Unlike traditional sensor networks, every node (vehicle) in VSN is equipped with various sensing (distance sensors, Global Positioning System GPS, and cameras), storage, and communicating capabilities, which can provide a wide range of applications including environmental surveillance and traffic monitoring, etc. VSN has the potential to improve transportation technology and the transportation environment due to its unlimited power supply and resulting in minimum energy constraints. However, VSN faces numerous challenges in terms of its design, implementation, network scalability, reliability, and deployment over large-scale networks, which need to be addressed before it is realised.

2. Contributions

In this special issue, we collected and compiled twelve outstanding contributions focusing on various aspects, including its modelling, security, trust management, test-bed implementation of vehicular networks, and VSN technology. In the following, a brief summary of each accepted paper is provided to encourage the readers.

In the first paper, the authors emphasize the importance of software-defined networks (SDNs) and cellular networks in the realization of vehicular networks [1]. The paper provides an overview of the existing cellular network-based solutions for vehicle-to-everything (V2X) communication. Furthermore, the paper also discusses the existing architectures for integrating cellular networks with vehicular networks. Based on the discussed architectures, the role of SDN and its features are discussed for realizing V2X communication. Without loss of generality, the primary focus of this paper is on software-defined vehicular networks (SDVNs). The authors took different architectures and their implementations and carried out a comparative analysis of these techniques to define elements that are essential for the design of SDVNs. Overall, the paper provides the features of different implementations pertaining to SDVNs.

Hyogon and Kim [2] cover a very important topic of content trust in the safety-critical applications where a vehicle receives a safety message which is then used by a decision-support system to trigger a designated action by the vehicle which could be, for instance, deceleration, emergency brake, and so on. In this regard, it is critically important to perform a plausibility check on the content of the received message. This paper discusses the existing plausibility-based mechanisms to provide content trust in vehicular networks. The paper proposes a beacon-based 'whispering' approach where low-power beacon messages are used to verify the neighbors and then decide whether to trust the content of the received message or not. This work is also closely related to the Sybil attack where illusion is created by creating fake nodes. The low-power in the beacon messages is an important contribution where the authors take into account the fact that using low power in the beacon could be beneficial for proving the proximity of the neighbors. Thus, they could be used to check the plausibility of the received message contents.

Salman et al. [3] addressed the problem of data dissemination in smart cities. In smart cities, a massive amount of data is generated by a huge number of data sources and there is a need for efficient mechanisms to collect data and send it to the control units for further processing. The vehicular network is one option to carry out such tasks where vehicles are used as data carriers. Instead of using dedicated mechanisms for data sharing with the control units, in this work, the authors use mobility patterns of the vehicles and leverage them for data dissemination as well. This phenomenon not only increases efficiency, but also reduces the carbon emission because of the massive data dissemination in smart cities. This paper develops a mathematical model to measure the degree of data offloading by taking into account the communication between vehicles and RSUs. The software then develops an algorithm to select the data dissemination nodes in an energy-efficient way to offload data to the control centers in smart cities. The paper also takes Auckland city as an example to validate the efficacy of the proposed data dissemination schemes.

Likewise, Hadiwardoyo et al. [4] have targeted another very interesting idea of bringing UAV communication to the connected vehicles domain. The rationale behind this idea is to bring connectivity and positioning services among cars which are non-line of sight due to the terrain and infrastructure hazards. In this paper, the authors have modelled UAV to act as a mobile roadside unit (RSU) and proposed an algorithm to achieve good visibility levels towards the current location of a target car. The positioning technique proposed optimizes the position of the UAV, defining its best altitude so that it can avoid terrain blockages.

On the other hand, works in [5,6] studied the cache management problem in vehicular networks over the new paradigm of informationcentric networking. In particular, Amadeo et al. [5] presented the benefits of tracking the content lifetime in named data networking (NDN) packets to prevent stale information from becoming disseminated in the vehicular network. Furthermore, they also proposed an efficient NDN-compliant caching strategy that accounts for the content lifetime for both replacement purposes and caching decisions.

Unlike the conventional case when all nodes store copies of the popular data, Meng et al. [6] proposed a new distributed caching strategy at the edge of the network in vehicular social networks environments to reduce the number of overall data dissemination problems. The proposed strategy called DCS is studied comparatively against a number of conventional caching strategies and the presented results show its efficiency in terms of memory consumption, path stretch ratio, cache hit ratio, and content eviction ratio.

As the VSNs have become popular over time, a massive increase in the data traffic has been observed from the connected vehicles. This data traffic is usually transferred via 5G mobile networks. Therefore, the device-to-device (D2D) communication mechanisms have also been studied recently to make the resultant communication performance better for vehicles within 5G-based VSN. However, D2D communications are prone to network interference. The interference is usually reduced via different interference management techniques including power controls and optimal mode controls. Hyebin and Lim [7] proposed a novel technique using joint power-control and optimal mode-selection via reinforcement learning which provides energy optimization within VSN. Extensive simulations are carried out to validate the proposals, which suggests that the proposed scheme performs best in terms of achievable data rate and system energy efficiency.

Recently, blockchain has been introduced as a novel mechanism to achieve security in the vehicular networks. In particular, Lewis et al. [8] proposed a novel blockchain-based event driven message protocol dissemination framework for vehicular networks using edge computing in the 5G cellular architecture. In this proposed architecture, the authors used a lightweight multi-receiver signcryption scheme without pairing to ensure low-time consuming operations, security, privacy and access control in the network. Further, the architecture uses a private blockchain system in the network for reliability and auditability purposes. The proposed architecture is validated, and the efficiency of the protocol is evaluated in terms of overall security, communication and computational costs.

On the other hand, Chuanyi and Li [9] have explored a completely different yet timely topic of resource-limited wireless sensor networks and cluster formation. Communication protocols in WSNs are very much in numbers, however, most of those schemes failed to consider the resource efficiency issue of the trusted computing itself. In this new study, the proposed cross-validation scheme computes trust values among cluster members and cluster heads. The proposed trust management scheme is believed to be fast and resource-saving as it enables the cooperation of nodes in an efficient way. Further, the trust model is effective against collaborative attacks as well. Through extensive simulations, a proof of concept is provided to further validate the scheme.

Geetanjali et al. [10] discussed the issue of malicious intruders in the vehicular networks. The main aim of these intruders is to mislead the overall communication by disseminating malicious content to both connected and autonomous vehicles in the network. To address these issues, the authors proposed a novel blockchain-based framework which can ensure the secrecy and transparency in the network as the information is stored and traced in the backend blockchain. This framework is validated across various security criteria including fake requests of the user, compromise of smart devices, probabilistic authentication scenarios and alteration in stored user's ratings. The proposed framework achieved the success rate of 79% over the baseline method which shows that this blockchain-based framework can be utilized to secure the connected and autonomous vehicles in the network.

Moving further, Sani et al. [11] have proposed a new MAC protocol for wireless sensor networks (WSN) that comprises a new Initial Control Frame Message, Traffic Estimation Function, Control Frame Message, and Adaptive Function. Using these four data structures, through different simulations in OMNET++, the protocol achieves higher latency and less energy consumption.

Farman et al. [12] proposed an efficient and accurate barrier control system to recognize the vehicle license plate using sensor platforms. As the license plate has various backgrounds, colors and fonts, it is extremely challenging to recognize the license plate of the vehicle accurately. In the proposed method, a vehicle is detected automatically using ultrasonic sensors and then image-based recognition is utilized with the aim to recognize a vehicle license plate. The authors implemented this mechanism on a PC running MATLAB and Raspberry Pi running Python and OpenCV. The results showed high accuracy where several license plates were used and nearly 93% of license plates were identified correctly.

Acknowledgments: We would like to acknowledge all the authors for their valuable contribution in making this SI successful. Further, we are thankful to the Sensors editorial team for their continuous cooperation throughout the SI. Lastly, we are grateful to the anonymous reviewers for their valuable input, comments, and suggestions for the submitted papers.

Conflicts of Interest: The authors declare no conflict of interest.

References

1. Lionel, N.; Nkenyereye, L.; Islam, S.M.; Choi, Y.; Bilal, M.; Jang, J. Software-defined network-based vehicular networks: A position paper on their modeling and implementation. *Sensors* **2019**, *19*, 3788.
2. Hyogon, K.; Kim, T. Vehicle-to-Vehicle (V2V) Message Content Plausibility Check for Platoons through Low-Power Beaconing. *Sensors* **2019**, *19*, 5493.
3. Salman, N.; Liu, W.; Sarkar, N.I. Energy-Efficient Massive Data Dissemination through Vehicle Mobility in Smart Cities. *Sensors* **2019**, *19*, 4735.
4. Hadiwardoyo, S.A.; Calafate, C.T.; Cano, J.; Krinkin, K.; Klionskiy, D.; Hernández-Orallo, E.; Manzoni, P. Three Dimensional UAV Positioning for Dynamic UAV-to-Car Communications. *Sensors* **2020**, *20*, 356. [CrossRef] [PubMed]
5. Marica, A.; Campolo, C.; Ruggeri, G.; Lia, G.; Molinaro, A. Caching Transient Contents in Vehicular Named Data Networking: A Performance Analysis. *Sensors* **2020**, *20*, 1985.
6. Yahui, M.; Naeem, M.A.; Ali, R.; Zikria, Y.B.; Kim, S.W. DCS: Distributed Caching Strategy at the Edge of Vehicular Sensor Networks in Information-Centric Networking. *Sensors* **2019**, *19*, 4407.
7. Hyebin, P.; Lim, Y. Reinforcement Learning for Energy Optimization with 5G Communications in Vehicular Social Networks. *Sensors* **2020**, *20*, 2361.
8. Lewis, N.; Tama, B.A.; Shahzad, M.K.; Choi, Y. Secure and Blockchain-Based Emergency Driven Message Protocol for 5G Enabled Vehicular Edge Computing. *Sensors* **2020**, *20*, 154.
9. Chuanyi, L.; Li, X. Fast, Resource-Saving, and Anti-Collaborative Attack Trust Computing Scheme Based on Cross-Validation for Clustered Wireless Sensor Networks. *Sensors* **2020**, *20*, 1592.
10. Geetanjali, R.; Sharma, A.; Iqbal, R.; Aloqaily, M.; Jaglan, N.; Kumar, R. A blockchain framework for securing connected and autonomous vehicles. *Sensors* **2019**, *19*, 3165.
11. Sani, A.M.; Yee, L.; Hussain, M.R.; Khan, N.; Ang, T.F.; Anisi, M.H.; Huang, Z.; Ali, I. An Adaptive Wake-Up-Interval to Enhance Receiver-Based Ps-Mac Protocol for Wireless Sensor Networks. *Sensors* **2019**, *19*, 3732.
12. Farman, U.; Anwar, H.; Shahzadi, I.; Rehman, A.U.; Mehmood, S.; Niaz, S.; Awan, K.M.; Khan, A.; Kwak, D. Barrier Access Control Using Sensors Platform and Vehicle License Plate Characters Recognition. *Sensors* **2019**, *19*, 3015.

Review

Software-Defined Network-Based Vehicular Networks: A Position Paper on Their Modeling and Implementation

Lionel Nkenyereye [1], Lewis Nkenyereye [2,*], S. M. Riazul Islam [3], Yoon-Ho Choi [4], Muhammad Bilal [5] and Jong-Wook Jang [1,*]

[1] Department of Computer Engineering, Dong-Eui University, Busan 614-714, Korea
[2] Department of Computer and Information Security, Sejong University, Seoul 05006, Korea
[3] Department of Computer Science and Engineering, Sejong University, Seoul 05006, Korea
[4] Division of Computer and Electronics Systems Engineering, Hankuk University of Foreign Studies, Yongin-si 17035, Korea
[5] School of Computer Science and Engineering, Pusan National University, Busan 46241, Korea
* Correspondence: nkenyele@sejong.ac.kr (L.N.); jwjang@deu.ac.kr (J.-W.J.)

Received: 15 August 2019; Accepted: 30 August 2019; Published: 31 August 2019

Abstract: There is a strong devotion in the automotive industry to be part of a wider progression towards the Fifth Generation (5G) era. In-vehicle integration costs between cellular and vehicle-to-vehicle networks using Dedicated Short Range Communication could be avoided by adopting Cellular Vehicle-to-Everything (C-V2X) technology with the possibility to re-use the existing mobile network infrastructure. More and more, with the emergence of Software Defined Networks, the flexibility and the programmability of the network have not only impacted the design of new vehicular network architectures but also the implementation of V2X services in future intelligent transportation systems. In this paper, we define the concepts that help evaluate software-defined-based vehicular network systems in the literature based on their modeling and implementation schemes. We first overview the current studies available in the literature on C-V2X technology in support of V2X applications. We then present the different architectures and their underlying system models for LTE-V2X communications. We later describe the key ideas of software-defined networks and their concepts for V2X services. Lastly, we provide a comparative analysis of existing SDN-based vehicular network system grouped according to their modeling and simulation concepts. We provide a discussion and highlight vehicular ad-hoc networks' challenges handled by SDN-based vehicular networks.

Keywords: software-defined vehicular network; vehicle-to-everything (V2X); modeling and implementation; software defined network

---

1. Introduction

Vehicle-to-everything (V2X) communications are definite technologies in vehicular networks to drastically reduce road accidents and enable a high-level of vehicle automation. For years, the technology of choice for V2X, on one hand, has been Dedicated Short Range Communication (DSRC) [1], which is based on IEEE 802.11p technology [1,2]. On the other hand, Cellular-V2X (C-V2X) technology is seen as a new communication standard supporting V2X services [3]. LTE-V2X technology is a derivative of the cellular uplink technology that maintains similarity with the current LTE systems [2]. Furthermore, the focus on V2X technology expands the availability of a wide range of services that include cloud-based vehicular services and edge computing [3]. Therefore, vehicles access these cloud-based services through road side units (RSUs). Thus, RSUs increase the reliability of disseminating critical safety messages to a large number of vehicles [4].

RSUs are communication nodes with the vehicular networks. This means that the vehicle needs to have access to road infrastructures through RSUs using infrastructure-based communications (hereafter V2I) [5]. For instance, RSUs forward received messages to intelligent transportations system (ITS) application servers by exploiting wide area networks [5]. Although communication capabilities between vehicles depend highly on the number of RSUs deployed and their coverage, RSUs are surely costly to deploy and to maintain. Consequently; there is a trade-off between full connectivity through RSUs and the deployment cost. To overcome the deployment cost of RSUs, road operators (ROs) can additionally leverage spectrum owned by mobile network operators (MNOs) to control traffic management services. In this situation, ROs are certainly expected to deploy and manage public-sector RSUs [6]. Following this, the ROs can enter into business arrangements with MNOs to surely deploy RSUs and run V2X services provided by ITS's authorities [6]. Therefore, MNOs should leverage existing cellular infrastructure to promote efficient deployment of V2X services.

Though the IEEE 802.11p was tested, automotive makers have manifested interest in C-V2X technology and question the applicability of the IEEE 802.11p for enabling many new V2X services. These doubts about the use of IEEE 802.11p coincides with the emerging of the fifth generation (5G) technology which aims to reduce network management through automation [7]. Furthermore, the commitment of automotive OEMs to test cellular communication for V2X motivated them to be part of a wider progression of 5G era [7]. The key technology of 5G design is mainly focused on the automation of network resources by using network slicing [8] which in turn is based on two new network technologies: network function virtualization (NFV) and software-defined networks (SDNs) [9]. The SDN concept together with edge computing could resolve most issues in vehicular networks such as irregular connectivity packet loss rate [8,10]. Therefore, software-defined-based vehicular network (SDVN) systems [8,10] improve resource utilization, selection of best routes, and facilitate network programming [9]. These SDVN architectures define local SDN domains through clustering in order to access the global intelligence of the network managed by the SDN controller [11,12].

There is a considerable amount of research work on SDVN [8–12] that focuses on different concepts, including the definition of SDN, software entities of the control plane, routing protocols using SDN-based VANET, etc. Some authors have proposed innovative architectures based on existing V2X scenarios that provide optimization results of their proposed architecture. There is also a number of surveys [13,14] that summarize the current work in the literature. However, it is quite challenging for most of the researchers to quickly decide which proposed solution could be suitable for their use case from schemes that propose modeling, architecture and optimization.

In this paper, we provide a review of published articles in the literature to comprehend the present state of research concerning software-defined networks-based vehicular networks with a particular focus on the articles whose contributions include modeling and implementation. Consequently, we performed a search on Google Scholar with the following keywords: software-defined networks, software-defined networks-based vehicular networks and modeling and implementation. In addition, we used the same keywords on other three research web engines, namely ScienceDirect, IEEE and ACM. Since SDN and VANETs are relatively new topics, we did not retrieve a huge number of papers that required an established protocol for evaluation and selection. Therefore, articles were manually selected or excluded if a given article provides clear modeling and implementation techniques. Other criteria were used in the selection such as significance, citation or rank of the publication venue.

In this work, we mainly focus on providing implicit literature that focuses on classifying existing SDVN solutions based on their modeling and implementation. To the best of our knowledge, it is the first work that groups SDVNs based on their modeling and implementation schemes. Therefore, in this paper the main contributions are summarized as follows:

- We first overview the current studies available in the literature on C-V2X technology in support of V2X applications.
- We then present the different architectures and their underlying system model for LTE-V2X communications.

- We also describe the keys ideas of software-defined networks and their concepts for V2X services.
- We define four elements that are considered for modeling and implementations of SDN for vehicular networks. We then present a comparative analysis for existing schemes grouped according to their modeling and simulation concepts.
- We provide a discussion and highlight vehicular adhoc network(VANET)' s challenges handled by SDN based vehicular network.

The remainder of the paper is organized as follows: the current studies and technologies for V2X services are detailed in Section 2. A comparative study of architectures and a system model of LTE-V2X communication in the implementation of V2X services are discussed in Section 3. The modeling and implementation of software-defined vehicular networks for V2X is detailed in Section 4, together with a definition of SDN, before briefly discussing findings on the comparative study of existing SDN based vehicular network in Section 5. Finally we conclude our work in Section 6.

2. Current Studies and Technologies for V2X Services

This section relates the evolution of vehicles equipped either with IEEE 802.11 p or C-V2X wireless communication technologies for deploying V2X services. This section describes the V2X and C-V2X communications modes. A comparative study of existing architectures and a system model of LTE-V2X communication in the implementation of V2X services are detailed.

2.1. V2X Communication Modes

A vehicle can interact with its environment through various types of communication as specified in [15]:

(1) Vehicle-to-Vehicle (V2V): A type of communication, in which User Equipements (UEs) (such as vehicles) communicate using V2V services.
(2) Vehicle-to-Pedestrian (V2P): A type of communication, in which both UEs (vehicle, pedestrian) communicate using V2P services.
(3) Vehicle-to-Infrastructure (V2I): A type of communication, in which one part is a vehicle- capable user equipement (VUE) and an RSU entity, both communicating using V2I services.
(4) Vehicle-to-Network (V2N): A type of communication, in which one part is vehicle-capable user equipment (VUE) and the other part is a V2X application server on the cloud for instance, both communicating using V2N services. As shown in Figure 1, V2N relates to any communication between vehicles and computing infrastructures such as RSU deployed either with eNodeB or like a standalone stationary UE [15].

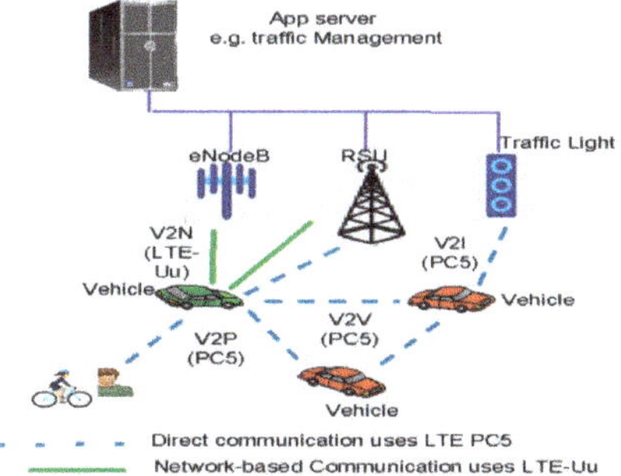

Figure 1. 3GPP Release 14 [16] for V2X services using direct communication over side link PC5 and LTE-Uu.

2.2. Evolution of Vehicles Using V2X Services

The study on the socio-economic benefits of cellular V2X [17] conducted by "The Analysys Mason" [17] specifies four (4) case scenarios to study the evolution of vehicles either equipped with IEEE 802.11 or C-V2X technologies for deploying V2X services. These case scenarios are numbered from one (1) to four (4). Scenario one (1) is the case adoption of C-V2X and IEEE 802.11p in the absence of any government measures. The second scenario is the case all new vehicles to support ITS services using IEEE 802.11 p in 2020; the third scenario is the case in 2023 all new vehicles are equipped with LTE PC5. The fourth scenario is the case the Equitable 5.9 GHz use is adopted for V2X communications.

Lessons learned from the study in [17] are described in Table 1, which summarizes case scenarios about V2X communications and relevant challenges. In the absence of any government regulations, V2V would use IEEE 802.11p or LTE-V2X PC5 [18]. This means that no-direct communication interoperability between IEEE 802.11p and PC5 exists. Therefore, V2V is possible via cellular LTE and vehicles without IEEE 802.11p or PC5 will use V2I and V2P via LTE-Uu of a smartphone brought in the vehicle. The case scenario 2 concerns all new vehicles that will use IEEE 802.11p in 2020 to support ITS services. Although vehicles without IEEE 802.11p would not communicate via V2V and V2I, vehicles equipped with IEEE 802.11p and LTE Uu could communicate via the cellular network. The challenge of the scenario case 3 (all new vehicles equipped with LTE PC5) would dictate ROs to add PC5-based RSU to existing RSUs potential. This means that vehicles without PC5 enabled would have to use V2I via smartphone. The case scenario 4 that predicts the use of Equitable 5.9GHz would allow automotive OEMs to use IEEE 802.11 p for V2V/V2I and Cellular (LTE-Uu) for V2N. In conclusion, the base case (case scenario 1) and equitable 5.9GHz RSU (case scenario 4) [19] deployment are thus suggested to be the most profitable way to deploy V2X services based on the net benefit perspective.

Table 1. Use Case scenarios to study the penetration of V2X services. The study was carried out by Analysys Mason [17].

Scenario#	Description	Vehicular Communication	Remarks
Base case	Adoption of C-V2X and IEEE 802.11p in the absence of any government measures	V2V using IEEE 802.11p or LTE-V2X PC5	V2V is possible via cellular LTE and V2I and V2P via LTE-Uu of a smartphone
Scenario 2	In 2020, all new vehicles to support ITS services via IEEE 2020	IEEE 802.11p for V2V and V2I	Road operators should install new RSUs or expand them to support V2I
Scenario 3	In 2023, all new vehicles equipped with LTE PC5	V2V and V2I via LTE PC5	Road operators add PC5-based RSU to existing RSUs
Scenario 4	Equitable 5.9GHz use	Division spectrum for V2V based PC5 and IEEE 802.11p	IEEE 802.11 p for V2V/V2I, Cellular(LTE-Uu) for V2N and others use PC5 for V2V/V2I

The number of vehicles equipped with embedded C-V2X communication technology is expected to increase as shown in Figure 2. Even though after a while we would expect C-V2X to be equipped in a greater number of vehicles, vehicles which do not have embedded C-V2X would use LTE PC5-based smartphones for accessing V2I and V2P services. In this context, the base case (scenario 1) seems to be the one to be adopted by many automotive makers. A key challenge in this scenario is predicted when different automotive OEMs would deploy different V2V communication solutions. Consequently; the inefficient use of the equitable 5.9 GHz spectrum could occur due to no direct-communication interoperability of the two technologies (C-V2X and IEEE 802.11p).

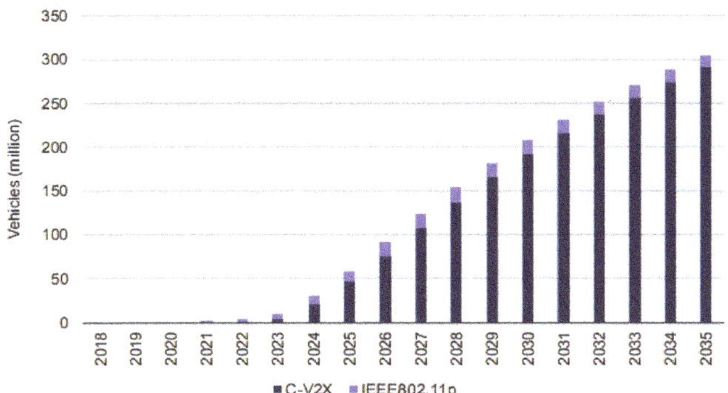

Figure 2. Evolution of the number of vehicles using V2X services in base case scenario 1 (Table 1) of vehicular technology [17].

2.3. 3rd Generation Partnership Project (3GPP) Cellular-V2X

3GPP Release 14 [16] for V2X services using direct communication over side link PC5 and LTE-Uu is shown in Figure 2. Direct communication uses links over the side link PC5 reference interface. In fact, side link PC5 defines features based on proximity service (ProSe) which is adapted for V2V communication scenarios [16]. PC5 communication mode enables V2I communication between vehicles and road infrastructures such as traffic control lights. In addition, V2N service uses LTE-Uu for allowing communication between vehicle and computing infrastructures, for example, an RSU implemented either with an eNodeB or as a standalone stationary UE, central cloud computing. A vehicular enabled UE exchanges data with deployed computing infrastructures over the LTE-Uu interface through RSU. The RSU broadcasts V2X messages towards multiple vehicles enabled UEs in a target area through the evolved multimedia broadcast multicast service (eMBMS) [16]. V2N serves VUEs in communication with an application server hosting ITS management applications, referred as a V2X Application Server (AS), which would provide a global state of traffic, the management of it, and service information [16,19–22].

Actually, cellular communication today represents the most embraced solution to collect data from vehicles and retransmit them to the network through RSUs. This avoids having to build new or set-ups expensive installations of RSUs [23,24]. To address admittedly V2X services use cases, the technical specification group (TSG), radio access networked (RAN) define V2V service using device-to-device (D2D) as specified in Release 12 [21]. Thus, a direct communication interface called sidelink (or PC5 interface) was thereafter specified in Release 14 [16] to allow direct communication link between devices. In addition, improvements to this interface have been added within Release 14 to study the V2V use cases in the ITS 5.9 GHz band and more specifically in [22].

3. A Comparative Study of Architectures and a System Model of LTE-V2X Communication in the Implementation of V2X Services

Important research on LTE-V2X communication in implementing V2X services has started to show relevant results. The relevant results of existing works focus mostly on the following network concepts: (i) long-term evolution-vehicle (LTE-V) standard supporting V2V communications using PC5 in LTE [25], (ii) methodical and assimilated V2X solution based on time-division LTE (TD-LTE) [26], (iii) multi-channel licensed-assisted access (LAA) schemes to enlarge multi-carrier Wi-Fi network [27]. We identify the following categories of work addressing system model of LTE-V2X communication in the implementation of V2X services:

(1)   Relevant use cases and requirements for V2X services
(2)   Design choices determining the performance of LTE-V2X communications

3.1. Relevant Use cases and Requirements for V2X Services

Boban et al. [28] describe the benefits of vehicles cooperating through V2X communication. They define descriptions and requirements of some relevant use cases which would be supported by future V2X communications systems. Among relevant use cases presented, some of them are bandwidth-demanding applications with high link reliability estimated to reach 99%. Considering latency, the authors mentioned that a low latency with a value below 10 ms is required for most of relevant uses cases. Therefore, these relevant uses cases require a high throughput of tens of Mb/s per vehicle. In addition, Seo et al. [4] provide a survey of the service flow and conditions of the V2X services based on LTE systems. They also discuss relevant scenarios suitable for an operational LTE-based V2X services system. Their work reveals some challenges such as high mobility and high density of vehicle which would bring a great impact in designing practical and technical solutions to satisfy the requirements of V2X services.

3.2. Design Choices Determining the Performance of LTE-V2X Communication

Masegosa et al. [25] put forward an overview of the long-term evolution-vehicle (LTE-V) standard supporting V2V communications using LTE's direct interface known as PC5 in LTE. The overview of physical layers changes presented under release 14 for LTE-V allows both communications modes 3 and 4 of the LTE-V. LTE-V is under study and its specifications would be published in Release 15 [6]. This Release 15 defines specifications on fifth-generation (5G) for supporting both V2X services and self-driving vehicles' applications. Indeed, the goal of Masegosa and al.'s work [25] was to review V2X Communications under mode 3 and mode 4 with LTE-V. In mode 3, the resources are assigned by the cellular network while mode 4 does not depend on cellular coverage, and vehicles autonomously take their radio resources using a relegated scheduling scheme supported by congestion control.

The results of the works in [25] discusses the performance achieved by the most major wireless technology IEEE 802.11p compared to LTE-V when vehicles transmit 10 packets per second (pps) to a distance of 160 m. In case the 802.11p data rate is increased to 18 Mb/s to a distance up to 160 m, IEEE 802.11p achieves a smaller packets data rate (PDR) than LTE-V thanks to the physical layer performance and the overriding effect of propagation. The authors analyzed also the performance of (LTE-V) standard when the channel load increases, this means when a vehicle transmits 50 packets per second (pps); the results show that the packet collisions become the primary source of errors.

Chen et al., [26] put forward a long-term evolution (LTE)-V model with a contribution on a methodical and assimilated V2X solution based on time-division LTE (TD-LTE). The main idea is the use of a centralized architecture that highlights features of TD-LTE and LTE-V-cell optimizes radio resource management for supporting better V2I. The results from their study are compared with the well-known wireless technology, IEEE 802.11p. The comparison reveals that LTE-V inherits the advantages of TD-LTE, including local features of TD-LTE and LTE-V-cell for supporting V2I communication implemented based on a centralized architecture. Therefore, they suggested that LTE-V would consort new features to overcome the challenges of V2V communications, such as congestion control.

Mukherjee et al. [27] studied the impact of unlicensed spectrum operation on the LTE physical layer architecture and the study of farther enhancements about licensed-assisted access (LAA). They present a brief survey of valuables enhancements for LAA for upcoming LTE releases. The experimental results of their proposed system expose clearly that from the synchronization point of analysis and the influence on the non-substitute Wi-Fi network, both classes of multi-channel LAA LBT schemes are realizable and can enlarge the performance of a multi-carrier Wi-Fi network assimilated when it is synchronizing with another Wi-Fi network.

Kawasaki et al. [29] proposed a performance evaluation between two methods of LTE-based V2X. The two methods are Uu-based LTE-V2X based and PC5-based LTE-V2X which is supported by device to device (D2D) communication [22]. The authors argue that queuing latency is significantly affected by bandwidth allocation, latency, parallel degree (PD) both in PC5-based and Uu-based. The authors reveal that the numbers of admissible parallel transfer are decided by different factors in Uu-based and PC5-based LTE V2X. However, in case the number of parallel transfer is equivalent to a larger logical bandwidth, queuing latency is estimated to remain smaller. The experimental evaluation results show that at PD=8, Uu-based was recorded to have the latency of 69.91msec and PC5-based LTE to have a latency of 11.82 msec. To sum up, the latency of PC5-based had only 16.9% of the latency in Uu-based. PC5-based LTE unveiled to retain a better performance than Uu-based while PC5-based requires additional functions compared to the existing LTE.

4. Modeling and Implementations of Software-Defined Vehicular Networks for V2X

4.1. Definition of Software-Defined Networks

Software-defined networks (SDNs) [30] are based on the separation of data and control planes. In SDNs, communication between the control layer and network layer takes place through the SDN control protocol because the control plane and forwarding plane are decoupled. Based on this principle of decoupling data and control plane, a standard protocol with multivendor support was needed for enabling communication between SDN's layers. As a result, OpenFlow was developed for this purpose [30]. OpenFlow was the first open-source control protocol for communicating between the SDN controller and the network devices. OpenFlow enables the implementation of a user application program to manipulate directly network devices without implementing various network protocols. Furthermore, OpenFlow maintains what it calls a flow table [31] on the network device (forwarding devices). The flow table contains information on how the data needs to be forwarded [30]. The SDN controller can then use OpenFlow to program the network devices of an OpenFlow-enabled switch by altering this flow table [32]. To program the forwarding information and set up the path across the network, the OpenFlow architecture supports two modes of operation, reactive and proactive [33]. The reactive mode is the default method of implementing SDN using OpenFlow and assumes that there is no intelligence of a control layer running on the network devices. In this mode, the first packet of the data traffic received on any of the forwarding nodes is sent to the SDN controller, and then the SDN controller uses this information to program the flow across of the whole network. In proactive mode, the SDN controller is preconfigured with some default flow values, and the traffic flow is programmed preemptively as soon as the switch is brought up. SDN controller and switches exchange the flow of information over the network using a secure channel such as Secure Socket Layer (SSL) or Transport Layer Security (TLS) while the OpenFlow manages communication between network layer and control layers [34,35].

4.2. Software-Defined Networks and their Concept in Vehicular Networks for Deploying V2X Services

The control layer plane is responsible for collecting and maintaining the status of all SDN cellular network devices, RSUs, and the vehicles [8]. An example of such SDN deployment in V2X services could be the route prediction on demand. The application could monitor vehicles on the roads and provides additional route prediction paths at a certain time of the day or when the vehicles are temporarily disconnected due to the high speed of the vehicles. The control layer would have to provide with the information about the vehicle's future route based on the Global Positioning System (GPS) or a navigation system [11]. The ability to deploy V2X services through SDN concepts is perhaps the most significant for automakers to solve the challenges of the no-direct interoperability of vehicle's wireless interface. Today, deployment of V2X services demands higher agility in network restoration, massive scalability, faster deployment, and operating expense optimization [36]. Therefore, V2X services cannot simply afford to be slowed down by the lack of speed in human-driven processes.

Automakers' onboard wireless communication interfaces have been traditionally specific to their vehicles. Automakers offer limited support for allowing external network devices to make decisions based on the logic and constraints across the vehicular networks. SDN offers a solution by linking V2X services to the vehicular network and bridging the challenge that existed with manual control and management processes. In addition, maximum use of automated tools and application have become a necessity to meet the V2X service demands. Automation and programmability capability are needed to support the provisioning of V2X services, the monitoring, and interpreting of V2X networks devices data. Therefore, automated tools implement run-time changes based on high mobility of vehicular networks, road traffic loads, and disconnection due to a high speed.

Since the SDN puts the intelligence of the vehicular networks in a central controlling software called SDN controller which conveys vehicular routing protocols to VANET's wireless nodes (such vehicles, RSUs). In fact, the vehicular routing protocols automatically react to the vehicle's mobility since the global view of the network is permanently available on SDN Controller. Therefore, the dissemination of routing path based on the vehicle's speed could be built directly into the SDN controller. Alternatively, the open protocols to manage the V2X applications can run on the top of the SDN controller using the northbound bound APIS [30] to proceed down the routing policies and rules to the controller and southbound APIS [30] to convey routing policies from SDN controller to the V2X forwarding devices. In conclusion, features of the SDN should handle the issue of high mobility and then improve V2X messages exchanged in a heterogeneous VANET architecture.

4.3. Architecture Overview of Software-Defined Vehicular Networks

Figure 3 depicts the components of various wireless communications in the software-defined vehicular network. To allow an SDN-based vehicular network (SDVN), simulation conducted on it leads to a certain number of SDN components. The SDN controller is the central logical intelligence of the SDN-based vehicular system. The SDN controller has a generalized and global view of the vehicular network and implements Openflow protocol to handle routing policies to eNB-type RSU controller on RSU. In fact, eNB-type RSU controller deployed on the edge of the vehicular network shortens the decision of generating new routing packets undefined in forwarding devices' flow tables. The SDN controller V2X network management conveys routing policies to UEs (vehicles) by implementing ITS's goals set up on the cloud or at the edge of the network for lowering processing decisions. The SDN controller is not only responsible to provide the whole performance but also provide routing rules for wireless devices (vehicles) selecting best routing paths to their destinations in VANET. OpenFlow enabled V2X-EU is the SDN wireless node and is responsible to control the data plane elements [12]. Data plane on vehicle implements OpenFlow protocol and is embedded in the OnBoard Diagnostic Unit (OBU). Furthermore, data plane elements are the VUEs that perform control message in term of routing policies from the eNB-type RSU controller to execute predefined actions which state ITS's goals once implemented in the application plane of the SDVN.

4.4. Modeling and Implementations of SDN for Vehicular Networks

The study of existing works on SDN-based vehicular networks was conducted based on the four (4) basics elements of modeling and simulation scheme [37]. First, we identified the targeted drawback that the researchers addressed. The second element is the classification of the existing SDVNs or VANETs system on which belong the addressed drawback. The third is the systems analysis which allows identifying parts of the SDVN system that are relevant to the problem. Finally, the model of the proposed solution, in turn, provides the implementation of the model related to the SDVN system in considering the outputs of its system analysis. Modeling and simulation scheme of existing work on SDVN was proposed to study the issues of several problems that originate from the complexity of ITS's applications understudy in VANETs and Internet of Vehicles. Thus, the models of software-defined vehicular networks contribute as a network technology to provide a solution to current VANETs' applications. In addition, SDVN is considered as a system because it is a part of VANET technology that will influence the design of future vehicular network architectures. The summary of the modeling and simulation schemes of existing works on SDN-based vehicular networks is described in Table 2.

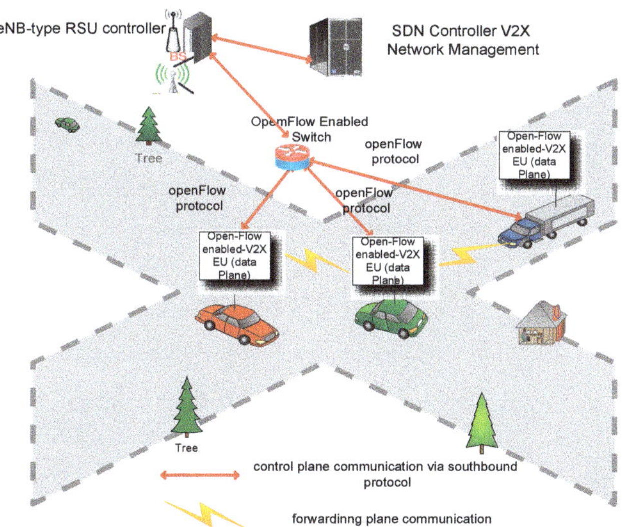

Figure 3. Software-Defined Vehicular Network. Data plane on vehicle implements OpenFlow protocol and is embedded in the OnBoard Diagnostic Unit (OBU). The SDN controller has a Generalized Vehicular cloud Openflow Controller on RSU. The SDN controller conveys routing policies to UEs (Vehicles) by implementing ITS's goals set up on the cloud.

Mainly, a VANET deals with systems in its objectives in a way of filling the separation between heterogeneity caused by communications interfaces equipped in vehicles or infrastructure-based communication. For instance, let us consider the fact that high mobility of vehicles causes dynamic topology change that in turn generates packet losses in the network, therefore routing protocols in mobile entities to effectively handle the short lifetime link are required. To this, the modeling and implementation of this issue conducted by the authors in [12] and summarized in Table 2 show that it is an SDVN's challenge and the research community suggests what could be done to solve the problem. The system analysis which represents the entities of the system that are relevant to the problem discloses trace of message overhead between vehicles (data planes entities) and the SDN controller. To this end, researchers should quickly decode that message overhead between vehicles (data plane) and SDN controller is the root cause, therefore, the implementation of the solution to a new problem related to routing protocols that could break link quality would start on message overhead on the SDN controller. Thus, the model proposed by the authors in [12] involves a new routing protocol that improves the packet delivery ratio by selecting stable routes with the lowest latency to control the overhead message on the SDN controller. Inalterability in protocol deployment due to the heterogeneity of wireless infrastructures prompted the authors in [38] to provide a system analysis that centers on abstracting heterogeneous wireless nodes as SDN switches enabled OpenFlow and designing SDN controller to manage dynamically network resources.

The output of the system analysis prompts the authors in [38] to propose a solution model that includes an adaptive protocol for heterogeneous multihop routing, a topology that enables SDN management overhead via the status of SDN switches and finally provide use cases of SDVN-enabled V2V, V2I, and V2N. To improve the performance in communication by mitigating the connectivity loss between vehicles and central SDN controller in [39], the authors suggested a system analysis based on selecting local SDN controller domains through clustering concepts. They proposed a hierarchical SDNV as the implementation model to decrease connectivity loss at the SDN controller, consequently; enhance the robustness of internetworking of data plane entities.

Table 2. Summary of related works on SDN based vehicular networks grouped according to the modeling and implementation scheme.

Description of the Problem	System	System Analysis	Model of the Proposed Architecture
Connectivity loss between vehicles and SDN controller [39]	SDVN	Local SDN controller domains through clustering	Hierarchical placement of SDN controllers decrease connectivity latency between them
Routing in mobile cloud [12]	SDN-based routing	Track message overhead between vehicles and controller	Control the overhead of the SDN controller and packet delivery ratio
Amount of data transfered for multimedia applications [1]	SDVN	Analyze throughput, end-to-end delay	RSU micro-datacenter, stochastic switching for reconfiguration overhead
Heterogeneity of wireless infrastructures and inalterable in protocol [38]	SDVANETs	Abstract heterogeneous wireless nodes as SDN switches enabled OpenFlow Allocate network resources through SDN controler	Deploy adaptive protocol for heterogeneous multihop routing; mitigate SDN management overhead via status of SDN switches; SDN enabled V2V, V2I and V2N.
Efficient resource utilization [11]	Software-defined Cloud/Fog network	SDN supports hybrid mode, Control plane is distributed between SDN controller, BS and RSU	Fog computing concept is adding to provide FSDN
Latency control [10]	Software-defined Mobile Edge computing	Software-defined cloud/edge vehicular networking	Latency control mechanisms: radio access steering at the base stations (BSs)
Latency control [40] on Multiple core network for autonomous driving vehicle	Software-defined VANET with 5G	Local knowledge of surroundings nodes, SDN controller, Broadcast beacon message	Cellular network integrated with network Model, SDN control eNB infrastructure, RSU controller controls RSU
Latency control and cost on cellular network [32]	Software-defined VANET with 5G	Control communication: VANET based, cellular network-based, hybrid-based	Optimize southbound communication via rebating mechanism, game equilibrium, two-stage leader-follower game for best decision between vehicle and controller
Dynamic resource management [14]	Software-Defined VANETs	Topology of SDN controller, Model of Node in Mininet-WiFi	Extend modeling of node car in mininet-WiFi
Control latency communication [13]	Vehicular networking; heterogeneity of radio access technologies	Vehicle network architecture for resource management, SDN controller, redesign of existing vehicular networks	Model SDHVNet architecture

ITS scenarios in future VANETS require quality of service (QoS)s and efficient utilization of network resources for enabling autonomous driving. The authors in [11], [14] addressed the challenge of efficient resource utilization. In [11], the system on which the problem is associated use the fog(edge) computing technology, thus the SDN-based fog network is evaluated to propose location-aware services with less communication latency. To this end, the system analysis centers on the deployment of the SDN to support hybrid mode (central-based and distributed-based configuration of SDN controller) and on the configuration of control plane (SDN controller) in distributed mode with both the base station (BS) and RSU. Considering the outputs of the SDN-based fog computing, the authors in [11] propose a model that combines edge(fog) computing services for allowing heterogeneous communication access for V2V, V2I, and V2N. The authors in [14] provide a system analysis that centers of the topology deployment of SDN controller and the possibility to model communication nodes (vehicles) as an SDN switch using the open-source simulation tool known as Mininet-WiFi [14]. The proposed model offers efficient utilization of network resource after modeling the vehicle as a node using Mininet-WiFi. Taking mobile edge computing step further, the authors in [10] investigate the possibility of deploying VANET's application with low-latency and high-reliability communication delay in software-defined mobile edge computing. The system analysis provided by the authors in [10] takes into consideration the edge vehicular network architecture which in turns provide a modeling solution on how to control the communication latency through radio access steering at the base station.

The advancement of 5G in the automotive field brings the integration of VANETS and 5G technology to construct 5G software-defined vehicular network with SDN technology as a primary key enabler. The authors in [32,40] investigated the challenge of communication latency and the cost on multiple core network for the autonomous driving vehicle. The modeling and implementation of [32] provide a systematic analysis based on the control of latency at VANET position, the cellular network- or hybrid-based (VANET and cellular position). The outputs of the system analysis prompt the authors in [32] to model their solution for decreasing communication latency by optimizing southbound communication via both rebating mechanism and the use of game equilibrium associated with the two-stage leader-follower game in order to select best routing paths between vehicle and controller. In [40], the modeling concepts centers on the system analysis based on broadcasting V2V beaconing messages so that the local knowledge of surroundings nodes and their topology are available at the SDN controller. After system analysis, the proposed solution provides a model that includes the integration of SDN controller, eNB infrastructures, RSU controller and 5G to design 5G based SDN concept.

The full transformation of VANET into SDVN requires to model SDVN solutions not only based on system architectures of SDVN but also based on mathematical analysis. Since the SDVN integrates the use of the SDN concept on the VANETs, a mathematically-based model is the natural modeling language to break up complexity problems and make VANETs' and SDVNs' challenges tractable. Mathematical-based theory applied to SDVN should bring further improvements and variations for allowing SDN to fully enhance VANETs, consequently, minimizing latency and cost, safety message delivery using heterogeneous communication interfaces [41]. A thorough mathematical model theory for all the above–analyzed articles that would lead to a new proposed concept that along with its implementation shall be addressed in future work.

5. Discussion

In this section, we summarize our findings from the classification of SDVNs based on the modeling and implementation schemes. The modeling strategy used in this paper to break up SDVNs' architecture helps to sort out existing VANETs' challenges addressed by integrating the SDN concept in VANETs. Some of the problems and system analysis in the process of modeling for problem-solving have been covered in Section 4.4, however, the rest of this section covers the summary of the four elements on which we centered the modeling of existing SDVN architecture in order to comprehend the current VANET's challenges solved by SDVN system. The simplicity of modeling proposed in this paper aims at encouraging a research combination towards SDVN with 5G and with edge computing as an alternative solution for future VANET's applications. Based on our study, we provide SDVNs' systems analysis of existing SDVN architecture.

The comparative study of existing SDVN based on the modeling and implementation scheme as shown in Table 2 provides a list of a number of VANET issues addressing the full transformation of VANETs to SDVN. To this end, the identified VANET issues handled over to SDVN systems are summarized in the following contributions: firstly, the contributions in [10,13,32,39,40] that address the issue of loss of connectivity by controlling the data plane latency, secondly, the contributions of authors in [12] that are related to routing protocols in the mobile cloud environment, thirdly, the contributions from authors in [11,14] which address the issue of resource utilization. Finally, the authors in [1] provide a study on the amount of data transferred for multimedia applications. Lastly, the contributions in [32,40] address the issue of communication latency and the cost of using multiple core network for autonomous vehicles.

Moreover, handover control and proper allocation of radio resource were analyzed in [42] to mitigate the challenge of mobility management and transmission delay. The mobility management in VANETs increases delays in the transmission where handover procedures are not properly implemented.

To this, SDVN with fog computing would allow meeting the requirements of low transmission delay by adopting a hybrid handover scheme, optimizing radio resource allocation through the Markov decision process [42]. However, inefficient control for high mobility that causes unsteady wireless channel for SDVN and latency on the distribution of commands from controllers and interworking breach through heterogeneous networks were among ongoing VANETs' challenges to contend with SDVN. In addition, network slicing and NFV [25] in SDVN introduce potential research opportunities. In fact, SDN allows operative network slicing in a dynamic topology. The NFV with the use of hypervisor has the task of adjusting OpenFlow in the way to enable heterogeneous network interworking.

The system analysis of SDVN systems identifies components, architectures, protocols directly linked to the SDVN challenge addressed. Note that there are six (6) architecture systems of SDN- based vehicular network proposed in the literature: SDVN [1,39], SDN-based routing [12], software-defined VANETs (SDVANETs) [14,38], SDN-based cloud/mobile (fog/edge) computing [10,11], software-defined VANET with 5G [32,40]. A comprehensive study of SDVN architecture, its benefits and services are described in [36]. Although six systems of SDVN architectures are currently implemented and simulated, system analysis provides insights to relevant components, architectures, protocols and simulation tools to be considered before providing a solution model to VANET's challenge. In fact, we can list a few of SDVN system's analysis as summarized in Table 2: placement of SDN controller [11,13,14,39,40], communication control VANET-based or cellular network-based [32], local knowledge of surrounding nodes via beacon or geo-broadcast messages [39,42], network simulator tools such as Mininet-WiFi [14], trace of overhead messages between vehicles and SDN controllers [1].

Comprehensive surveys on the software-defined networks in [41,43] lack a comparative study on the system analysis of existing SDVNs to point out SDVN components, architectures and algorithms investigated to tackle SDVN drawbacks. Authors in [41,43] investigate SDVN architectures to identify their benefits and challenges against the VANETs regarding communication in [41], and security in [43]. Within the existing SDVN solutions, technology for SDN controllers, implementation tool for the OpenFlow protocol have been proposed, yet a comprehensive study on SDVN architectures based on the modeling will provide the required insights on the components needed for further enhancements. Since the system analysis of SDVN systems provide key enabling technologies for investigating SDVENs' challenges, the solution model proposed in the implementation based the system analysis entities in SDVN shows potential research opportunities towards an efficient SDVN that could allow a huge number of next-generation VANET applications.

6. Conclusions

Software-defined networks (SDNs) are a network technology based on the separation of data and control planes. This paper mainly focuses on discussing implicit literature that concentrates on classifying existing SDVN solutions based on their modeling and implementation. In addition, this work provides an overview of the current studies available in the literature on C-V2X applications in support of V2X applications. The keys ideas of software-defined networks and their concepts for V2X services were also presented. We show that the simplicity of modeling that was proposed provides a detailed analysis of known solutions including SDVN or SDVN with 5G, SDVN-based cloud/mobile edge computing in order to solve current VANET issues in most cases. Loss of connectivity between vehicles and SDN controllers, routing in mobile (edge) cloud computing, were among the issues tacked by existing solutions such as SDVN, software-defined edge computing, SDVN with 5G and SDN-based routing that are currently implemented in order to solve current and ongoing VANETs challenges. Lastly, we discussed some guidelines for future research work.

Author Contributions: Conceptualization, writing—original draft preparation, L.N.; methodology, investigation, writing—review and editing, L.N.; formal analysis, validation, editing S.M.R.I.; supervision Y.H.C, visualization, M.B.; writing—review and editing, ressources, funding acquisition, J.J.W.

Funding: This research received no external funding.

Acknowledgments: This research was supported the BB21+ project in 2019.

Conflicts of Interest: The authors declare no conflict of interest.

References

1. Mohammad, A.S.; Ala, A.F.; Mohsen, G. Software-Defined Networking for RSU clouds in support of the Internet of Vehicles. *IEEE Internet Things J.* **2015**, *2*, 133–144.
2. Ke, Z.; Yuming, M.; Supeng, L.; Yejun, H.; Yan, Z. Mobile-Edge Computing for Vehicular networks. *IEEE Vehic. Technol. Mag.* **2017**, *12*, 36–44.
3. Campolo, C.; Molinaro, A.; Menichella, F. 5G Network Slicing for Vehicle-to-Everything Services. *IEEE Wirel. Commun.* **2017**, *24*, 38–45. [CrossRef]
4. Seo, H.; Lee, K.D.; Yasukawa, S.; Peng, Y.; Sartori, P. LTE Evolution for Vehicle-to-Everything Services. *IEEE Commun. Mag.* **2016**, *6*, 22–28. [CrossRef]
5. Silva, M.C.; Masini, M.B.; Ferrari, G.; Thibault, I. A Survey on Infrastructure-Based Vehicular Networks. *Mob. Inf. Syst.* **2017**, *2017*, 1–28. [CrossRef]
6. 5G Americas V2X Cellular Solutions. Available online: http://www.5gamericas.org/files/2914/7769/1296/5GA_V2X_Report_FINAL_for_upload.pdf (accessed on 31 August 2019).
7. 5GAA, White Papers, Toward Fully Connected Vehicles: Edge Computing for Advanced Automotive Communications. Available online: http://5gaa.org/wp-content/uploads/2017/12/5GAA_T-170219-whitepaper-EdgeComputing_5GAA.pdf (accessed on 31 August 2019).
8. Jianqi, L.; Jiafu, W.; Bi, Z.; Qinruo, W.; Houbing, S.; Meikang, Q. A Scalable and Quick-Response Software Defined Vehicular Network Assisted by Mobile Edge Computing. *IEEE Commun. Mag.* **2017**, *55*, 94–100.
9. He, X.; Ren, Z.; Shi, C.; Fang, J. A Novel Load Balancing Strategy of Software-Defined Cloud/Fog Networking on the Internet of Vehicles. *China Commun.* **2016**, *13*, 140–149. [CrossRef]
10. Deng, J.D.; Lien, S.Y.; Lin, C.C.; Hung, C.S.; Chen, W.B. Latency Control in Software-Defined Mobile-Edge vehicular Networking. *IEEE Commun. Mag.* **2017**, *55*, 87–93. [CrossRef]
11. Truong, B.N.; Lee, M.G.; Doudane, G.Y. Software Defined Networking-based Vehicular Adhoc Network with Fog Computing. In Proceedings of the 2015 IFIP/IEEE International Symposium on Integrated Network Management (IM), Ottawa, ON, Canada, 11–15 May 2015.
12. Ku, I.; Lu, Y.; Gerla, M.; Ongaro, F.; Gomes, L.R.; Cerqueira, E. Towards Software-Defined VANET: Architecture and Services. In Proceedings of the 13th Annual Mediterranean AdHoc Networking Workshop (MED-HOC-NET), Piran, Slovenia, 2–4 June 2014; pp. 103–110.
13. Adnan, M.; Wei, E.Z.; Quan, Z.S. Software-Defined Heterogeneous Vehicular Networking: The Architectural Design and Open Challenges. *Future Internet* **2019**, *11*, 1–17.
14. Ramon, D.R.F.; Claudia, C.; Christian, E.R.; Antonella, M. From Theory to Experimental Evaluation: Resource Management in Software-Defined Vehicular Networks. *IEEE* **2017**, *5*, 1–8.
15. Wang, X.; Mao, S.; Gong, M.X. An overview of 3GPP cellular vehicle-to-everything standards. *Get Mobile.* **2017**, *21*, 19–25. [CrossRef]
16. 3GPP. Study on LTE Support for Vehicle to Everything(V2X) Services. Available online: https://portal.3gpp.org/desktopmodules/Specifications/SpecificationDetails.aspx?specificationId=2898 (accessed on 31 August 2019).
17. Rebbeck, T.; Stewart, J.; Lacour, H.A.; Lillen, A.; McClure, D.; Dunoyer, A. Socio-Economic Benefits of Cellular V2X, Final Report for 5GAA. Available online: http://www.analysysmason.com/contentassets/b1bd66c1baf443be9678b483619f2f3d/analysys-mason-report-for-5gaa-on-socio-economic-benefits-of-cellular-v2x.pdf (accessed on 31 August 2019).
18. US Department of transportation. The smart/Connected City and Its Implications for Connected Transportation. Available online: www.its.dot.gov/index.htm (accessed on 31 August 2019).
19. 3GPP. LTE-Based V2X Services. Available online: http://www.tech-invite.com/3m36/tinv-3gpp-36-885.html (accessed on 31 August 2019).
20. 3GPP. Liaison Statement from 3GPP RAN on LTE-Based Vehicle-to-Vehicle Communication. Available online: http://www.3gpp.org/news-events/3gpp-news/1798-v2x_r14 (accessed on 31 August 2019).
21. 3GPP. Evolved Universal Terrestrial Radio. Available online: https://www.etsi.org/deliver/etsi_ts/136300_136399/136300/09.04.00_60/ts_136300v090400p.pdf (accessed on 31 August 2019).

22. 3GPP. Study on the Enhancement of 3GPP Support 5G V2X Services. TR 22.886. 2016. Available online: https://www.3gpp.org/DynaReport/22-series.htm (accessed on 31 August 2019).
23. Wang, X.; Mao, S.; Gong, M.X. An Overview of 3GPP Cellular Vehicle-to-Everything Standards. Available online: https://www.sigmobile.org/pubs/getmobile/articles/Vol21Issue3_2.pdf (accessed on 31 August 2019).
24. 3GPP. Service Requirements for V2X Services. Available online: https://portal.3gpp.org/desktopmodules/Specifications/SpecificationDetails.aspx?specificationId=2989 (accessed on 31 August 2019).
25. Masegosa, M.R.; Gozalvez, J. A New 5G Technology for Short-Range Vehicle-to-Everything Communications. *IEEE Vehicular Tech.* **2017**, *10*, 30–39. [CrossRef]
26. Chen, A.; Hu, J.; Shi, Y.; Zhao, L. LTE-V: A TD-LTE-Based V2X Solution for Future Vehicular Network. *IEEE Internet of Things J.* **2016**, *3*, 907–1005. [CrossRef]
27. Mukherjee, A.; Cheng, J.; Falahati, S.; Koorapaty, H.; Kang, D.H.; Karaki, R.; Falconetti, L.; Larsson, D. Licensed-Assisted Access LTE: Coexistence with IEEE 802.11 and the Evolution toward 5G. *IEEE Commun. Mag* **2016**, *54*, 50–57. [CrossRef]
28. Boban, M.; Kousaridas, A.; Manolakis, K.; Eichinger, J.; Xu, W. Use Cases, Requirements, and Design Considerations for 5G V2X. Available online: https://arxiv.org/pdf/1712.01754.pdf (accessed on 31 August 2019).
29. Kawasaki, R.; Onishi, H.; Murase, T. Performance Evaluation on V2X Communication with PC5-based and uu-based LTE in Crash Warning Application. In Proceedings of the 2017 IEEE 6th Global Conferenceon Consumer Electronics (GCCE2017), Nagoya, Japan, 24–27 October 2017.
30. Kreutz, D.; Ramos, M.V.F.; Veissimo, E.P.; Rothenberg, E.C.; Azodolmplky, S.; Uhlig, S. Software-Defined Networking: A Comprehensive Survey. *Proc. IEEE* **2015**, *103*, 14–76. [CrossRef]
31. Fei, H. *Network Innovation through OpenFlow and SDN: Principles and Design*; CRC Publishing: Boca Raton, FL, USA, 2016.
32. Li, H.; Dong, M.; Ota, K. Control Plane Optimization in Software-Defined Vehicular Ad Hoc Networks. *IEEE T.Veh.Technol.* **2016**, *65*, 7895–7904. [CrossRef]
33. Liu, Y.; Ding, Y.A.; Tarkoma, S. *Software-Defined Networking in Mobile Access Networks*; Department of Computer Science, University of Helsinki: Helsinki, Finland, 2013 19 September; pp. 1–29.
34. Software-Defined Network. What Is It, How Does It Work, and What Is Good for. Available online: http://www.cs.tau.ac.il/~{}msagiv/courses/rsdn/SDN-TAU.pdf (accessed on 31 August 2019).
35. Chayapathi, R.; Hassan, F.S.; Shah, P. Software defined Networking (SDN). In *Network Functions Virtualization (NFV) with A Touch of SD*; Addison-Wesley Professional: Boston, MA, USA, 2016; ISBN 0-13-446305-6.
36. Toufga, S.; Owezarski, P.; Abdellatif, S.; Villemur, T. A SDN hybrid architecture for vehicular networks: Applications to Intelligent Transport System. In Proceedings of the 9th Europen Congress on Embedded Real Time Software and Systems (ERTS), Toulouse, France, 31 Januray–2 February 2018; pp. 1–8.
37. James, E.C. An Introduction to the Use of Modelig and Simulation throughout the Systems Engineering Process. Available online: https://ndiastorage.blob.core.usgovcloudapi.net/ndia/2012/systemtutorial/14907.pdf (accessed on 31 August 2019).
38. Zongjian, H.; Jiannong, C.; Xuefeng, L. Enabling rapid innovation for heterogeneous vehicular communications. *IEEE* **2016**, *30*, 10–15.
39. Sergio, C.; Azzedine, B.; Rodolfo, I.M. An Architecture for Hierarchical Software-Defined vehicular Networks. *IEEE Commun. Mag.* **2017**, *55*, 80–86.
40. Piyush, D.; Mohsin, R.; Hoa, L.; Nauman, A. Software-Defined Approach for Communication in Autonomous Transportation Systems. *Energy Web* **2017**, *4*, 1–9.
41. Manisha, C.; Sandeep, H.; Krishn, K.M.; Arun, K.S.; Zhigao, Z. A survey on Software-defined networking in vehicular ad hoc networks: Challenges, applications and use cases. *Sustain. Cities Soc.* **2017**, *35*, 830–840.
42. Yeomin, Z.; Haijun, Z.; Keping, L.; Qiang, Z.; Xiaoming, X. Software-Defined and Fog-Computing-Based Next Generation Vehicular Networks. *IEEE commun. Mag.* **2018**, *56*, 34–41.
43. Wafa, B.J.; Mauro, C.; Chhagan, L. Software-Defined VANETs: Benefits, Challenges, and Future Directions. Available online: https://arxiv.org/abs/1904.04577 (accessed on 31 August 2019).

Article

Vehicle-to-Vehicle (V2V) Message Content Plausibility Check for Platoons through Low-Power Beaconing †

Hyogon Kim [1,*] **and Taeho Kim** [2]

[1] Department of Computer Science and Engineering, Korea University, Anam-Dong, Sungbuk-Gu, Seoul 02841, Korea

[2] Department of Computer Science, University of Colorado Boulder, 1111 Engineering Drive ECOT 717, 430 UCB, Boulder, CO 80309-0430, USA; taeho.kim@colorado.edu

* Correspondence: hyogon@korea.ac.kr; Tel.: +82-2-3290-3204

† This paper is an extended version of the conference paper. Kim, T.; Kim, H. Vehicle-to-Vehicle Message Content Plausibility Check through Low-Power Beaconing, 2017 IEEE 86th Vehicular Technology Conference, Toronto, ON, Canada, 24–27 September 2017.

Received: 28 October 2019; Accepted: 10 December 2019; Published: 12 December 2019

Abstract: Although the IEEE Wireless Access in Vehicular Environment (WAVE) and 3GPP Cellular V2X deployments are imminent, their standards do not yet cover an important security aspect; the message content plausibility check. In safety-critical driving situations, vehicles cannot blindly trust the content of received safety messages, because an attacker may have forged false values in it in order to cause unsafe response from the receiving vehicles. In particular, the attacks mounted from remote, well-hidden positions around roads are considered the most apparent danger. So far, there have been three approaches to validating V2X message content: checking based on sensor fusion, behavior analysis, and communication constraints. This paper discusses the three existing approaches. In addition, it discusses a communication-based checking scheme that supplements the existing approaches. It uses low-power transmission of vehicle identifiers to identify remote attackers. We demonstrate its potential address in the case of an autonomous vehicle platooning application.

Keywords: V2V communication; message contents plausibility; power control

---

1. Introduction

In vehicle-to-everything (V2X) communications, various threats exist. They range from physical types such as signal jamming and Global Positioning System (GPS) spoofing to more logical types such as higher level protocol attacks or cryptanalyses. Unless thoroughly addressed, they can pose grave safety problems for human lives, not to mention less serious losses. Unfortunately, there is not a single overarching solution for all, due to the aforementioned diversity of the potential attacks. Each threat may require a different set of countermeasures.

The most notable security feature of V2X communication is the use of security credentials, a.k.a. certificates, commonly assumed by the Wireless Access in Vehicular Environment (WAVE) in IEEE 1609.2 [1] and the 3G Partnership Project (3GPP) Cellular V2X framework in 3GPP TR 36.885 [2]. Every safety message, the Society of Automotive Engineers (SAE) J2735 basic safety message (BSM) [3] broadcast from each vehicle in particular, is required to carry the certificate or its digest to provide the message source authentication and integrity check [1]. Also, it is used to provide privacy, by requiring frequent and continuous change [4].

However, one obvious threat that the certificate-based V2X security framework is not addressing is the message content fabrication. Unless appropriately filtered, vehicles can utilize received BSMs without knowing whether the contents of the messages sent by the nearby vehicles are genuine or fake. In case an attacker has access to the device with a valid credential, she could forge an erroneous message and transmit it, which will not be filtered by the certificate check. Then, the attacker could cause the receiver vehicles to inappropriately react to the faulty message content and possibly face critical consequences such as collisions. Since tampering with the sensory input data such as GPS position information to be transferred by communication protocols is easier than breaking the protocols or the cryptographic system, this type of attack may be more beneficial for the attackers.

Although the content fabrication attacks from a close distance can easily be refuted by the multitude of on-board sensors such as radars, cameras, and lidars on the vehicle, remote attacks out of the line of sight (LoS) of these sensors cannot be adequately filtered. A notable case would be roadside attacks, where the attacker can broadcast false information to the passing vehicles. For example, the attacker can fabricate false positional messages of a "ghost vehicle" to cause a multi-car pile-up in a long platoon [5]. Therefore, to obtain the users' confidence in this technology, the credibility of the V2X messages from remote locations out of the range of the LoS sensors should be checked.

Below, we propose a solution where vehicles employ low-power beaconing messages (called "whispers") to judge whether the vehicles can trust the contents of the received vehicle messages or not. This paper particularly focuses on the data fabrication attacks on the position information, a crucial and endangered subpart of the system [6]. We demonstrate the value of the proposed approach by considering a promising use case of V2X, platooning, so that the platoons can maintain traffic efficiency in the situation that the roadside attacker transmits fake messages. Although studies applying vehicular communication to truck platooning have dealt with these issues theoretically and practically Ref [7,8], the fact that platooning vehicles can receive vehicle messages containing false contents was not addressed in them. As the most effective attack target is the platoons, which is considered one of the promising applications of cooperative autonomous driving [9]. Finally, our proposal can be implemented only by utilizing the existing power control functionality that is already used for the congestion control in the V2X communication.

The contributions of this paper can be summarized as follows.

- By narrowing the spatial attack window to a small distance to the potential victims on the road, the proposed scheme can force the attackers into the detection range of the line-of-sight (LoS) hardware sensors. Thus, the scheme creates an opportunity for the cooperation between the LoS sensors and the communication for the message contents validation.
- By applying the proposed scheme to platooning, potentially the most vulnerable application of V2X to the false position attack, we show that it can mitigate not only the collision risk but also the discomfort from the unnecessary responses to the false position attack.

2. Related Work

The problem of message forgery has been noticed from the early days of V2X communication [10], to have dire safety consequences [11]. It is even considered the most serious security threat for its ease of implementation and execution [6]. Although the certifying authority (CA) is supposed to revoke the certificates on misbehaving devices as per the IEEE 1609.2 [1], there is an issue of how the authority can obtain and accumulate sufficient evidence that the devices are faulty or compromised [12]. Even if it could, it will take time during which the attacker has a time window of attacks to exploit. Especially if the attacker misbehaves in a discontinuous manner, namely evading detection intelligently, the task of identifying the misbehaving device and revoking its certificates will not be easy in the first place. So, the message correctness check is considered the primary focus of misbehavior detection [13].

There have been mainly three approaches to solving the problem, i.e., behavior analysis [11,14,15], sensor-fusion [16–18] and communication-based constraint check [5,12,19]. These approaches are not mutually exclusive, so they can be used alone, or in combination. The first exploits physical constraints of vehicle movement dynamics. The second uses multiple sensor inputs to verify the position information in the received message. The third uses maximum communication range constraints.

Below, we will briefly discuss these three that check the plausibility of the position information conveyed in a vehicle-to-vehicle (V2V) message. We note that because many of the existing approaches have been around for a long time, they do not reflect the recent standards development. It is one of the reasons that we need a new solution approach that is based on the standardized framework for the V2X communication.

2.1. Vehicle Dynamics-Based Validation

This approach relies on models of plausibility for vehicle behaviors. Laws of physics or driver behavior model can be used to judge if a claimed movement of a suspected vehicle is plausible. Stübing et al. [11] used Kalman filter to analyze the path taken by neighbor vehicles. By comparing the Kalman prediction of the path of a neighbor vehicle with the mobility data claimed in its periodic safety messages such as position, speed, and heading, this method judged if the deviation of the reported mobility data from the Kalman prediction is in an acceptable range; otherwise, it raises an alert. Finding that the Kalman filter cannot keep up with highly dynamic maneuvers such as sudden overtaking and hard braking, though, it additionally employed a hidden Markov model (HMM) to verify such movements. Sun et al. [20] also used Kalman filter on the received signal from the transmitter. Yavvari et al. [14] more extensively utilized the information contained in the BSM [3] to check the plausibility of the claimed movement (location and kinematics) by the message sender. Since the BSM has the lateral and longitudinal acceleration, speed, position, heading angle, and vehicle length and width, these data can be used to check the plausibility of the claimed movement given the past few transmissions from the message sender. If a claimed value exceeds the error range of the given dynamics model of the vehicle, the checking algorithm flags anomaly. Leinmüller et al. [19] tried to check if the V2X message has false position information by the fact that vehicles can move only at a well-defined maximum speed such as the general speed limit on streets. It also exploited the fact that only a restricted number of vehicles can reside in a certain area, whereby it can prevent Sybil attacks. It also used maps to check if a vehicle can navigate through the claimed position. Ghaleb et al. [15] used neural networks to find misbehavior in the communicated information. The local dynamic map (LDM) is constructed from the shared information, and each message is determined legitimate or malicious based on the historical behavior of the model.

2.2. Sensor-Based Cross-Checking

Bißmeyer et al. [16] used sensor fusion and particle filters to check the plausibility of the position information in the incoming messages and assign a trust level to the message sending vehicle. A separate particle filter was used for each tracked neighbor vehicle. The particle filter combines all available different position information from a variety of input sources such as V2X message, radar, road map, etc., to detect inconsistencies among them, introduced by faulty nodes or malicious attackers. For example, a radar, lidar, or camera could detect no neighbor vehicle at a position claimed through V2X. Not only each V2X message is evaluated in terms of the trustworthiness using the particle filter framework, but also the trust level of the message sending vehicle is computed based on the message trust rating. Schmidt et al. [17] presented a similar heuristic framework to combine the plausibility ratings from different sensor modules such as radars, lidars, and ultrasonic to check the message plausibility. The sensor-based check is part of a battery of tests using the movement analysis, sensor-proofed position check, minimum distance moved check, map-proofed position check, etc. Kim et al. [18] also combined various information sources such as sensors, maps, and input from other vehicles. Yan et al. [19] checked the position information in the received message through radar sensors. Each vehicle first monitors the neighboring vehicles using the radar in its range. Then it propagates the information to other vehicles to share the information globally. LeBlanc et al. [21] relies on road-side units (RSUs) to provide GPS reference values to defeat false position attacks.

2.3. Communication-Based Validation

The last approach to evaluate the trustworthiness of V2X reported position information uses the message conveying technology itself. First of all, a vehicle could compute the angle [22] or distance to the message sender using the received signal strength, the time-of-flight, and the angle-of-arrival of the V2X message. However, highly dynamic V2X channel conditions such as shadowing by blocking vehicles or multipaths created by metal-hulled vehicles make it hard to obtain reliable measurements. Moreover, the strict time synchronization between vehicles that is required for the computation would be hard to satisfy on the on-board units (OBUs) as these devices are not created for such purpose. Therefore, in this section, we will discuss only the schemes that rely on the most conservative constraints, i.e., the maximum possible communication distances.

Parno et al. [5] proposed a scheme where a vehicle's relative location is defined by its entanglement with other vehicles. Each vehicle regularly broadcasts its identity (a public key) along with its signature of a current timestamp. When a vehicle A receives such a broadcast from another vehicle B, it signs B's ID and rebroadcasts it. If a vehicle C on the opposite lane rebroadcast A's identity before it rebroadcasts B's identity, then B can conclude that A is ahead of him/her. As the authors noted, however, this scheme was a sketch of a possible solution and had strong assumptions to make it work. For one, it depends on the opposite-side traffic, a condition we cannot always rely on. However, the idea of entanglement within the communication range is viable, and we borrow it for our own proposal discussed later. Raya et al. [12] proposed a method to evict a misbehaving vehicle from the trusted set of neighbors even before the certifying authority (CA) includes the attacker's certificates in the revocation list. In particular, if the attacker reports an implausible position that stands out from the observation of the honest majority, i.e., if messages are received beyond their expected area of propagation, the honest neighbors in the communication range of the attacker notice it. Then the honest majority begins to warn any new vehicle that comes into the communication range of the attacker to watch out for the potentially fabricated information from the suspected vehicle. Furthermore, if the accusation in the warning message collects enough supporting signatures from the honest majority, it promotes to the disregard message. When the disregard message with enough signatures is picked up by the roadside unit (RSU), it is forwarded to the CA so that it can revoke the certificates of the attacker. Leinmüller et al. [23] also exploited the maximum communication range limitation in addition to other behavioral anomalies of the vehicle in question to verify neighbors' position information. In case an observing vehicle M overhears communication between N and A, it compares their positions and check if a possible attacker A can be within the maximum communication range of N. If the previously claimed position of A contradicts the condition, M considers A sending out false position information when it is actually at a position closer to N. Since this work is for multi-hop forwarding situation where the identity of the next forwarding node is known, we cannot directly apply it to the one-hop broadcast situation. Moreover, this checking method cannot be used against the remote attacker pretending to be close when it is actually far away.

Schmidt et al. [17] used a checking method that requires the neighboring vehicle should be heard during at least twice the maximum communication range d_{TX}, which is called the minimum distance moved (MDM), in order to be trusted. This check is to cope with stationary attackers whose transmission range is d_{TX}. Unfortunately, however, the current standards typically allow the adaptation of the transmit (Tx) power for the purpose of congestion control. In particular, the SAE J2945/1 stipulates that a Wireless Access in Vehicular Environment (WAVE) device can transmit at a Tx power ranging from 10 dBm to 23 dBm, depending on the channel congestion condition [24]. Consequently, a factor of 20 power difference obviously wildly affects the communication range of BSM. Moreover, the range can be also highly dependent on the given scenario and buildings in the vicinity, and be anywhere between 100 m and 500 m [13]. Under these circumstances, the position plausibility check based on the estimated maximum communication distance constraint cannot be reliable. Received signal strength indicator (RSSI)-based plausibility check mechanisms are subject to the same problem. Ruj et al. [25] exploits the time of flight to check the distance to the transmitter. But this scheme requires extremely precise time synchronization between vehicles.

Lastly, we stress that our paper is not about Sybil attacks. In Sybil attacks, the attacker can forge many false identities [26,27]. In the current V2X standards, however, the vehicle identity in each message must be proven by the attached security credential based on public key infrastructrure (PKI) Ref [1]. Our solution is based on the current standards, so we assume that even the attack should use its certificate to mount the attack. In our threat model, the attacker is not capable of tampering with the PKI. It can only tamper with the sensor input (e.g., Global Positioning System (GPS) coordinates) that is provided to the IEEE 1609.2 security module. In fact, it is the focus of this paper. When an authenticated attacker tries to propagate false information, we can narrow the spatial attack window to a short distance to the potential victims so that the hardware sensors can double-check the received false information.

In Section 3, we will introduce a complementary scheme in the third category, based on physical constraints of low-power communication, to defend against remote roadside attackers. Although it does not preclude the possible employment of the other two approaches in combination, an independent check based on the communication constraints has certain advantages over them. First, it can work even when the sensor-based position check does not work, e.g., the in non-line-of-sight (NLoS) condition. This is because most vehicle sensors such as camera, radar, and lidar are LoS devices. Second, it can work when the behavioral analysis may fail. For one, the behavioral analysis cannot be applied to a stopped vehicle, since it does not exhibit any movement behavior. Since V2X is expected to become a regulation-enforced safety feature in the near future [28], we believe the communication-based position verification should be included irrespective of other additional checks. In particular, our scheme can be easily implemented in the WAVE or the cellular V2X (C-V2X) framework.

3. Neighbor Verification through Low-Power Beacons

The core idea of our proposal builds on the physical communication constraint that messages from neighbor vehicles transmitted at a small transmit (Tx) power reach only the immediate proximity of the sender [29]. It contrasts with existing communication-based approaches that exploit the maximum communication range (e.g., 300 m) of the attacker as the constraint to be checked against [6,17]. In this section, we discuss our proposal to use low-power beaconing for proximity proving purpose, in order to defend against position data forging attack from stationary roadside attackers. The proposed scheme has a few desirable properties. First, it does not require a new hardware component beyond existing wireless access in vehicular environment (WAVE) on-board unit (OBU). Second, it does not excessively increase channel utilization as to hamper normal beaconing activities using BSMs [3]. Third, it works where sensor fusion is not applicable, and against a stationary attacker to which the behavior-based checking is not applicable. Below, we sketch the solution approach.

3.1. Sketch of Solution Approach

See Figure 1 that depicts the movements of two honest vehicles U and V on a road strip, and two attackers K and K' on the roadsides. An attacker K' is less than d_B apart from a potential victim V, where d_B is the maximum distance that a BSM beacon can reach. Note that the attacker can use the standard Tx power (e.g., 23 dBm) or increase d_B by using an amplifier or directional transmission. Without the proposed low-power beacon check scheme, K' can inject false positions or other safety-critical information in its BSM, $BSM_{K'}$, and coax V into believing it and elicit a dangerous and unnecessary reaction (e.g., hard braking). But with the proposed low-power beaconing, each vehicle is designed to trust only those vehicles that are within the low-power beacon range $d_W \ll d_B$. It checks the condition not by the location information in neighbors' BSMs, but per the physical constraints of wireless communication. Notice that it does not require the addition of new hardware or modification of the standard WAVE OBU, because the message Tx power can be explicitly controlled by application in WAVE short message protocol (WSMP) [30].

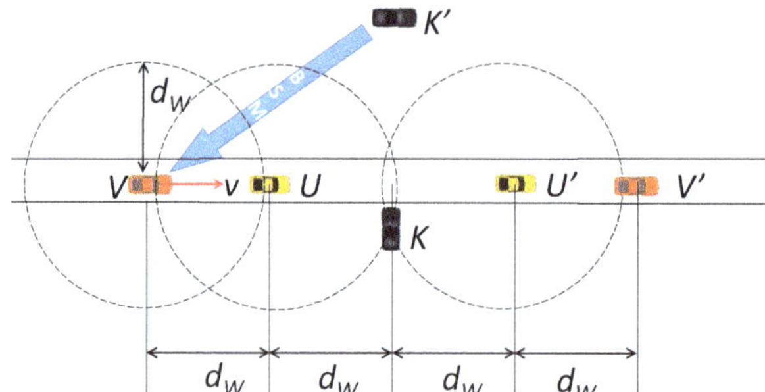

Figure 1. Low-power beacon checking setup.

In our solution approach, each vehicle broadcasts special beacons carrying the sender's randomly chosen identifier at much lower Tx power, in addition to BSM beacons. For convenience, we will call this special beacon "whisper" in the rest of this paper. Only if a neighbor echoes the received whisper identifier in its own whispers, the neighbor is trusted. In Figure 1, the remote attacker K' does not hear these whispers of V's, as it is not within d_W from V. Thus K' cannot include V's identifier in its whispers, so V rejects the BSMs from K' as suspicious. For an attacker to mount the false position attack, therefore, it must come within the distance d_W to the roadside. Moreover, it should exchange whispers with a passing vehicle (e.g., U) to enable an attack. For example, the attacker K in Figure 1 has a chance to mount the attack. Even in this case, the window of attack is limited to $\pm 2\, d_W$ in principle. So, with a sufficiently small d_W, we can drastically reduce the attack window for K. Furthermore, we can introduce an extension to this baseline scheme to further limit the attackers that come within d_W from the roadside, which we will discuss in Section 3.6. Recollect that the purpose of limiting the Tx power of the whispers is to let only the neighbor vehicles in close proximity, but not a remote attacker, hear them. Finally, a desirable fallout from using a small Tx power for whispers is that it helps suppress the increase of the channel utilization due to this additional security measure to a small value. Even with this added security mechanism, therefore, we can still keep a larger chunk of the channel bandwidth for ordinary BSMs. Below, we discuss the details of the sketched idea.

3.2. Attacker Model

Before delving into the details of the proposed scheme, we specify the adversary model as follows.

- The attacker is stationary and located on the roadside [6,11], which will be the most frequent attack scene. The stationary attacker is identified in Leinmüller et al. [6] as the most threatening attack for its low complexity of implementation and execution.
- Before transmission, the attacker can tweak the data obtained from sensors or from the in-vehicle networks such as the controller area network (CAN) bus.
- The attacker has a valid security credential [11]. So, the attacker can correctly encode invalid position data.
- The attacker can increase the Tx power or use directional transmission to affect farther vehicles.

Note that the neighbor check using whispers is focused on the safety-related events that take place in the local neighborhood of each vehicle. To be precise, the local neighborhood is the two-hop range of the whispers ($= 2\, d_W$). Thus the check completely prevents the attackers from placing "ghost vehicles" unless the attacker comes into the two-hop distance of the whispers. Then for the attackers that are indeed within the two-hop distance, we can develop an extension of the whisper scheme to further remove the attacks. For most safety-critical events, the attack prevention within $2\, d_W$ will be sufficient, as they take place in close proximity, e.g., forward collision. There will be safety applications where vehicles need to heed messages from long distances, but they are beyond the scope of this paper.

3.3. Neighbor Check through Low-Power Beaconing

In the WAVE framework, every vehicle periodically transmits a beacon called BSM, at up to 10 Hz. Such safety beaconing is similarly performed in C-V2X as well, with two more higher rates (20 Hz and 50 Hz) [31]. In this paper, we additionally require that every vehicle V transmit special beacons called whispers, denoted by W_V, that carry the following information:

- I_V: whisper identifier (WID) of V
- $L_V = \{I_x | x \in N_V^t\}$: list of WIDs heard by V, where N_V^t is one-hop neighbors that passed the neighborhood check using whispers
- $dig(C_V)$: digest of V's certificate [1]

The whisper identifier (WID) I_V is randomly chosen by each vehicle, and changed every update interval t_u, much more frequently than the pseudonym change [24]. (the WID should also change upon the pseudonym replacement as stipulated by SAE J2945 [24], as well as every t_u.) Otherwise, once the attacker learns of I_V, it would be able to prove itself as a close neighbor of V by using I_V. Then the attacker could attack V with false information as long as its forged BSM can reach V. By changing I_V frequently, however, the vehicle V can make it hard for the attackers that once learned of I_V to attack V later in time.

Every vehicle V puts in L_V all whispered neighbor identifiers I_x that heard and passed its "whisper check", and rebroadcasts them in its own whisper W_V. The whisper check refers to the following test: if a receiver U of W_V finds its whisper identifier $I_U \in L_V$, U trusts the content in subsequent beacons BSM_V to come from a close neighbor V. If a neighbor vehicle fails the whisper check, on the other hand, its whisper identifier (WID) is neither stored nor rebroadcast. The one-hop whisper neighborhood N_V of V is defined by the maximum distance from which a whisper reaches V. $N_V^t \subseteq N_V$ is the subset of the one-hop neighborhood that passed the whisper check.

Each vehicle V stores the binding $(I_x, dig(C_x))$ for each neighbor $x \in N_V^t$. This is to prevent impersonation attacks. Suppose V has received a whisper from U and confirmed $I_V \in L_U$. At this moment, the binding $(I_U, dig(C_U))$ is made at V. Once the binding is made, K cannot impersonate U because it does not have the private key of U to sign the BSM that is validated only by C_U. On the other hand, suppose that K hears the whisper and the BSM from U before the binding $(I_U, dig(C_U))$ is made at V. In this case, K can attempt to pretend to be U and make a biding $(I_U, dig(C_K))$ at V. However, U is not (yet) a trusted neighbor of V, so the whisper W_U does not have I_V. Therefore, K could not prove that it is a one-hop whisper neighbor of V. When BSM_K reaches V, V will reject this forged BSM.

Algorithm 1 shows the description of the proposed whisper checking logic. Line 3 is the whisper check. If the received whisper fails this test, the sender is determined to be unverifiable as a neighbor (line 16). If the whisper check succeeds, the whisper ID is used as a key to find the bound certificate (line 4). If none, we create one to prevent the impersonation attack (line 5). If there is a binding, however, the WID-certificate binding is checked against the values in the received whisper (line 9). If they match, it is a confirmation of both the neighbor relation and the WID-certificate mapping. The trust is strengthened as the credit for the received whisper is incremented (line 10). This information will be later used in an enhanced version of the base algorithm discussed here. If the binding and the whisper say otherwise, the whisper is ignored (line 12). Note that the binding should be newly created when either the pseudonym change or WID change takes place.

Algorithm 1 Whisper check at U.

1: **procedure** WHISPER-CHECK(W_V)   ▷ U received whisper W_V from V
2:   Extract L_V, I_V, and $dig(C_V)$ from W_V
3:   **if** $I_U \in L_V$ **then**   ▷ Did V hear my whisper?
4:    **if** $B_U(I_V) == \varnothing$ **then**   ▷ B is the WID-certificate binding set; No binding exists for V yet
5:     $B_U(I_V) \leftarrow dig(C_V)$ ▷ Make one for V; C_V is the link to BSM from V
6:     $L_U \leftarrow L_U \cup I_V$ ▷ Store V's whisper ID (WID) for rebroadcast
7:
8:    **else**   ▷ Binding exists for I_V
9:     **if** $B_U(I_V) == dig(C_V)$ **then**   ▷ Binding confirmed?
10:      $C(I_V)$++   ▷ Credit up
11:     **else**   ▷ Conflict
12:      return   ▷ Ignore this whisper
13:     **end if**
14:    **end if**
15:   **else**
16:    return   ▷ Ignore this unverifiable neighbor
17:   **end if**
18: **end procedure**

3.4. Range Extension

One obvious question will be about the appropriate value of the whisper range d_W. It should depend on applications that require contents plausibility checks. What if an application requires a wider checking range? One may argue that we could increase the Tx power of whispers to extend the range, but it would increase the channel utilization as well. As a consequence, it can trigger the BSM congestion control [24] so that fewer BSMs are transmitted and the vehicle position tracking error increases. In essence, it can have safety ramifications. In order to extend the range of whisper check to two hops without such side effects, we could take an approach similar to Parno et al. [5]. Namely, when W_V passes the whisper check at U, we can let U not only trust BSMs from V, but also those from the vehicles in L_V. So U can identify two-hop trusted neighbors $N_U^{t2} = \bigcup L_x$, where $x \in N_U^t$.

Unfortunately, under the two-hop whisper check, when the distance d_K between the attacker's true position and the message receiving vehicles is less than d_W, the attacker can also exploit the overheard whisper identifiers (WIDs) to extend the attack range. Figure 2 depicts the situation. Here, the attacker K can obtain I_V from L_A, for $A \in N_V^t$ and $d(A, K) \leq d_W$. Echoing this overheard I_V in its whisper W_K, the attacker can extend its attack range from d_W to $2d_W$. Not only that, after V passes K, a similar situation can happen if there is another intermediate vehicle (e.g., B) between the attacker and the victim.

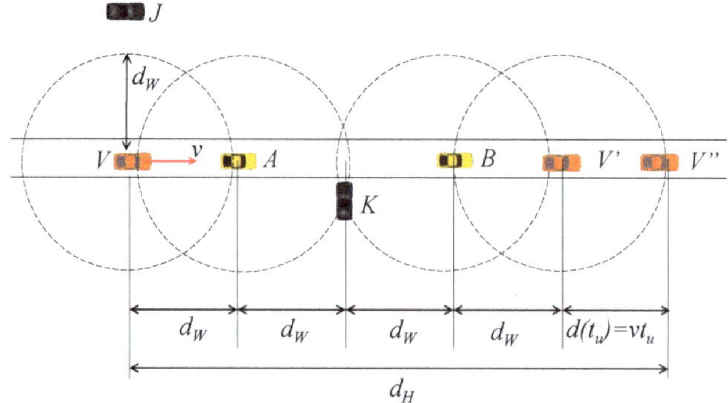

Figure 2. Distances in credit-based whisper.

Furthermore, there are two more factors that can extend the attacker's reach even beyond $2d_W$. The first is the whisper ID update. Recollect that the WID is changed every t_u (see Section 3.3). If vehicle V moves at speed v, it can add as much as vt_u to d_K because V can continue to use the same I_v even after moving more than $2\,d_W$ away from K, before the next update instant. The second factor is the message gap between consecutive whispers. Suppose V changes its WID from I_v to I_v'. Since the update will be reflected in the next whisper to transmit, a neighbor U can send its BSM containing the old WID I_v. If V performs the whisper check against its new WID I_v', this legitimate BSM from U will be filtered. Therefore, before whispering with the new WID, each vehicle should accept BSMs with its old WID. Assuming the whispering rate of c Hz, neighbors BSM with the old WID can arrive as late as $1/c$ second after the update, which we reflect to the attacker's overhearing range calculation. In total, the attacker can attempt to deceive a vehicle at distances

$$d_X \leq 4 \cdot d_W + v \cdot (t_u + 1/c),, \tag{1}$$

where d_X is the maximum distance that the attacker can exploit V's WID. We stress that the extended attack range d_X is not where the attacker can mount the attack from. The attacker should still be physically within d_W of the road to overhear the whispers. In this paper, however, we shun away from the multi-hop whispering possibilities due to the potential to extend the attacker's capability. We leave it as future work and focus on exploring the efficacy of the single-hop whispering.

Note that the changed mobility model can affect the attack distance d_X. In particular, the increased vehicle speed v affects Equation (1), extending the attack distance. In this case, according to the same equation, more rapidly changing the whisper ID (namely, smaller t_u) can be used to counter it.

3.5. Whispering Rate and Tx Power

Lowering the Tx power of whispers helps filter the messages from remote roadside attackers and reduce the channel utilization incurred by the whispers. But the Tx power set too low can render the BMS from even close neighbors not trusted. Therefore, we should find an appropriate Tx power. As to the channel utilization, another parameter that affects it is the whispering rate. To find appropriate whispering Tx power and frequency to be used for the rest of this paper, let us consider two highway driving scenarios depicted in Figure 3. There are four lanes on the simulated road. The BSM Tx power is set to 23 dBm, and the messaging rate to 10 Hz. On the other hand, the Tx power of whispers is varied between 7 and 10 dBm, and the messaging rate between 6 and 8 Hz. The channel model is set to two-ray ground. The physical layer transmission rate is set to 6 Mbps for the most robust delivery [32].

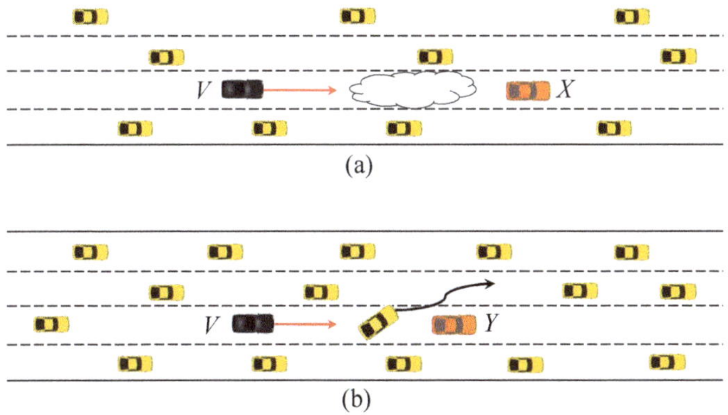

Figure 3. Highway driving simulation scenarios (**a**) poor visibility (**b**) blocked line of sight.

In (a), vehicles move at 80 km/h in the rainy road condition, where the road surface friction coefficient is 0.3 [33]. The headway distance between vehicles on the same lane is set to 100 m. In this case, the host vehicle V cannot detect the stopped vehicle X due to poor visibility. In (b), vehicles move at 120 km/h on dry road where the friction coefficient is 0.8. The headway distance between vehicles on the same lane is set to 50 m. There is a vehicle between V and the stopped vehicle X. The stopped vehicle warns approaching vehicles through BSMs. Other vehicles check the distance to the stationary vehicle at the moment of receiving the first BSM that passes the whisper check. In (b), the intervening vehicle changes lanes to evade Y. The human reaction time typically ranges from 0.7 s to 1.5 s [34]. In this paper, we set it to 1 s. When vehicles move at 80 km/h on a rainy road or 120 km/h on a dry road, the braking distances based on our assumptions are 104.5 m and 102.8 m, respectively. Given these parameters, we compute the distance d between the host vehicle and the stopped vehicle when a whisper checked the first BSM from the stopped vehicle arrives at the host vehicle. Based on d, Figure 4 shows the probabilities in the two scenarios that the host vehicle collides with the stopped vehicle as it could not stop in time. The collision depends on the message delivery loss ratio and latency experienced by the BSM from the stopped vehicle. Now, given the success probability of the forged message delivery, the vehicle collision takes place if

$$d + v_f \cdot \frac{v}{|a|} < v \cdot \left(1 + \frac{v}{|a|}\right) + \frac{v^2}{2a},$$

where d is the headway distance to the front vehicle U, v_f is the speed of U, v is the speed of the host vehicle V, a is the maximum deceleration.

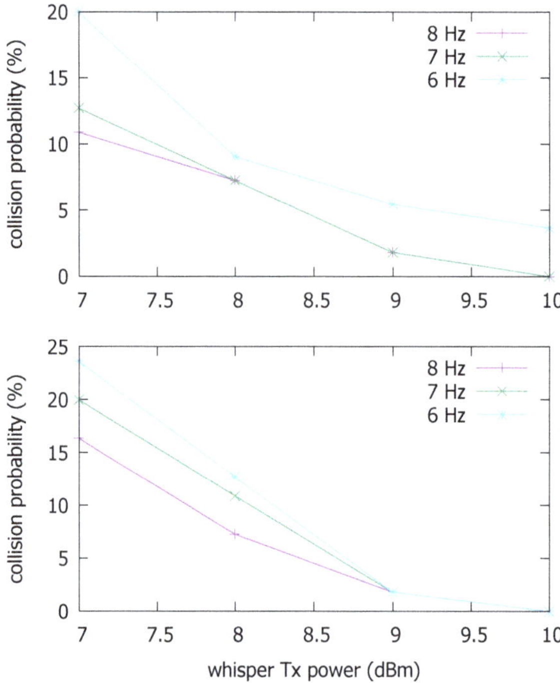

Figure 4. Vehicle collision probabilities as functions of messaging rate and Tx power

Let d_{ca} denote the distance between the BSM sender and the receiver at which the collision probability is less than 5%. In (a), the message reception is possible at distances $d > d_{ca}$ at 9 dBm and 7 or 8 Hz. In (b), all three messaging rates qualify at 9 dBm. For this reason, we will assume in the rest of this paper that the whisper Tx power is set to 9 dBm, and the whispering frequency to 7 Hz. Note that these numbers are only used as rough guidelines to demonstrate the potential of the proposed approach. If need be, they will be refined in future work.

3.6. Further Check Based on Credit

An attacker within d_W of the roadside can still hear the whispers of the target vehicles and mount the attack. In order to cope with such attackers, we can extend the whisper scheme by incorporating the notion of credit. In this extension, each vehicle V tracks the credit $C_V(U)$ for each neighbor vehicle U. Without further sophistication, we simply let the credit increase by one point per each passed whisper test and decrease by one point each second. In Algorithm 1, we showed how the credit for each neighbor is increased. To prevent the attacker from arbitrarily inflating the credit, the increase for each neighbor can be bounded by the whispering frequency f_W, so that the reception of whispers beyond f_W does not add to the credit. Assuming the minimum whisper rate cannot fall below 1 Hz, legitimate neighbors within d_W of V will maintain a non-zero credit at V. Given the whispering frequency of $f_W = 7$ Hz, for instance, the maximum credit that one neighbor can accumulate per second is $7 - 1 = 6$. Based on this observation, we can set the credit threshold over which we can trust a given neighbor vehicle at V as

$$\theta_V = (f_W - 1) \cdot d_X / v(V),, \tag{2}$$

where $v(\cdot)$ is the speed of the given vehicle. The threshold θ is essentially the credit that a roadside attacker can maximally accumulate at V while V travels a distance d_X at v, or equivalently, during $d_X/v(V)$ seconds. Here, d_X is the maximum distance that a vehicle can use the same whisper ID, as in Equation (1). To show how long a neighbor vehicle should travel with a host vehicle to be considered credible, Figure 5 plots $d_X/v(V)$ as a function of t_u and v for $d_W = 170$ m.

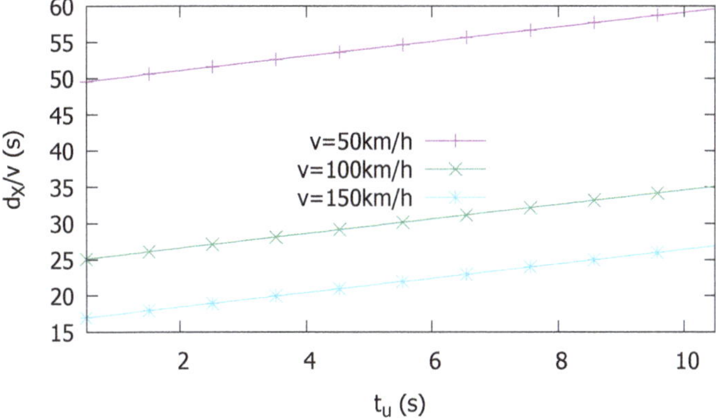

Figure 5. Elapsed time required for trust vs. whisper time allowed for the attacker, $d_W = 170$ m.

A single-hop whisper check will roughly halve θ. Therefore, in reality, the time that neighbor vehicles should be in the communication range of a host vehicle to be trusted can be small. The credit-based enhancement ascertains $C_V(U) > \theta_V$ for V to trust the message from U. Namely, if the neighboring vehicle U can accumulate more than a roadside attacker maximally can, V trusts U.

Obviously, there is a risk of completely ignoring the legitimate vehicles whose credit falls short of θ, e.g., due to message losses arising from adverse channel conditions. Or, a new vehicle can join the traffic from a junction. But Figure 5 is a very conservative estimate considering that we define d_W to be the maximum distance at which the whisper reception probability is non-zero. The whisper ID can change much faster, as long as the binding with the certificate digest is maintained. Using small t_u reduces θ, helping vehicles within the whisper range d_W quickly exceed the threshold for each other.

Using smaller d_W would reduce the threshold even further, especially when v is small, reducing d_W by using smaller Tx powers can be useful because the smaller distance is driven in a given time than in high-speed driving so that credibility checking may be more focused on closer neighborhood.

4. Attack Filtering Performance

In this section, we investigate the attack filtering performance of the proposed scheme, by simulating a highway driving scenario. Figure 6 depicts the attack situation used in simulation. Vehicles are moving on a 4-lane highway at the same speed of 120 km/h. The headway distance between vehicles on the same lane is 33.3 m, or equivalently, a 1 s gap at the given speed. The whisper size is the sum of the whisper ID (2 bytes), the digest of the certificate (8 bytes) [1], and the list of received whisper IDs (at most 90 × 2 bytes). The BSM size 80 bytes for the message and 125 bytes for the certificate, so it is 201 bytes. The whisper and BSM are transmitted as a WSMP message in IEEE 802.11 frame, and the lower layer overhead is an additional 80 bytes. In this paper, we assume that BSMs are transmitted at 10 Hz, and whispers at 7 Hz. Note that at such whispering rate, legitimate vehicles that fail to deliver some of the whispers due to the poor channel condition will succeed within seconds at most, and assert their neighborship without being suspected for long. The power of BSMs is 23 dBm, and whispers, 9 dBm. The path loss model is two-ray ground, and the fading model is Rician with $k = 3$. We perform the simulation using the Qualnet simulator, and the simulation configuration is summarized in Table 1. In the given situation, we consider the attack successful at a vehicle V if V hears BSM_K with a false position information (i.e., that of $K' \neq K$) and BSM_K passes V's whisper test. Then, we count through simulation the number of vehicles that are successfully attacked for varying distances of attacker to the road, d_K.

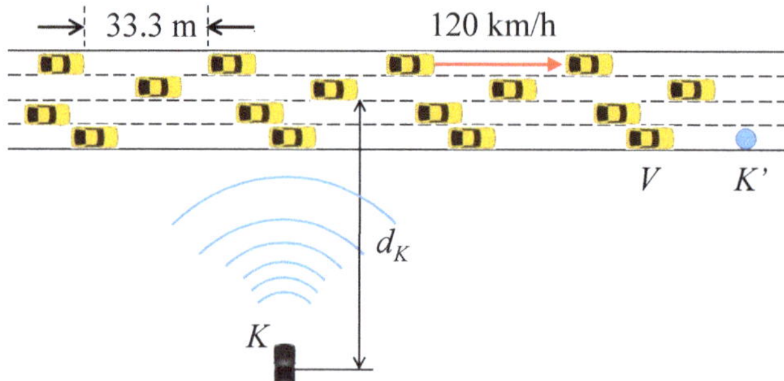

Figure 6. Attack scenario.

Table 1. Simulation parameters.

Parameter	Value	Explanation
P_W	9 dBm	Whisper Tx power
f_W	7 Hz	Whisper frequency
L_W	≤194 B	Whisper size
f_B	10 Hz	BSM frequency
P_B	23 dBm	BSM Tx power
L_B	205 B	BSM size
v	120 km/h	Vehicle speed
d_I	33.3 m	Headway distance
Path loss	Two-ray ground	
Fading	Rician ($K = 3$)	

4.1. Attack Mitigation Performance

As discussed above, if $d_K > d_W$, then $I_v \in L_K$, so V can filter K's BSM. This holds true even if K increases the BSM transmission power or uses directional transmission to extend its transmission range towards farther victim vehicles. But as V comes closer so that its whisper can be decoded by K, then K's BSM can pass V's test. Figure 7 shows the number of successful attacks as a function of d_K. Namely, it plots how far from the road center the attacker can successfully mount the attack. Although the number of vehicles that fall to the attack is also a function of the vehicle traffic density, vehicle speed, and the duration of the attack, we fix them as in the previous section and focus on the effect of d_K. Without the whisper check, the attacker can successfully achieve the attack from as far as $d_K > 600$ m at the BSM Tx power of 23 dBm. With Tx power-boosting or directional transmission, d_K could be even larger. This result suggests that with good line-of-sight (LoS), the attacker may well position himself safely apart from the highway, avoid visual detection by the passing victims, and still pose a significant threat. The reason that the number of vehicles exposed to the attack is approximately 3.5 inside the attack range is because we have a headway distance of 33.3 m between vehicles. The length of a vehicle is 5 m, so with four lanes, approximately 3.5 vehicles enter the attack range every second. With the whisper check, however, the attack enabling distance d_K is significantly reduced to $d_W \approx 170$ m. The result confirms that the whisper check is effective in narrowing the attack range to d_W. So, at least, the attacker should come near to the roadside in order to mount the false position attack under the whisper scheme.

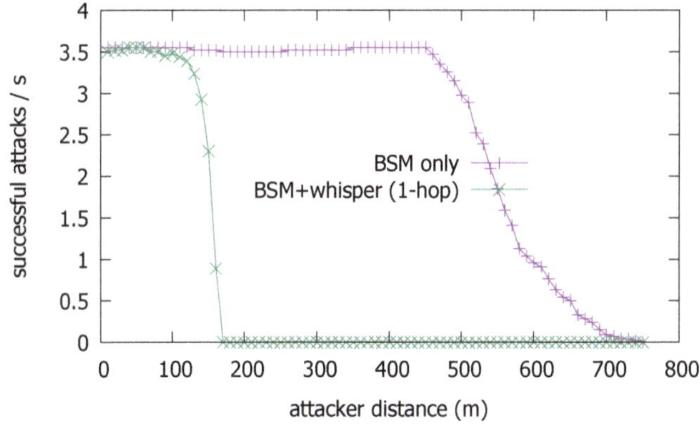

Figure 7. Successful attacks with basic safety message (BSM) only and with BSM + whisper.

4.2. Channel Utilization Increase

Additional whispering activity inevitably increases channel utilization. If excessive, it could even jeopardize the more important safety message exchanges such as BSM. So, we also measure the channel used in the simulation to check on this possibility. Figure 8 shows the channel busy percentage (CBP) as a result of using whispers. We notice that the increase in CBP remains at 2% to 3% in each of the considered variations of Tx power and frequency. Even if the whisper messaging rate is 60% to 80% of the BSM beaconing rate, the increase in CBP is not as significant due to the smaller power at which the whisper messages are transmitted. So, the low-power beaconing scheme only slightly increases the CBP, and it can be done without excessively disturbing the ordinary beacon exchanges.

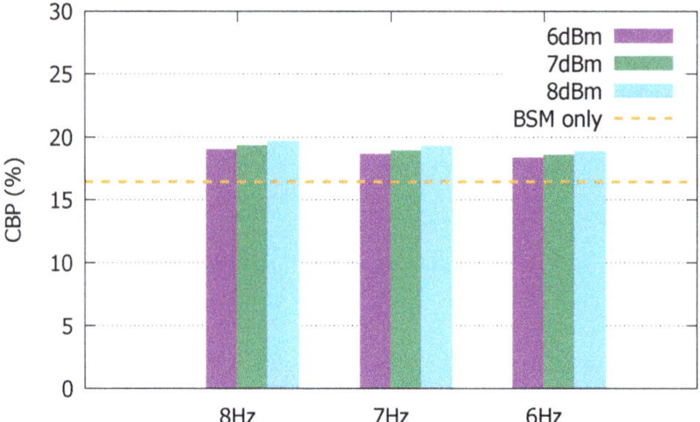

Figure 8. Channel busy percentage (CBP) increase due to whispers.

4.3. Effectiveness of Credit-Based Enhancement

To check the efficacy of the credit-based enhancement, we repeat the simulation for Figure 6 with the employment of the credit-based check. Figure 9 shows the result.

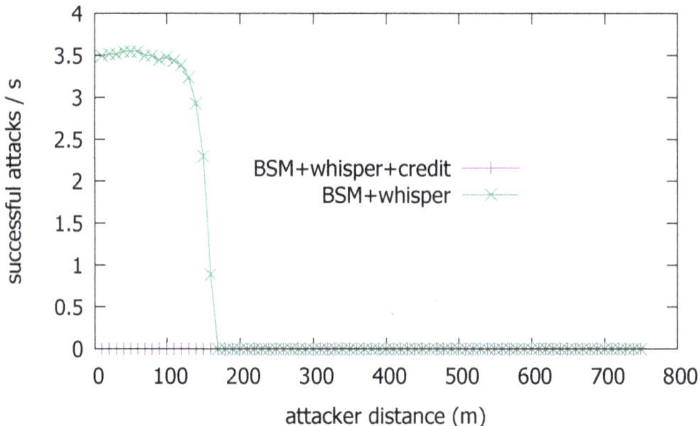

Figure 9. Attack success probability with additional credit-based check.

We see from the figure that the additional credit-based check completely shuts out the road-side attacker, as it fails to accumulate enough credit, just as intended by Equation (2). A crucial observation in the issue of position plausibility check is that the risk of vehicle collisions is higher between those that are in close proximity. Therefore, it is imperative that the nearby vehicle positions should be ascertained more than those of farther vehicles. It is why the SAE J2945/1 standard stipulates that only the vehicles within 100 m of the ego vehicle be tracked [24]. From this viewpoint, the credit-based double check is highly recommended due to its usefulness in checking the positions within short, safety-critical distances, d_W in particular.

We can summarize the significance of the proposed scheme as follows. In connected/autonomous cars, V2X is no longer an option, but a mandatory component that allows vehicles to sense larger distances and non-line-of-sight (NLoS) situations. Without the proposed solution, the attackers can safely position themselves at a significantly larger distance from the victims (connected/autonomous cars). The hardware sensors such as cameras, radars, and lidar are all line-of-sight (LoS) sensors, and their coverage is limited. Therefore, with only the sensors, we could not cope with the attacks that use longer range and non-LoS technology that is the V2X. By using a smaller Tx power to make d_W narrow, we can create an opportunity for the cooperation between the LoS sensors and the communication for the message contents validation.

5. Application to Platoon Protection

As traffic and cargo volume increase worldwide, many researchers have studied how to reduce traffic congestion and to carry cargo efficiently through truck platooning. In 2011, California PATH [35] team conducted experiments with three heavy trucks that have only an automated longitudinal control and confirmed the improvement of fuel consumption by about 10% on the average. Energy ITS [36] team performed tests with three automated heavy trucks and one light truck with the gaps of 10 m and 4.7 m at 80 km/h in 2013. Currently, there are numerous others seeking even higher efficiency goals [37]. These project teams conducted their research using vehicle sensors and vehicle-to-vehicle (V2V) communication, and they demonstrated that platooning can improve energy [38] and traffic efficiency [39]. Like this, as platooning is one of the most promising applications of V2X [9], and one of the prominent target for roadside remote attackers when V2X becomes available [5], it is imperative to explore how the proposed scheme can help mitigate attacks against platoons.

5.1. Problem Formulation

In platoons, each vehicle can utilize V2V communication to obtain information about vehicles that are in non-line-of-sight (NLoS) points that cannot be detected by its sensors. This is also essential for all platooning vehicles to use the platoon leader's driving information. For our purpose, it is considerably difficult to define the platoon formation and the surrounding traffic in a completely generic manner. Thus in this paper, we consider the scenario as depicted in Figure 10, where the autonomous platooning and non-platooning vehicles move together on a road, and they are subjected to a roadside attack from K.

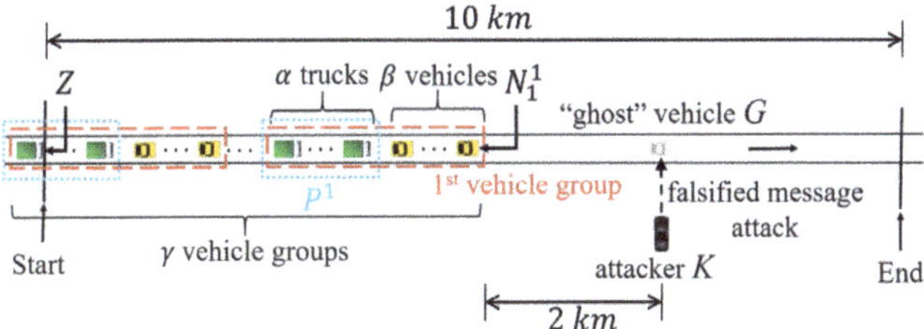

Figure 10. Simulated platooning scenario.

Since the platooning is dangerous to perform in urban roads, the highway environment is usually assumed. Figure 10 is the platooning scenario in the highway environment. Following the 3GPP TR 36.885 (Annex A) suggests for evaluation scenarios [2], we used a straight road longer than 2 km. Unlike the 36.885 that suggests 2.5 s gap between vehicles that run at 70 km/h, however, all non-platoon vehicles in our paper try to keep at 80 km/h with 1 s gap (approximately 22 m) and platoon vehicles, 10 m.

In this formulation, there are γ vehicle groups on a 10 km stretch of a lane. Each vehicle group is composed of $\alpha + \beta$ vehicles, where α and β are the numbers of platooning vehicles and non-platooning vehicles, respectively. It models the driving situation from the viewpoint of platoons where it is behind a non-platoon vehicle population in a lane. As the platoon uses a smaller inter-vehicle gap than non-platoon vehicles, the false position attack from K is a more grave safety threat to the platoon vehicles. As above, all vehicles exchange BSM at 23 dBm, but for whispers, we will assume here that they use the Tx power of 9 dBm, slightly higher than in the previous sections. As a result, d_W is 170 m in our simulation setting while BSM propagation dwindles in power till 600 m beyond which the packet delivery is hardly possible.

Let N_j^i be the ith non-platooning vehicle of the jth group, and P^k denote the kth platoon. For the string stability of the platoons, we model into the simulator the Rajamani controller [40] that computes the desired acceleration of each vehicle in the platoon. For this, we assume that the platoon vehicles receive the movement data of both the preceding vehicle and the platoon leader through V2V communication, whereas non-platoon vehicles receive only those of the preceding vehicle. We set the intra-platoon safety gap, namely the required minimum distance between two consecutive vehicles in the platoon, to 10 m as assumed in many works of literature. The length of a vehicle is 5 m, and the safety gap between non-platooning vehicles or between a non-platooning vehicle and the immediately following platoon leader is one second headway time at 80 km/h, which is approximately 22 m. Although presenting the case where collisions occur will obviously be more dramatic, we set the safety gap to a large value because we want to show that even if there is no collision, the attack can still affect the comfort and the efficiency of the vehicle traffic running with platoons.

We assume that all platoon vehicles try to keep their speed at 80 km/h. In contrast, non-platoon vehicles can use higher speeds when it wants to close the gap with a preceding vehicle. The maximum acceleration and deceleration of both types of vehicles are assumed to be 3 m/s^2 and -5 m/s^2, respectively [41]. All vehicles can use their sensors to detect the obstacles within 150 m of the line-of-sight (LoS). Notice that N_1^1 has no obstruction in front, so it can easily recognize obstacles in the front. However, the front view of its followers are blocked by the preceding vehicles, making it difficult to detect obstacles. An attacker K is situated at d_W from the roadside. Although a very remote attacker can be more easily rejected as explained above, by putting the attacker within the hearing range of the whisper, we want to stress-test the proposed scheme. K broadcasts forged messages announcing a ghost vehicle G, which is not actually present, at a distance of 2 km ahead of N_1^1. K continuously broadcasts forged messages, causing the vehicles to brake when the vehicle groups come in the range of its BSM transmission.

5.2. Effect of Whisper Check With Platoons

We first measure the time it takes for the last vehicle Z of the entire vehicle groups to pass through the endpoint of the 10 km stretch, while varying α, β and γ ($5 \leq \alpha \leq 10, 1 \leq \beta, \gamma \leq 10$). It will reveal any accumulated delay effect if the vehicles in front are affected by the forged message attacks. Specifically, without the whisper check, each vehicle with a block front view will believe that G is actually present at the forged coordinate. If the vehicle determines that there is a risk of colliding with G, it will activate its brake. For this, we assume that each vehicle brakes at 30% of the maximum deceleration when the time to collision (TTC) with the preceding vehicle is 2.6 s, and at the maximum deceleration when the TTC is 1 s to prevent collision [42]. Each vehicle activates the brake until its LoS sensors finally perceive that G is not actually present.

When the whisper check is not employed, Figure 11 shows the velocity changes of some of the non-platoon vehicles in the first vehicle group under the forged message attack that a ghost vehicle is at G. We assume $\alpha = \beta = 10$ in this example.

Since N_1^1 has the LoS for the ghost vehicle G, it can use its sensors to perceive that K's messages are forged and keeps going at the same velocity as before without braking. But as for the other vehicles in the group, they brake to slow down due to the absence of LoS to G. So, N_1^2 begins to brake at 30% of the maximum braking when its TTC to G reaches 2.6 s. But as soon as N_1^1 passes the forged position, the LoS to the forged coordinate is clear for N_1^2. Its sensors detect that G does not exist, so accelerates again to reach the target safe distance with N_1^1 (at around $t = 90$ s in Figure 11). Notice that the velocity of the following vehicles more severely decreases towards the end of the non-platoon vehicle column. Due to the slowed vehicles in front, the latter vehicles spend more time until they reach G, when their sensors find the absence of the ghost vehicle. Some middle vehicles are not shown for readability, but this accumulated slowdown effect is amplified to a very large and elongated velocity instability as it propagates to the last non-platoon vehicle N_1^{10}. From the velocity changes of N_1^{10}, we observe in the simulation data that its acceleration and deceleration occurs as many as 60 times, potentially causing discomfort to the passengers.

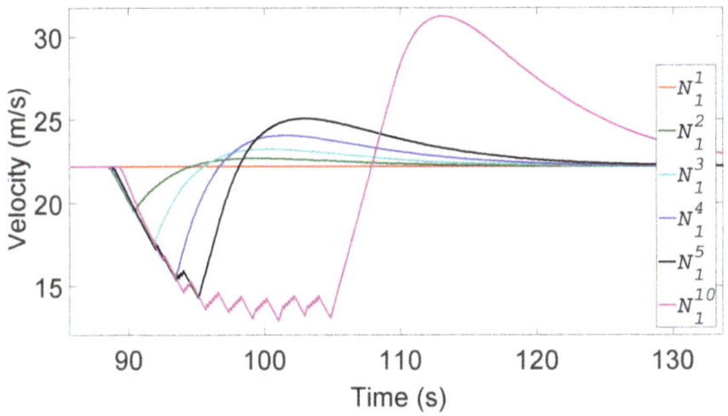

Figure 11. Velocity changes of the non-platoon vehicles in the first vehicle group without forged message filtering.

The velocity fluctuation in the non-platoon vehicles directly affects the immediately following platoon vehicles. The leader of the platoon P^1 that immediately follows N_1^{10} has to slow down due to the attack and then increases its speed to the original speed only after its other sensors check the absence of G. However, unlike the non-platoon vehicles, the platoon members of P^1 are controlled by the leader, so at the command of the leader, they move at the same speed. Recollect that the platoon members are prohibited from accelerating to more than 80 km/h. The other platoons $P^2, \ldots, P^{10}$ experience similar dynamics. Furthermore, the speed of P^i decreases more than that of P^{i-1} as the effect accumulates. Figure 12 shows the velocity changes of some of the platoons. In general, P^i suffers i episodes of significant velocity fluctuation due to the remote attack. We observe that the latter platoons, most notably P^{10}, suffer from elongated and repeated instability.

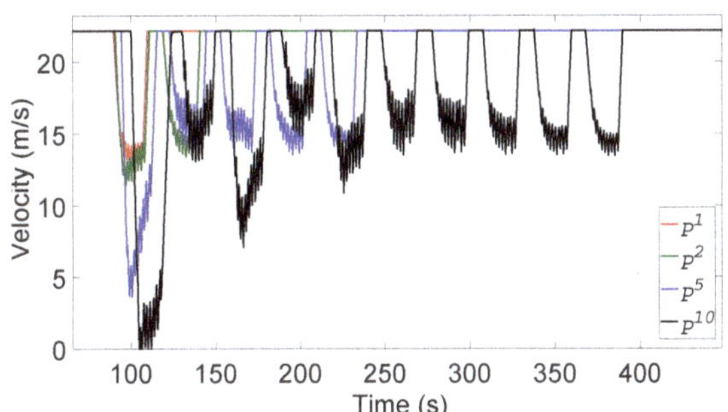

Figure 12. Velocity changes of platoons without forged message filtering.

Using the whisper check can drastically reduce the undesirable impacts of the remote forged position attack such as exposed by Figures 11 and 12. Figure 13 sheds light on these impacts of whisper checks from yet another angle, the distance loss. In particular, it shows the distance lost by the last vehicle Z compared with the case in which vehicles employ the whisper checks, under various values of β and γ and $\alpha = 10$. We fix the value of α because the platoon followers travel at the same speed as the platoon leader and had little effect on the velocity change of Z. When all vehicles use the whisper check, Z travels 10 km at a velocity of 80 km/h without any distance loss, regardless of β and γ values, because the vehicles successfully filtered out forged messages. In contrast, in the absence of the whisper check, the average velocity of Z decreases as β and γ increase, and there is lost distance. For instance, for $\alpha = \beta = \gamma = 10$, the average velocity of Z while moving 10 km dropped to about 52 km/h, far below 80 km/h with the whisper check in play.

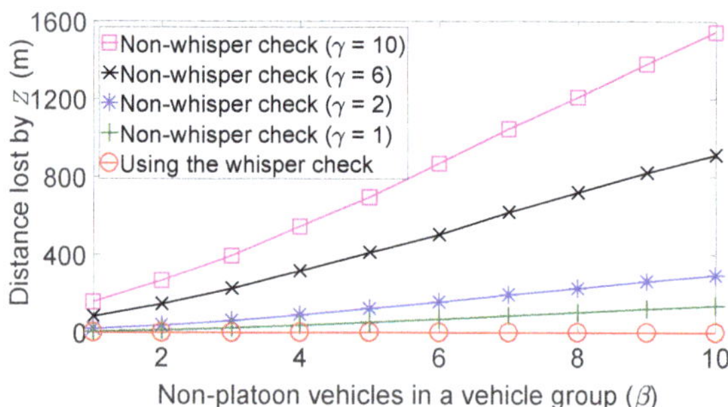

Figure 13. Velocity changes of platoons without forged message filtering.

Although we set the inter-vehicle distance to sufficiently large values so that the roadside attack did not cause direct collisions, the loss of average speed and the potential discomfort due to repeated and wild fluctuations of the vehicle speed would be enough to annoy the vehicle riders. If the inter-vehicle distance were lower for higher road throughput or fuel-efficiency, it would increase the probability of more serious events.

6. Conclusions

In this paper, we explore a Tx power-based communication constraint check that can filter remote attacks that aims to disseminate false position information to running vehicles. By using low-power beacons, vehicles can mutually check if the BSM hence the position information therein indeed comes from a physically close neighbor. At least, it would pressure an attacker to come close within the low-power transmission range of the victim vehicles to mount an effective attack. In case the attacker indeed comes in close range, the on-board hardware sensors such as radars, lidar, and cameras with a typically smaller range than V2X can kick in to validate the claimed position.

Through extensive simulation, we demonstrated that there is value in using the low-power beacon exchanges between vehicles in preventing the harmful impacts from remote false position attacks through V2X communication. Specifically, we confirm that traffic efficiency and comfort of platooning may be decreased due to the remote attack. We show, however, that if we employ the low-power beaconing message check to platooning, we can successfully cope with forged message attacks and can overcome the problem. The additional bandwidth cost is small thanks to the low Tx power, and the Tx power reduction for the additional beacons is easily implementable within the current V2X standard frameworks.

Author Contributions: H.K. conceived and designed the experiments; T.K. performed the experiments and analyzed the data; H.K. wrote the paper.

Funding: This work is supported by the Korea Agency for Infrastructure Technology Advancement(KAIA) grant funded by the Ministry of Land, Infrastructure and Transport (Grant 19CTAP-C151975-01).

Conflicts of Interest: The authors declare no conflict of interest. The founding sponsors had no role in the design of the study; in the collection, analyses, or interpretation of data; in the writing of the manuscript, and in the decision to publish the results.

References

1. IEEE 1609 WG. *IEEE Standard for Wireless Access in Vehicular Environments (WAVE) —Security Services for Applications and Management Messages*; IEEE Std 1609.2-2016; IEEE: Piscataway, NJ, USA, 2016.

2. 3G Partnership Project. *Study on LTE-Based V2X Services, TR 36.885 v14.0.0*; 3GPP: Sophia Antipolis, France, 2016.

3. SAE International. *Dedicated Short Range Communications (DSRC) Message Set Dictionary*; SAE: Warrendale, PA, USA, 2016.

4. IEEE 1609 WG. *IEEE Standard for Wireless Access in Vehicular Environments (WAVE)—Multi-Channel Operation*; IEEE Std 1609.4-2010; IEEE: Piscataway, NJ, USA, 2011.

5. Parno, B.; Perrig, A. Challenges in Securing Vehicular Networks. In Proceedings of the ACM HotNets, College Park, MD, USA, 14–15 November 2005.

6. Leinmüller, T.; Schmidt, R.; Held, A. Cooperative Position Verification—Defending Against Roadside Attackers 2.0. In Proceedings of the 17th ITS World Congress, Busan, Korea, 25–29 October 2010.

7. Gehring, O.; Fritz, H. Practical results of a longitudinal control concept for truck platooning with vehicle to vehicle communication. In Proceedings of the IEEE Conference on Intelligent Transportation System (ITSC), Boston, MA, USA, 9–12 November 1997.

8. Bergenhem, C.; Hedin, E.; Skarin, D. Vehicle-to-vehicle communication for a platooning system. *Procedia-Soc. Behav. Sci.* **2012**, *48*, 1222–1233. [CrossRef]

9. 3G Partnership Project. *5G; Service Requirements for Enhanced V2X Scenarios 3GPP TS 22.186 Version 16.2.0 Release 16*; 3GPP: Sophia Antipolis, France, 2019.

10. Blum, J.; Eskandarian, Z. The Threat of Intelligent Collisions. *IEEE IT Prof.* **2004**, *6*, 24–29. [CrossRef]

11. Stübing, H.; Firl, J.; Huss, S.A. A two-stage verification process for car-to-X mobility data based on path prediction and probabilistic maneuver recognition. In Proceedings of the IEEE Vehicular Networking Conference, Amsterdam, The Netherlands, 14–16 November 2011.

12. Raya, M.; Papadimitratos, P.; Aad, I.; Jungels, D.; Hubaux, J.-P. Eviction of misbehaving and faulty nodes in vehicular networks. *IEEE J. Sel. Areas Commun.* **2007**, *25*, 1557–1568. [CrossRef]

13. Van der Heijden, R.W.; Dietzel, S.; Leinmüller, T.; Kargl, F. Survey on Misbehavior Detection in Cooperative Intelligent Transportation Systems. *IEEE Commun. Surv. Tutor.* **2019**, *21*, 779–811. [CrossRef]

14. Yavvari, C.; Duric, Z.; Wijesekera, D. Vehicular dynamics based plausibility checking. In Proceedings of the IEEE ITSC, Yokohama, Japan, 16–19 October 2017.

15. Ghaleb, F.A.; Zainal, A.; Rassam, M.A.; Mohammed, F. An effective misbehavior detection model using artificial neural network for vehicular ad hoc network applications. In Proceedings of the IEEE Conference on Application, Information and Network Security (AINS), Sarawak, Malaysia, 13–14 November 2017.

16. Bißmeyer, N.; Mauthofer, S.; Bayarou, K.M.; Kargl, F. Assessment of node trustworthiness in vanets using data plausibility checks with particle filters. In Proceedings of the Vehicular Networking Conference (VNC), Seoul, Korea, 14–16 November 2012.

17. Schmidt, R.K.; Leinmüller, T.; Schoch, E.; Held, A.; Schäfer, G. Vehicle Behavior Analysis to Enhance Security in VANETs. In Proceedings of the V2VCOM, Eindhoven, The Netherlands, 3 June 2008.

18. Kim, T.H.J.; Studer, A.; Dubey, R.; Zhang, X.; Perrig, A.; Bai, F.; Bellur, B.; Iyer, A. VANet alert endorsement using multi-source filters. In Proceedings of the ACM VANET, Chicago, IL, USA, 24 September 2010.

19. Yan, G.; Olariu, S.; Weigle, M.C. Providing VANET security through active position detection. *Comput. Commun.* **2008**, *31*, 2883–2897. [CrossRef]

20. Sun, M.; Li, M.; Gerdes, R. A data trust framework for VANETs enabling false data detection and secure vehicle tracking. In Proceedings of the IEEE Conference on Communications and Network Security (CNS), Las Vegas, NV, USA, 9–11 October 2017.

21. LeBlanc, H.J.; Hassan, F.; Gomez, E.; Alsbou, N. Inter-vehicle communication assisted localization with resilience to false data injection attacks. In Proceedings of the ACM CarSYS, New York, NY, USA, 3–7 October 2016.

22. Kuk, S.; Kim, H.; Park, Y. Detecting False Position Attack in Vehicular Communications Using Angular Check. In Proceedings of the ACM Carsys, Snowbird, UT, USA, 16–20 October 2017.

23. Leinmüller, T.; Schoch, E.; Kargl, F.; Maihöfer, C. Improved security in geographic ad hoc routing through autonomous position verification. In Proceedings of the ACM VANET, Los Angeles, CA, USA, 29 September 2006.

24. SAE International. *Dedicated Short Range Communications (DSRC) Common Performance Requirements*; SAE: Warrendale, PA, USA, 2017.

25. Ruj, S.; Cavenaghi, M.A.; Huang, Z.; Nayak, A.; Stojmenovic, I. On Data-Centric Misbehavior Detection in VANETs. In Proceedings of the IEEE Vehicular Technology Conference (VTC Fall), San Francisco, CA, USA, 5–8 September 2011.

26. Jin, D.; Song, J. A traffic flow theory aided physical measurementbased sybil nodes detection mechanism in vehicular ad-hoc networks. In Proceedings of the IEEE/ACIS 13th International Conference on Computer and Information Science (ICIS), Taiyuan, China, 4–6 June 2014.

27. Bouassida, M.S.; Guette, G.; Shawky, M.; Ducourthial, B. Sybil nodes detection based on received signal strength variations within vanet. *Int. J. Netw. Secur.* **2009**, *9*, 22–33.

28. National Highway Traffic Safety Administration (NHTSA). *US DoT Advances Connected Vehicle Technology to Prevent Hundreds of Thousands of Crashes*; NHTSA: Washington, DC, USA, 2016.

29. Kim, T.; Kim, H. Vehicle-to-Vehicle Message Content Plausibility Check through Low-Power Beaconing. In Proceedings of the IEEE 86th Vehicular Technology Conference: VTC2017-Fall, Toronto, ON, Canada, 24–27 September 2017.

30. IEEE 1609 WG. *IEEE Standard for Wireless Access in Vehicular Environments (WAVE)—Networking Services*; IEEE Std 1609.3-2016; IEEE: Piscataway, NJ, USA, 2016.

31. 3G Partnership Project. *Evolved Universal Terrestrial Radio Access (E-UTRA); Medium Access Control (MAC) Protocol Specification (v15.7.0, Release 15)*; Technical Report 36.321; 3GPP: Sophia Antipolis, France, 2019.

32. Bai, F.; Stancil, D.D.; Krishnan, H. Toward understanding characteristics of dedicated short range communications (DSRC) from a perspective of vehicular network engineers. In Proceedings of the ACM MobiCom, Chicago, IL, USA, 20–24 September 2010.

33. Fricke, L.B. *Traffic Accident Reconstruction*, 2nd ed.; Northwestern University Traffic Institute: Kenosha: WI, USA, 1990.

34. Green, M. How long does it take to stop? Methodological analysis of driver perception-brake times. *Transp. Hum. Factors* **2000**, *2*, 195–216. [CrossRef]

35. Lu, X.-Y.; Shladover, S.E. *Automated Truck Platoon Control*; California PATH Research Report UCB-ITS-PRR-2011-1; University of California: Berkeley, CA, USA, 2011.

36. Tsugawa, S. Results and issues of an automated truck platoon within the energy ITS project. In Proceedings of the IEEE Intelligent Vehicles Symposium, Ypsilanti, MI, USA, 8–11 June 2014.

37. Eckhardt, J. *European Truck Platooning Challenge 2016-Creating Next Generation Mobility*; Storybook: The Hague, The Netherlands, 2016.

38. Liang, K.Y.; Mårtensson, J.; Johansson, K.H. Fuel-Saving Potentials of Platooning Evaluated through Sparse Heavy-Duty Vehicle Position Data. In Proceedings of the IEEE Intelligent Vehicles Symposium, Ypsilanti, MI, USA, 8–11 June 2014.

39. Ren, W.; Green, D. Continuous Platooning: A New Evolutionary Operating Concept for Automated highway Systems. In Proceedings of the IEEE American Control Conference, Baltimore, MD, USA, 29 June–1 July 1994.

40. Rajamani, R. *Vehicle Dynamics and Control*; Springer: New York, NY, USA, 2006.

41. Fernandes, P.; Nunes, U. Platooning with IVC-enabled autonomous vehicles: Strategies to mitigate communication delays, improve safety and traffic flow. *IEEE Trans. Intell. Transp. Syst.* **2012**, *13*, 91–106. [CrossRef]

42. Grover, C.; Knight, I.; Okoro, F.; Simmons, I.; Couper, G.; Massie, P.; Smith, B. *Automated Emergency Brake Systems: Technical Requirements, Costs and Benefits (TRL Published Project Report PPR 227)*; Transportation Research Library: Crowthorne, UK, 2008.

Article

Energy-Efficient Massive Data Dissemination through Vehicle Mobility in Smart Cities

Salman Naseer [1,2,*], **William Liu** [1] **and Nurul I Sarkar** [1]

[1] Department of Information Technology and Software Engineering, Auckland University of Technology, Auckland 1010, New Zealand; william.liu@aut.ac.nz (W.L.); nurul.sarkar@aut.ac.nz (N.I.S.)
[2] Department of Information Technology, University of the Punjab Gujranwala Campus, Gujranwala 52250, Pakistan
* Correspondence: salman.naseer@aut.ac.nz; Tel.: +64-22-689-7027

Received: 18 September 2019; Accepted: 29 October 2019; Published: 31 October 2019

Abstract: One of the main challenges of operating a smart city (SC) is collecting the massive data generated from multiple data sources (DSs) and transmitting them to the control units (CUs) for further data processing and analysis. These ever-increasing data demands require not only more and more capacity of the transmission channels but also results in resource over-provision to meet the resilience requirements, thus the unavoidable waste as a result of the data fluctuations throughout the day. In addition, the high energy consumption (EC) and carbon discharge from these data transmissions posing serious issues to the environment we live in. Therefore, to overcome the issues of intensive EC and carbon emission (CE) of massive data dissemination in SCs, we propose an energy-efficient and carbon reduction approach by using the daily mobility of the existing vehicles as an alternative communications channel to accommodate the data dissemination in SCs. To illustrate the effectiveness and efficiency of our approach, we take the Auckland City in New Zealand as an example, assuming massive data generated by various sources geographically scattered throughout the Auckland region, to the control centres located in the city. Results obtained show that our proposed approach can provide up to four times faster transferring the large volume of data by using the existing daily vehicles' mobility, than the conventional transmission network. Moreover, our proposed approach offers about 32% less EC and CE than that of conventional network transmission approach.

Keywords: smart city; delay tolerant network; infrastructure offloading; opportunistic network; vehicular mobility; energy consumption; carbon emission

---

1. Introduction

In the near future, SCs are envisioned to provide services such as road lights conversing with the smart grids, urban parks associating with administrations, seasides conveying cautions on pollution levels, and flood alerts to disaster management. This results in generation of huge data, and they are stored in the data clouds or some data control centers for processing and analysis. A huge number of sensors, installed in each city will continually generate a tremendous amount of data [1]. For instance, Westminster City Council has introduced solar waste bins that can speak to city council workers and inform them how much full they are. The framework used infrared and telemetry sensors and prompted a 60% decrease in cost for waste collection. Smart homes are now turning into reality, and it is expected that they will be available commercially by next year. Splunk predicts that one smart home today can create as much as 1 GB of data in a week. This means that all the UK smart homes may generate more than 26 million GBs of information [2]. Furthermore, video surveillance of entire city, smart sensors, and smart girds also generate big volume of data on each day for processing and analysis [3].

The transmission of a large amount of SCs data is going to swamp existing infrastructure networks, and it presents some unwieldy challenges. The operator's concern is to enhance the performance of their network by adding more capacity to their networks or by efficiently using the existing network resources. To address the issues of this explosion of data traffic we have different solutions. One of the expensive solutions is to elevate the existing networks to the next generation networks. Another expensive solution is to enhance the network capacity. However, the problem with these solutions is that it requires an enormous amount of cost for operational expense (OPEX) and capital expenditure (CAPEX) [4].

To transmit huge information in SCs, one of the candidate network is power line communication (PLC), it may provide LAN connectivity, WAN access, and some command and control capabilities [5]. However, interoperability problems, low dispersion and short-range communication of PLC networks become weak points for its success in the market of data transmission system [6]. Other energy-efficient solutions may include use of Zig-Bee, Bluetooth and WiFi along with PLC for smart metering and smart homes. Due to capacity and wireless range limitations, these solutions may be suitable for a short-range communication rather than long-distance communications [7]. Furthermore, fiber-optic in SC for huge data volume dissemination might be a good competitor. However, the city-wide deployment of optical-fiber system requires high budget [8].

Lately, Cellular Networks (CNs) appear to be an alternative solution for big data communication in SCs. However, huge mobile data demands have already created the issues of devastating CNs [9]. Figure 1 shows that mobile data demands will grow up to 38 PB per month by 2021 in New Zealand, which is 380% more than 2010 [10]. Broadband Internet and traditional core networks could also be possible candidates for SC data transmissions, but these networks are also now congested networks, traffic on the Internet has increased more rapidly than its existing capacity [11,12]. As a result, the problem of data dissemination between data sources and control units in SC ought to be solved by using some other types of hybrid networks instead of the using only Wi-Fi, 3G, LTE, Internet and so forth [13]. Under this topic, vehicular networks by using the existing routine rides in the city could be a possible solution to disseminate big data in SCs.

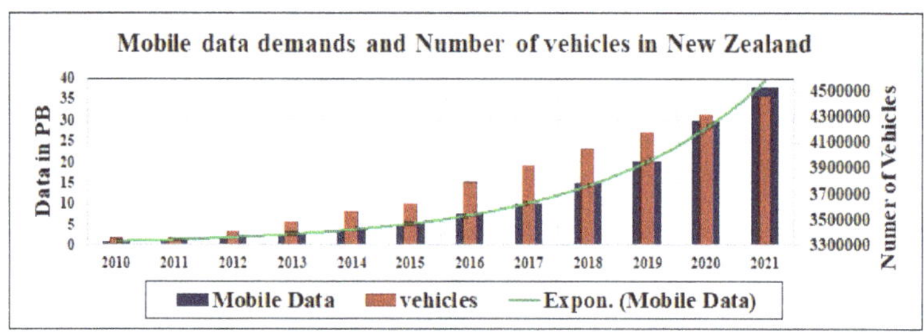

Figure 1. Mobile data demands and number of vehicles in New Zealand (NZ).

On the other hand there is one more big challenge in energy consumption in information communication technology (ICT). Andrey gives his analysis on global EC, with the best average and worst-case scenarios. International Energy Agency (IEA) estimated that Global electricity consumption would grow by 2.8 to 3.4% per year and ICT EC will be a big ratio (51%) of global EC in 2030. The electricity demand of communication technology is increasing and will be 30,715 TWh of a worst-case scenario in 2030. Andrey predicts that depending on different scenarios of ICT, EC in 2010 is 8% to 14%, in 2020 6% to 21% and in 2030 8% to 51%, respectively. Hence, it is also a critical issue that needs to address [14,15].

These factors of congested networks and EC in ICT create interest in alternate energy-efficient solutions to reduce the pressure of data traffic on traditional core networks. In this direction, Cho and Gupta [16] enhanced the data transfer performance by suggesting the concurrent use of conventional network and postal system. Some portion of the data is forwarded using the conventional network while some other delay-tolerant data is forwarded on hard disks through the postal system and it requires comprehensive data scheduling. Munjal et al. proposed a centralized Software-Defined Vehicular Connectivity procedure that enables scalable and adaptive control of the scheduled vehicles to offload traffic [17].

Thus, it is useful to introduce an energy-efficient option by using the existing vehicular networks to address this problem of big data transmission and to reduce the load on existing infrastructure networks. In this paper, we are proposing to use daily routine rides of SC's vehicles and annual average daily traffic (AADT) of Auckland city for delay-tolerant data transmission. The main contributions of this paper are as follows:

- We conduct a study to observe the correlation of current and predicted data demands of NZ's main cities on system model.
- We develop a mathematical model to measure the degree of data offloading by considering the Poisson arrival of smart vehicles and RSUs of the road network.
- A case of Auckland City is evaluated by using annual average daily traffic (AADT) of Auckland City in our proposed model to reduce the end-to-end delay for big data transmission.
- We design new algorithms using multi-commodity flow problem to select an energy-efficient network for data transmissions.

The rest of the paper is organized as follows. A correlation study is reported in Section 2. Section 3 present the related work. The system overview is presented in section IV along with mathematical models. We present our network flow model used to minimize EC and CE in Section 4. Analysis is performed in Section 5, along with two case scenarios. Finally, the paper is concluded in Section 6 along with future work.

2. A Possible Alternate Channel

Like rest of the world, the data demands in New Zealand are increasing each year, as depicted in Figure 2. There is an increase of 40% data demand in 2018 and an increase of 54% in the number of fiber connections for faster data communication, to make up 32% of all broadband connections. On the other hand, a significant decrease of 25% is shown on the slow Internet and dial-up connections. More than 70% of all Internet broadband connections with no data limits and cap. These data demands of broadband connections include the use of 281,615,000 GB of data that is equal to 90 million hours of streaming, or 10,700 years, high definition online shows of TV. Mobile phone users consumed 10,089,000 GB data in June 2018, which is up to 56% from 2017 [18]. The use of mobile data has been grown by 600% in the past four years and the expected growth for the next four years is 50% per year [10].

Cisco predicts that a SC with a populace of 1 million could produce 180 million GB data volume for each day or 42.3 ZB/month [19]. If we assume NZ main cities as smart cities, where data sources of different applications of smart cities like smart buildings, smart water, smart public services, smart lighting, smart mobility, smart waste management, smart meters, smart energy, smart sensors, and smart grids etc., will generate a massive volume of data and we need to transmit this big-data to data cloud, data centers or data control units. These expected data demands of 724 PB per day, will exceed the data usage beyond the available limit of existing infrastructure. There will be a massive gap between existing data usage and required data usage of these cities, as shown in Figure 3.

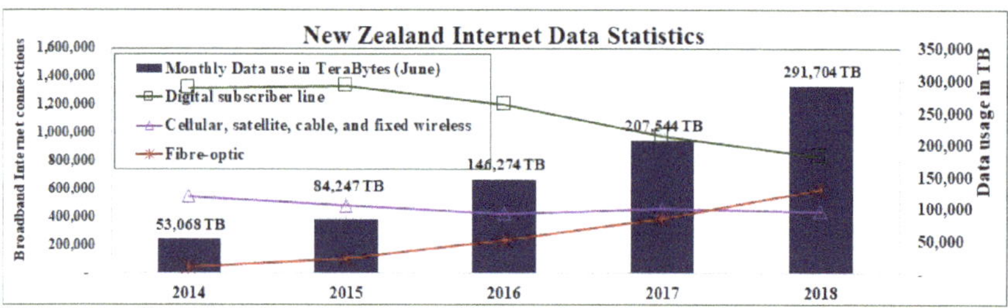

Figure 2. Past five years broadband data usage in NZ: Comparison of Fiber optic, Cellular/satellite/ and fixed wireless, DSL, and Monthly data use.

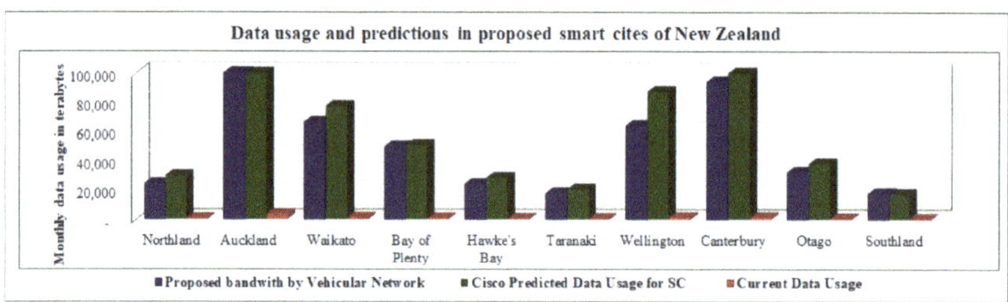

Figure 3. Data usage statistics of various NZ cities. Comparison of Cisco predicted, current, and the proposed data usage.

We propose a complimentary network, the vehicular network as a hybrid network, along with existing infrastructure, and mobile networks. Business Insider (BI), predicted that in 2021, 82% of all vehicles would be shipped as smart and connected vehicles [20]. These smart vehicles will have wireless interfaces, global positioning system, storage capacity, and processing capacity. We assume NZ's main cities as SCs. Ministry of transport NZ predicted that NZ has 0.79 per capita vehicles in 2016. This value is keep on increasing with an average increase of 2.12% in per capita vehicles since the past five years [21] as shown in Table 1. Figure 3, also predicts the parallel growth in mobile data demands and vehicle volume in NZ. Hence, it is forecasted that this significant volume of connected vehicles can produce huge bandwidth on the road network of NZ's smart cities. By using the data set of the ministry of transport NZ [21,22] as shown in Table 1, our proposed system predicts the vehicular network in SCs of NZ could produce a massive bandwidth up to 607 PB, if only 20% smart vehicles in NZ will transfer only one data assignment on a daily basis, as shown in Figure 3.

Table 1. Average percentage increase in per capita vehicles.

City Name	Population	Per capita Vehicle	Number of Vehicles	Yearly %Age Increase
Northland	168,300	0.71	119,493	1.14
Auckland	1,569,900	0.71	111,4629	2.03
Waikato	439,100	0.75	329,325	1.37
Bay of Plenty	287,100	0.86	246,906	1.66
Gisborne	47,400	0.66	31,284	0.62
Hawke's Bay	160,000	0.74	118,400	1.64
Taranaki	115,700	0.76	87,932	0.8
Wellington	496,900	0.64	318,016	1.28
West Coast	32,700	0.90	29,430	1.61
Canterbury	586,400	0.80	469,120	0.25
Otago	215,000	0.74	159,100	1.38
Southland	97,300	0.89	86,597	1.37

3. Related Work

Presently, different data clouds offer a data storage space on their web servers to their clients. Normally, numerous duplicates of data are stored on these servers [23]. The best example to transfer data from a client location to these servers is the data delivery services of Amazon Web Services (AWS). AWS offers data movement by using their data vehicles. For this purpose, AWS sends their vehicle to the client, the client loads required data onto the vehicle, and the vehicle is driven back to AWS for loading the client's data onto their web servers [24]. A digital map provider company Digital Globe used snowmobile services of AWS to transfer 70PB data onto the cloud of AWS [25]. This type of data transmission requires some fiscal cost because it needs a dedicated vehicle for this purpose. Moreover, it consumes energy and emits carbon in the environment.

A big data of delay-tolerant applications can be transferred by efficiently using the energy of daily routine trips of vehicles. Vehicular delay-tolerant network (VDTN) [26], is built on the theory of delay-tolerant network (DTN). It handles delay-tolerant applications at low cost and on unpredictable network conditions. Vehicles can be used as data carriers between terminal nodes either in rural areas or in emergency scenarios. This strategy can be helpful in both V2V and V2I communication. In this direction, Kashihara et al. [27], proposed a data offloading scheme to offload data, particularly to scheduled vehicles by using short-range and high-speed wireless communication. The motivation was to reduce traffic congestion because of the high data volume on traditional networks. Similarly, Hunjet et al. [28] and Usbeck et al. [29] uses VDTN and implement data forwarding schemes by using data farriers. A vehicular data dissemination project is implemented in France to reduce load on conventional networks [30]. Dessler et al. [31] used parked vehicle for data communication among other vehicles and proposed a protocol called vehicle cord where parked vehicles are used as RSUs.

In another work [32], Cho et al. uses hybrid network to upgrade the performance by proposing the concurrent use of traditional core network and postal framework. A portion of the time-critical data is exchanged by using the conventional core network while postal system is used to transfer delay-tolerant data with the help of hard drives. It requires complex data scheduling for data forwarding. Marincic et al. upgrade a similar work to minimize the EC and reduce the CE [33].

Moving Vehicles on the road and roadside APs of WiFi Network can also be used as a practical solution for cellular data offloading in SCs, called vehicular Wi-Fi offloading. The research in this direction has an objective to enhance the cellular data offloading performance, particularly for delay-tolerant and non-interactive applications [34,35]. At the same time, the advantages of opportunistic communication and D2D revolution are sound for some vehicular related use cases. It can empower location-based peer-to-peer applications and services. For example, thinking about the huge number of connected smart vehicles, a new software update of the smart vehicles can put a critical load on the cellular network, and cost huge money for vehicles owners. In this way, the new update can be downloaded by some designated vehicles by using D2D communication with RSUs, and then this update can be exchanged to other vehicles by D2D transmission. Along these lines, the majority of the cellular load can be shifted to V-D2D communication and accordingly cellular bandwidth, energy, and cost can be saved [36].

A network architecture is explained in [12] for smart city data offloading by using smart vehicles, and numerical analysis of this proposed architecture is demonstrated in [37]. In this architecture, the authors apply D2D communication, the daily vehicle count of Auckland roads and calculate the delay, throughput, and energy consumption. The proposed system outperforms the Internet with dedicated links of 100 Mbps and 1 Gbps. Also, the system can offload traffic by consuming less energy than the Internet. To transfer 20 TB data from a data location to the control unit, it consumes 40% less energy as compared to the traditional core network. In this paper, we are trying to describe how much data can be uploaded/downloaded to/from RSU while vehicle is on the move and how proposed algorithm can be used to select an energy-efficient network mode to transfer big data between two points.

4. The System Model

The proposed system model is shown in Figure 4. The wireless coverage of cellular network is available everywhere in the city, and smart vehicles are moving on the roads. These smart vehicles have GPS, WiFi, and cellular interfaces, also have storage and processing capacity. They can form a network with base station (BS) by using cellular interface, with roadside units (RSU) or with other vehicles by using WiFi interface. A central controller (CC) server is installed at cellular network. This controller will govern overall communication between data sources and control units.

For our proposed work, we assume that various SC devices and sensors generate huge volume of data in a city and these devices are the data sources. We need to transfer this data to different destinations in this case CUs are the destinations. CC selects cellular network or traditional core network to forward delay-sensitive data. For delay-tolerant data, CC selects cellular network, core network, or proposed vehicular network. Energy-efficient network mode selection algorithm is placed on CC server. When an information source needs to send information to a destination CU, it sends an information request packet to CC on cellular control channel. CC will choose the most appropriate alternative, to exchange information by means of core network, cellular network or vehicular network. This decision is made by considering the information of vehicle count on the road, delay-tolerant interval, history of vehicle's trajectory and energy cost. Based on these results, CC selects the suitable network, if infrastructure network is more appropriate for the given set of demands then CC guides the source to send data on path (a) or (b) of Figure 4, otherwise data can be forwarded by using vehicular transport network on path (c). For each given set of data transfer requirements, CC chooses optimal vehicles by using their trajectory's history and informs the DS to send data by using these particular vehicles.

On the receiver side, these vehicles upload the data whenever they encounter with an RSU by using their wireless interface. The RSUs are connected to the backbone network of the city and send the data packets to CU. After receiving the data, CU send the acknowledgments to the CC server. If a particular data packet is not transmitted to the CU before the delay-tolerant indicator expires, then CC informs to particular data source for re-transmission of missing data packets, by using the conventional network.

To calculate the delay and EC, following are the models, that are used in this paper for mathematical calculations. Notations used in the mathematical model are listed in Table 2.

Table 2. Notations used in this paper.

Symbol	Meaning
d	Effective distance
d_0	Reference distance
n	Path loss factor
Ψ	Gaussian random variable
ρ	Vehicle density
λ	Poisson arrival process parameter
B_v	Data offloading by a vehicle
B_{RSU}	Bandwidth of roadside unit
$\bar{s}$	Average speed of the vehicle
r	Radius of RSU's coverage area
D	Diameter of RSU's coverage area
d_{ac}	Distance between two consecutive RSUs with no wireless coverage
B	Core network bandwidth
D_{Vol}	Data volume
μ	Probability of vehicles to participate
E_{inc}	Incremental Energy cost

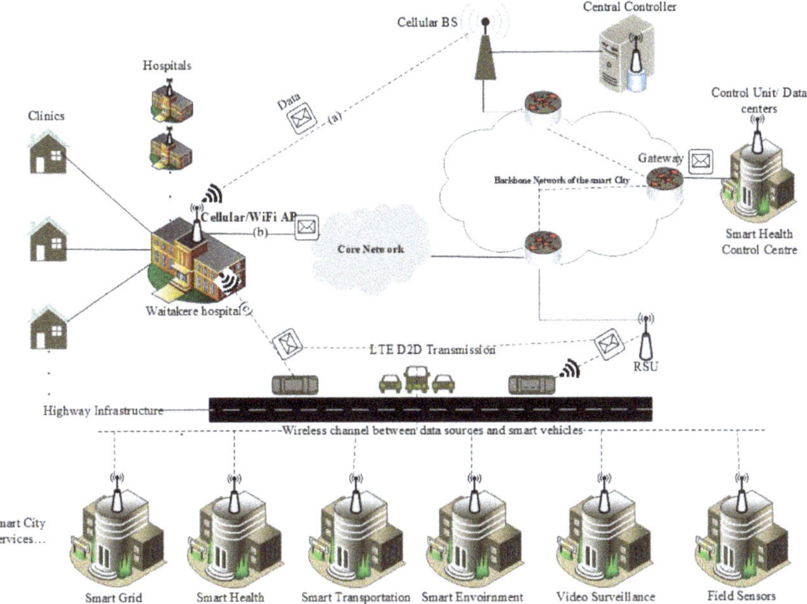

Figure 4. System model overview.

4.1. Delay Model

To transfer the data from source to destination, following are the two models of end-to-end delay for transport network and core network.

4.1.1. Delay Model for Transport Network

Effective distance: We have to identify the suitable region for vehicle in the wireless coverage area of an RSU. Normally, the packet loss rate depends upon the received power. By increasing the received power we can reduce the packet loss rate. Received power also depends upon the distance from the RSU. Therefore, to calculate the packet loss rate we can use the distance from RSU. Hence, to reduce the packet loss, it is essential to keep a vehicle in the appropriate coverage area of the RSU. We can calculate the path loss $PL(d)$ at a distance d by using path loss formula.

$$PL(d)[dB] = PL_F(d_0) + 10n \log\left(\frac{d}{d_0}\right) + \Psi \tag{1}$$

where d_0 is the reference distance where path loss inherits the characteristics of free-space loss PL_F, n is path loss exponent, depending upon the propagation environment. Ψ is the Gaussian random variable.

From Equation (1) we can find the effective distance from the vehicle to RSU.

$$d = d_0 \times 10^{\frac{PL_F(d_0) + \Psi - PL}{10(n)}} \tag{2}$$

d is the distance from the RSU to the vehicle and effective distance can be $2d$, the diameter of the wireless coverage area of RSU. In other words, each vehicle can have a chance to transmit the data packets to RSU at a distance from *0 to 2d* meters. In fact, the effective distance is restricted by each road, and it depends upon vehicle density on the road. Vehicle density is very high in urban areas and low in rural areas.

Delivery probability with vehicle density: The other parameters that affect the effective region of RSU are delay and bandwidth of WiFi is a key factor that can control the delay. RSU has a fixed bandwidth that can be shared among all vehicles inside the coverage area. If vehicle volume on the road is greater, then smaller will be the bandwidth share for vehicles. We assume a fixed number of vehicle m in the coverage area of RSU with bandwidth B_{RSU}. As we know that vehicles that connect to RSU follow Poisson arrival process {f(t), t $\geq$ 0} with parameter $\lambda > 0$ and vehicle volume on the road varies at different intervals of the day [38]. We can convert a day into various intervals, according to vehicle volume on the road, and these intervals follow the Poisson distribution. Suppose that there are n intervals $(I_1, I_1 \dots I_n)$ with parameters $(\lambda_1, \lambda_2 \dots \lambda_n)$. The probability of vehicles in the coverage area of an RSU at any time t can be calculated as follows.

$$P\{f(t+i) - f(i)\} = e^{-\lambda t}\frac{(\lambda t)^m}{m!} \tag{3}$$

where i is the start time of each interval, $(t+i) \in (I_1, I_1 \dots I_n)$, and $\lambda \in (\lambda_1, \lambda_2 \dots \lambda_n)$.

The vehicle density in the effective region of an access point depends upon the average speed $\bar{s}$ and Poisson arrival parameter lambda λ. By using different value of lambda $(\lambda_1, \lambda_2 \dots \lambda_n)$ in different intervals $(I_1, I_1 \dots I_n)$, we calculate the vehicle density ρ in the effective distance d of access point.

$$\rho = \frac{\lambda}{\pi \bar{s} d} \tag{4}$$

Suppose B_{RSU} is the bandwidth of RSU and B_v is the shared bandwidth of each vehicle when there are m vehicle in the coverage of RSU then

$$m = \frac{B_{RSU}}{B_v} \tag{5}$$

Transmission probability Φ, with m number of vehicles can be calculated as follows.

$$\Phi = \sum_{v=1}^{m}\left(e^{-\lambda t}\frac{(\lambda t)^v}{v!}\right) \tag{6}$$

Degree of data offloading: By using (2),(3), and (6) we can calculate the data transmitted/offloaded by a vehicle to RSU as follows.

$$B_v = B_{RSU} \times \Phi \times \frac{d}{\bar{s}} \tag{7}$$

Equation (7) can be represented in term of vehicle density (4) as follows.

$$B_v = B_{RSU} \times \Phi \times \frac{d}{\frac{\lambda}{\pi \rho d}} \tag{8}$$

where d is the distance of the vehicle from RSU and $\bar{s}$ is the speed of the vehicle.

End-to-end delay: The time T_d required to transfer a big data D_{vol} from source A to destination B depends upon two things. First travel time between A and B T_{AB}, 2nd the data loading time T_L.

$$T_d = T_{AB} + T_L \tag{9}$$

If d_{AB} is the distance between these two points and $\bar{s}$ is the average speed then T_{AB} can be calculated as follows.

$$T_{AB} = \frac{d_{AB}}{\bar{s}} \tag{10}$$

If N_v is the total number of vehicle required to transfer data D_{vol} then data loading time T_L can be calculated as follows.

$$T_L = \frac{D_{vol}}{N_v} \tag{11}$$

N_v depends upon vehicle count V_c on the road, bandwidth share of each vehicle B_v and probability μ of the drivers or vehicles willing to participate in this proposed system, then N_v can be calculated as follows.

$$N_v = V_c \times B_v \times \mu \tag{12}$$

By applying the value of N_v, T_{AB} and T_L in (9).

$$T_d = \frac{d_{AB}}{s} + \frac{D_{vol}}{V_c \times B_v \times \mu} \tag{13}$$

By using the value of B_v from (7).

$$T_d = \frac{d_{AB}}{s} + \frac{D_{Vol}}{V_c \times B_{RSU} \times \Phi \times \frac{d}{s} \times \mu} \tag{14}$$

4.1.2. Delay Model for Traditional Core Network

For core network case, the total delay depends upon the number of packets and number of intermediate nodes. Suppose L is the length of each packet then the total number of packets N can be calculated as follows.

$$N = \frac{D_{Vol}}{L} \tag{15}$$

Let Tp be the processing delay, Tq be the queuing delay, and Tt be the transmission delay and N_{Nodes} is the total number of nodes between sender and receiver then total end to end delay can be calculated as follows.

$$T_{Core} = N_{Nodes} \times 1^{st} Packet Delay + (N-1) \times (T_p + T_q + T_t) \tag{16}$$

We assume in case of a dedicated link when a packet arrives at a node; it will find no packet ahead in the queue. Then queuing delay becomes zero. In such case parallel processing will be performed at each node then the final end to end delay can be calculated as follows.

$$T_{Core} = N_{Nodes} \times 1^{st} Packet Delay + (N-1) \times T_t \tag{17}$$

If B is the transmission bandwidth then Equation (17) can be written as.

$$T_{Core} = N_{Nodes} \times 1^{st} Packet Delay + (N-1) \times \frac{L}{B} \tag{18}$$

4.2. Energy Model

To transfer the data from source to destination, following are the two models of energy consumption for transport network and traditional core network.

4.2.1. Energy Model for Transport Network

EC for delay-tolerant information transporting between two points by using vehicles depends on two things. 1st EC for data loading and 2nd EC for transporting the weight of storage device from point A to B.

$$E_{Veh} = E_{DataLoad} + E_{Transport} \tag{19}$$

If k is the total number of offloading points between A and B then $E_{DataLoad}$ can be calculated as follows

$$E_{DataLoad} = 2 \times \sum_{i=1}^{k} E_{D2D_i} \tag{20}$$

$E_{transport}$ depends upon the distance between the sender and receiver, gross weight of the vehicle, and fuel economy of the vehicle. If n vehicles are used to transport total data between them and $W_{Storage}$ is the storage device weight then we can calculate the EC for whole data shipment as follows.

$$E_{Transport} = C_{fuel} \times W_{Storage} \times \sum_{i=1}^{n} E_{Shipment_i} \tag{21}$$

C_{fuel} is fuel constant, and it converts the volume of fuel into energy consumed i.e., from litters to joules. If FE is the fuel economy then $E_{Shipment}$ can be calculated as follows.

$$E_{Shipment} = \frac{d_{AB}}{FE \times V_{Load}} \tag{22}$$

If W_{Total} is the gross weight of the fully-loaded vehicle then V_{Load} can be calculated as follows.

$$V_{Load} = W_{Total} - W_{Empty} \tag{23}$$

4.2.2. Energy Model for Core Network

Data transmission begins from the 1st datagram at the source and ends when the last datagram is delivered at the destination. EC for conventional network depends upon the energy consumed at the sender, receiver, and all intermediate devices. These devices contain switches and routers. If D_{Vol} is the total data volume, E_{SR} is the EC at source and destination, and we consider the incremental EC at all k intermediate devices, then the EC for conventional network case can be calculated by using the following equation.

$$CnEc = E_{SR} + \sum_{i=1}^{k} E_{inc_i} \tag{24}$$

If B_{up} is the uploading and B_{down} is downloading bandwidth at sender and receiver respectively, $\triangle P$ is power change at sender/receiver while transmitting/receiving the data, then E_{SR} can be calculated as follows.

$$E_{SR} = max\left(\frac{D_{Vol}}{B_{up}}, \frac{D_{Vol}}{B_{down}}\right)(\triangle P_{Sender} + \triangle P_{Receiver}) \tag{25}$$

For intermediate devices, if E_{bit} is the energy cost per bit then energy consumption for total D_{Vol} data is calculated as follows.

$$E_{inc} = D_{Vol} \times E_{bit} \tag{26}$$

Energy consumed per bit is calculated as a fraction of the max power P_{max} and available bandwidth B.

$$E_{bit} = \frac{P_{max}}{B} \tag{27}$$

5. Minimizing Energy Cost

The goal of this work is to minimize the total cost of sending the data requirements across the complex road network or traditional core network, while satisfying all demands/supplies, and respecting arc capacities. We solve the energy optimization problem for road network by using multi-commodity flow problem and calculate the energy cost for optimal paths. Similarly we calculate the energy cost for core network and finally we find that which network is suitable for data transmission with respect to energy consumption.

5.1. Multi-Commodity Flow Problem

Let $G = (V, E, C, A)$ be a capacitated undirected graph, where V is set of data sources or data centers locations, E is set of road links between data sources and destinations, C is the capacity of each road w.r.t vehicle count, and A is a set of cost per unit flow for a commodity b_i on each link $(i, j) \in E$. $R = \{(s_i, t_i, b_i)\}$ be a set of requirements, $s_i \in V$ is data source and $t_i \in V$ is destination data center for commodity b_i. For each edge $(i, j) \in E$ and each commodity r, associates a cost per unit of flow, designated by a_{ij}^r. The demand (or supply) at each node $i \in V$ for commodity r is designated as b_i^r, where $b_i^r > 0$ denotes a supply node and $b_i^r < 0$ denotes a demand node. We define decision variables x_{ij}^r that denote the amount of commodity r need to send from node i to node j. The amount of total flow, for all commodities, that can be sent across each link is bounded above by c_{ij}. We need to minimize the transport network energy consumption $TnEc$ by using the vehicle mobility of complex road network.

Minimize :

$$TnEc = \sum_{(i,j) \in A} \sum_{r \in R} a_{ij}^r . x_{ij}^r$$

Subject to :

$$\sum_{r \in R} x_{ij}^r \leq c_{ij} \qquad (i, j) \in E \qquad\qquad (Capacity)$$

$$\sum_{i,j \in E} x_{ij}^r - \sum_{i,j \in E} x_{ji}^r = b_i^r \quad i, j \in V, r \in R \qquad (Balance)$$

$$x_{i,j}^r \geq 0 \qquad\qquad (i, j) \in E, r \in R$$

$$\sum_{i,j \in E} x_{ij}^r(n) - \sum_{i,j \in E} x_{ji}^r(n) = \begin{cases} b_i^r & if\ n = s^r \\ -b_i^r & if\ n = t^r \quad (Flow\ conservation) \\ 0 & otherwise \end{cases}$$

$$\forall n \in V\ and\ r \in R$$

We use Algorithm 1, to solve the above energy optimization problem of multi-commodity flow, for multiple commodities of data transmission across the road network and calculate the energy-efficient paths.

Algorithm 1: Minimum cost multi-commodity flow

Input: A graph $G = (V, E, C, A)$, and set of requirements $R = \{(s_i, t_i, b_i)\}$
Output: Set of paths that meet the requirements
1 $G_{Temp_1} \leftarrow G_{Temp_2} \leftarrow G$, and $P = \phi$
2 **while** $|P| < |R|$ **do**
3   **for** $r_i \in R$ **do**
4      $p_i = minimum\_cost\_flow(G_{Temp_1}, s_i, t_i, b_i)$
5   **if** $p_i == \phi$ **then**
6        $p_i' = minimum\_cost\_flow(G_{Temp_1}, s_i, t_i, b_i)$
7   **for** $(e, c_{ij}) \in p_i'$ **do**
8   **if** $c_{ij} > Available\_Capacity\_in\_G_{Temp_1}$ **then**
9   Increase cost of e in G_{Temp_1} and G_{Temp_2}
10   **for** $p \in P$ **do**
11   **if** $e \in p$ **then**
12                $P \leftarrow P - p$
13   Return the capacity used by p to G_{Temp_1}
14   **end**
15   **end**
16   **end**
17   **end**
18   **else**
19   Reduce the capacity of edges in p_i from G_{Temp_1} $P \leftarrow P \cup p_i$
20   **end**
21   **end**
22 **end**
23 **return** P

5.2. Energy-Efficient Network Mode Selection

Let $G' = (V', E', A')$ be a capacitated undirected graph, where V' is a set of intermediate nodes (routers and switches), E' is a set of the links between these intermediate nodes, and A' is a set of unit energy cost to transfer a data commodity b_i on these links. We calculate the energy cost $CnEc$ for data transfer between data source and data center by using traditional core network from Equation (24). For transport network energy cost $TnEc$ can be calculated from Algorithm 1. The central controller apply Algorithm 2 to take the decision which network is suitable for a given set of data demands. This decision is forwarded to all data sources and intermediate nodes. Finally they select appropriate energy-efficient network interface to forward the data.

Algorithm 2: Energy-efficient network mode selection

Input: Graphs $G = (V, E, C, A)$, $G' = (V', E', A')$ and set of requirements $R = \{(s_i, t_i, b_i)\}$
Output: Energy-efficient network mode selection
1 $CnEc \leftarrow f(G', R)$
2 $TnEc \leftarrow f(P) \leftarrow multi\_commodity\_flow(G, R)$
3 **for** $r_i \in R$ **do**
4   **if** $CnEc_i > TnEc_i$ **then**
5      $mode_i$ = proposed network
6   **else**
7      $mode_i$ = traditional network
8   **end**
9 **end**
10 **return** $mode$

---

6. Performance Analysis

This section describes the numerical analysis of our proposed system with conventional network system. In first part, we set the value of various parameters and then we present two case scenarios to compare both systems.

6.1. Parameters Setting

To evaluate the proposed system model, we consider a straight expressway . Vehicles are moving with some predefined average speed $\bar{s}$ along the straight road. We are assuming that the vehicles are moving only in one direction for simplicity. We calculate the effective distance and data transfer probability in the wireless coverage area of an RSU. We also measure the vehicle density with different speeds of vehicles. Finally, by applying these parameters, we calculate the degree of data offloading by using different speeds and vehicle density.

Effective distance: To calculate the effective distance, we set reference distance $d_0 = 100$ m and carrier $f_c = 2.4$ GHz and apply the different path loss exponent values, according to Equation (2). The relationship between path loss [dB] and distance is shown in Figure 5a. Path loss increases with the distance between vehicle and RSU and approaches to 80 dB for 50 to 100 m. For the remaining analysis we set effective distance $d = 100$ m.

Delivery probability: The fixed bandwidth of RSU is divided among all the vehicles in the wireless coverage area of WiFi. Data transmission probability increases exponentially with different densities of vehicles as shown in Figure 5b. We vary the Poisson arrival parameter λ from 500 to 2000 by using Equation (6). The trend shows that the data transfer probability reaches to 0.95 in all cases when the number of vehicles approaches to 15.

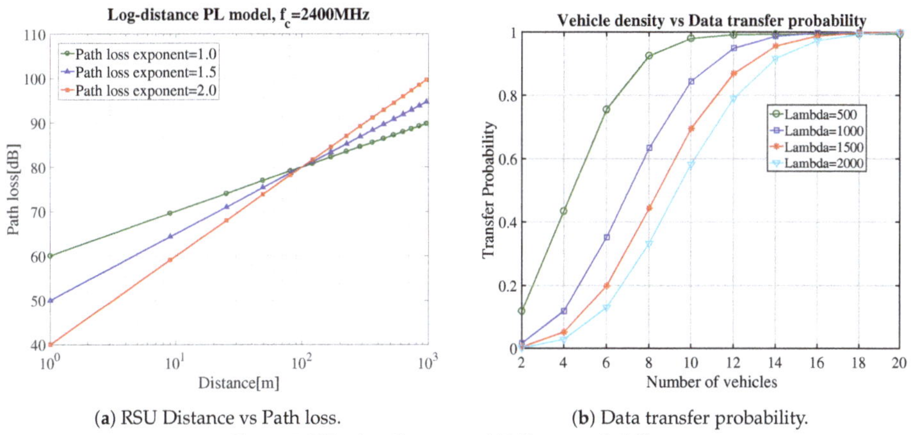

(a) RSU Distance vs Path loss. (b) Data transfer probability.

Figure 5. Effective distance and Delivery probability.

Vehicle density: ρ is evaluated with different speeds of vehicles by using effective distance d and Poisson arrival parameter λ in Equation (4). We vary the value of λ from 500 to 2500 and set effective distance $d = 100$. Vehicle speed is used to get the vehicle density in a specific area as shown in Figure 6a. If we increase the vehicle speed, then vehicle density decreases exponentially.

Degree of data offloading: By considering the above evaluation, the data offloading can be calculated by using Equation (7) in the wireless coverage area of an RSU. According to Figure 5a, we set the effective distance of the vehicle from RSU $d = 100$, and we set the transmission probability $\rho = 0.95$ according to Figure 5b. A vehicle can transfer various amount of data to RSU by using different speeds. The time in which a vehicle stayed connected with an RSU is inversely proportional to vehicle speed. According to Equation (7), the data transfer rate varies with the different speeds of vehicle and connection time duration of a vehicle decreases with increase in the speed of the vehicle. Moreover,

Figure 6b, shows that if we increase the vehicle density, i.e., there are more vehicles that want to send the data then the bandwidth will be divided among all vehicles.

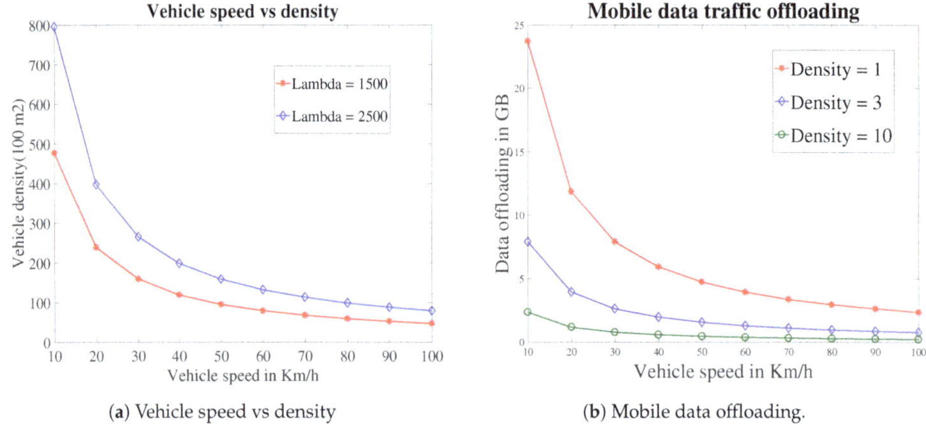

(**a**) Vehicle speed vs density   (**b**) Mobile data offloading.

Figure 6. Results for degree of data offloading.

In the following two subsections, we present two case scenarios for comparing the performance in terms of delay and EC.

6.2. Case Scenario I—Auckland City Case Scenario for Delay Tolerant Study

The proposed case of Auckland City is shown in Figure 7. We propose two cases for delay-tolerant information transmission, the core network, and the vehicle transport network. Assume a source at SH16 Royal Rd Off Ramp to Hobsonville Rd/SH18 Off Ramp WB with reference ID:01610016 in Auckland City needs to send some delay-tolerant data D_{Vol} to a destination in the city center. For the vehicular system, we find the distance d between these two points is 23 Km by using Google maps and vehicle count $AADT$ on the said link is 27857 [39]. The average speed $\bar{s}$ of the vehicle on this link is 50 Km/h. Assume each vehicle has IEEE 802.11ay wireless interface and disk capacity is 256 GB. We assume that if only 20% ($\mu = 0.2$) drivers take part in this proposed framework, then by applying Equation (14) we can estimate the delay value of the proposed network. In the core network case, a cellular network, a wired network, or a wireless network can be used for data forwarding, and its delay can be estimated by using the Equation (18). We exchange the different information volume between these two locations and get the outcomes as appeared in Figure 8a. For comparisons, we use the bandwidth 512 Mbps and 1 Gbps in traditional core network case. The outcome demonstrates that our proposed vehicular network outperforms the traditional core network. We enhance the data size and look at its impact on delay for data transmission. The results demonstrate that for huge information volume, the delay increases significantly for core networks as compared to our proposed vehicular network.

Figure 7. Auckland City Case Scenario for Delay Tolerant Study.

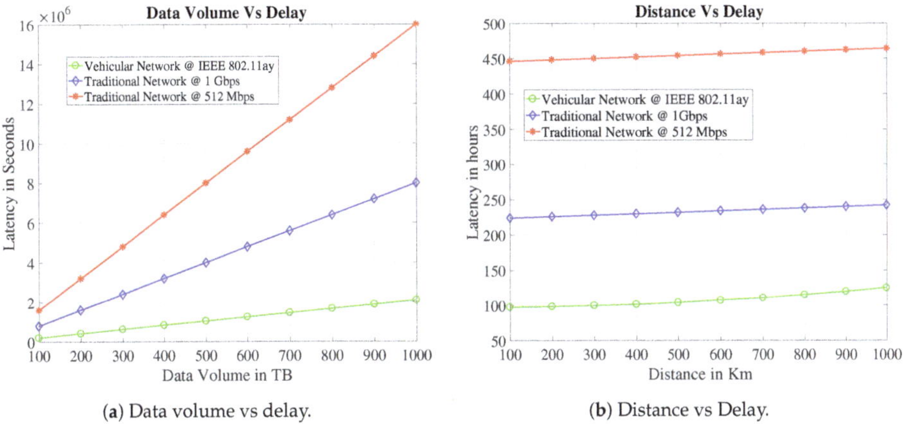

(**a**) Data volume vs delay.   (**b**) Distance vs Delay.

Figure 8. Delay performance study for Auckland City case scenario.

In the next case as shown in Figure 8b, we compare data transmission delay with distance for both cases. We transfer 100 TB data between two points by varying the distances. In this case, more routers and intermediate nodes will be used in core network, and delay depends upon the distance between two points as well. This transmission delay depends upon queuing delay, propagation delay, transmission delay, and processing delay. We assume that there is no queuing delay on the core network when a packet arrives in at the router, there is no inbound and outbound queue. Also, assume that the core network is built on fiber. Hence the propagation delay can also be ignored. We consider only transmission and processing delay and assume that there is a router after every 100 Km. Vehicular network delay also increases as the distance increases to the destination because each vehicle takes more time to reach the destination when we increase the distance. Similarly Figure 8b, also shows that the delay in the case of core networks also increases as the number of routing devices keeps on increasing. The results show that our system outperforms the traditional core networks when we increase the distance.

Vehicle Volume: Equation (14) shows that throughput of the vehicular network also depends upon vehicle volume on the road. To get the impact of vehicle count, we forward a massive size of data 1 PB between the same two points in Auckland. We vary the vehicle count on the road from 0 to 1000, By keeping constant the average speed and disk capacity, 50 Km/h and 256 GB respectively. Figure 9, demonstrates that at the beginning delay is very high in our proposed case as compared to traditional network, and this latency keeps on decreasing when we increase the vehicle volume on the road. In this scenario our system outperforms the conventional network with a dedicated bandwidth of 1 Gbps when vehicle volume is higher than 170. Our proposed scheme reduces the delay up to 47% when vehicle count is 850 in between these two points.

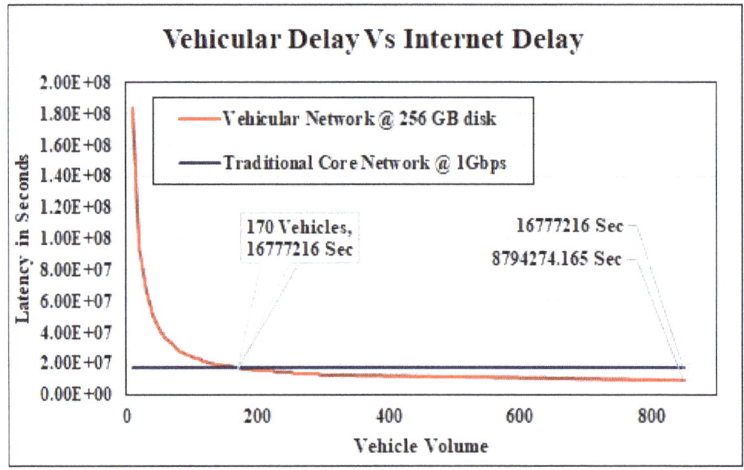

Figure 9. Latency in Transport Network Vs Traditional Network.

Energy Consumption: Our proposed EC model demonstrates that the EC relies on the information volume and the distance between sender and receiver. For core network, we assume that uploading and downloading bandwidth is 0.1 Gbps at sender and receiver. The intermediate nodes between sender and receiver are eleven LAN switches, two core and edge router. To get the impact of data volume, we forward different sizes of data between the same sender and receiver. The results in Figure 10, show that when we keep on increasing the data size, the EC also increases proportionally in both vehicular and conventional network cases. Vehicular system outperforms the traditional core network. It consumes only 66MJ energy to forward 100 tera byte data whereas the core network consumes 227 MJ of energy.

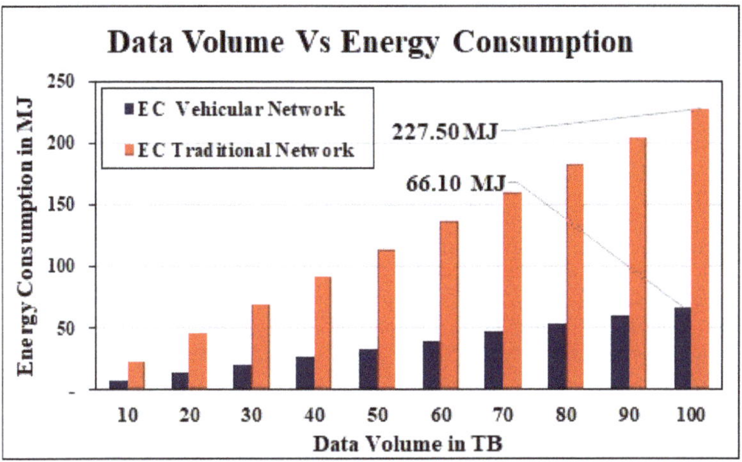

Figure 10. Data volume Vs Energy Consumption.

6.3. Case Scenario II- Finding the Best Routes

In this case scenario, we proposed a solution, where a road network is investigated for data transfer assignments as shown in Figure 11. Here, nodes represent the data offloading locations and links represent the roads. Each road has the traffic count as the capacity of the road. We use an embedded storage device of weight 0.95 Kg in the vehicle and calculate the energy cost to transfer 1TB of data for each road. In these calculations, we use Toyota Prius as a relaying vehicle, and the value of fuel constant is assumed as Cfuel = 37,624,722.29 J/L [40] in Equation (21). These calculations are presented in Table 3. For each offloading request; we apply our Algorithm 1 to find the best route by solving the data transfer assignment as a traditional multi-commodity flow problem.

The data set in Table 3, provides the cost a_{ij}^k of sending a unit of commodity r along arc (i, j), with distance and speed of that particular road. To calculate the EC of the traditional core network we use Table 4.

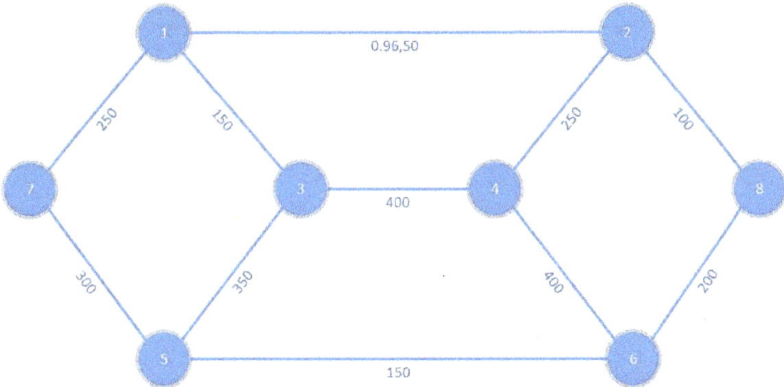

Figure 11. Graph of building network flow.

Table 3. Energy Cost per link in MJ/TB

From	To	Cost / TB	Distance (Km)	Speed (Km/h)
1	2	0.96	40	50
1	3	2.41	100	70
5	3	0.24	10	50
5	6	1.45	60	60
3	4	1.2	50	50
4	2	2.18	90	80
4	6	0.48	20	50
7	1	2.53	105	100
7	5	1.81	75	80
2	8	2.30	95	80
6	8	1.57	65	60

6.3.1. Energy Consumption

By using this multi-commodity flow model and our energy cost model, we evaluate the EC cost for a given set of five commodities (Figure 12) $R = \{c1, c2, ..., c5\}$ of delay-tolerant data, with a total data flow of 580 TB. We execute Algorithm 1 on SaS optimization tool [41] to find a set P of energy-efficient paths and optimal solutions for a given set of commodities to transfer data by using road network. To calculate the EC of the traditional core network, we use graph $G' = (V', E', A')$ by applying Equation (24), and unit cost for EC to transfer a data demand on each edge in MJ/TB is calculated in Table 4. Calculated results for the core network and VDTN are shown in Figure 12. It shows that our proposed data transport model outperforms the traditional core network, for all the commodities other than the commodity number 4. In this case the distance between the data source and destination is high. That is why vehicles consume more energy to transfer data at the final destination. Moreover, for this demand, a huge amount of energy is also consumed at intermediate nodes for data offloading from vehicle to intermediate nodes and uploading from intermediate nodes to vehicles. To transfer 580 TB data, the traditional core network consumes 1950 MJ energy, and our proposed vehicular network consumes 1495 MJ energy. In this way, our proposed vehicular network can save up to 24% of energy costs for these data assignments.

Table 4. Core network cost to transfer 1 TB on each link.

From	To	Lan L3 Switches	Edge Routers	Core Routers	Up BW	Down BW	MJ/TB
1	2	9	2	15	0.1	0.1	5.095
1	3	6	2	3	0.1	10.0	4.355
5	3	8	2	14	1.0	1.0	2.03
5	6	11	2	3	1.0	1.0	2.275
3	4	6	2	5	0.1	1.0	4.4
4	2	9	2	6	1.0	10.0	2.02
4	6	8	2	6	0.1	0.1	4.74
7	1	11	2	7	1.0	1.0	2.36
7	5	6	2	7	0.1	0.1	4.44
2	8	11	2	5	1.0	1.0	2.32
6	8	13	2	14	0.1	0.1	5.71
1	6	8	2	9	1.0	10.0	1.925
2	6	14	2	6	1.0	1.0	2.635
3	6	9	2	15	1.0	1.0	2.215
4	8	11	2	2	0.1	0.1	5.135

6.3.2. Energy Efficient Network Mode Selection

On the basis of the above calculated energy costs for traditional core network $CnEc$ and for transport network $TnEc$, the central controller apply Algorithm 2 to take the decision which network is suitable for each commodity in a given set of data demands. Central controller forward this decision to concerned data sources and intermediate nodes. The data sources and intermediate nodes select appropriate energy-efficient network interfaces to forward the data. With this optimal network mode selection, we can save the energy cost more than 32% for this given set of commodities.

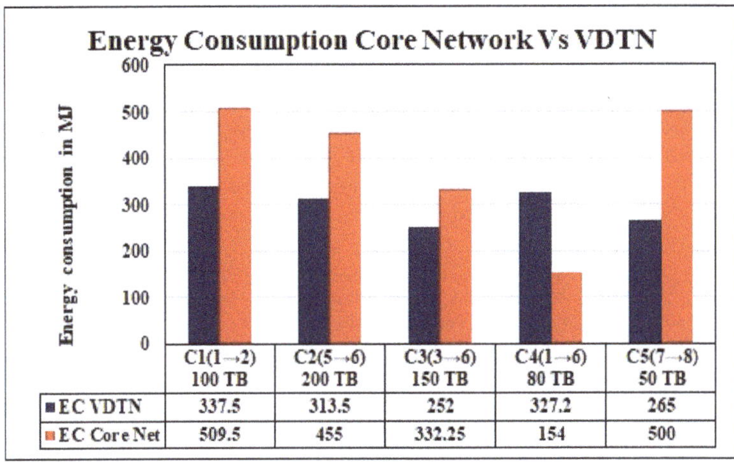

Figure 12. EC: A comparison of the core network and VDTN.

6.3.3. Carbon Emission

By applying energy-efficient network mode selection, we can further reduce the energy cost and carbon consumption. Information exchange for both core network and VDTN consumes energy from various means like fuel and power. This usage of energy transmits carbon into the environment. Figure 13 demonstrates that the CE, is less for our proposed optimal model as compared to traditional core networks and VDTN. To evaluate these outcomes, we used the carbon emission $0.703\ CO_2$ Kg/kWh [42] and the conversion unit 1 MJ = 0.2777778 kWh [43].

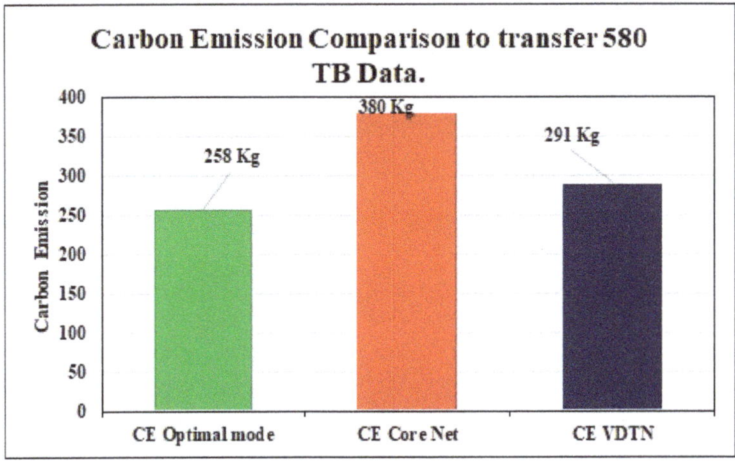

Figure 13. CE: A comparison of the core network, VDTN and optimal network mode.

7. Conclusions

In this paper, we developed both delay and energy models for sustainable data dissemination to be used to tackle typical SC daily data traffic. As a proof-of-concept of our algorithms, we presented Auckland City case scenarios where delay tolerant data is delivered to data centres. We found that our proposed system can effectively use the daily vehicle mobility of Auckland City for enormous information transmission to reduce the cost of EC and CE. The results obtained show that the proposed system can offer up to four times better data transfer rate than the dedicated core network with a data rate of 1 Gbps. For data transmission scenarios, our proposed approach can offer about 32% better EC and CE than the traditional network. The main conclusion is that our proposed system is suitable for delay-tolerant data delivery applications as it can reduce the network load further by sharing the burden of congested networks. Developing a more flexible and mature data offloading communication system by optimal mode selection on the basis of multi-objective optimization for delay-tolerant interval as well as energy cost, for data transmission is suggested as our future work.

Author Contributions: S.N. Modeled the end-to-end delay and energy consumption for both transport and conventional networks, implemented the case studies, and analysed the data under the supervision of W.L. and N.I.S. The manuscript was drafted by S.N., revised and proofread by W.L. and N.I.S.

Funding: This research was funded by University of the Punjab, Lahore, Pakistan (Notification No. D/117/Est.1) to Salman Naseer for his PhD studies at Auckland University of Technology, Auckland, New Zealand, under faculty development program.

Acknowledgments: The authors would like to thanks to all the reviewers who helped us in the review process of our work. Moreover special thanks to Auckland transport agency for providing the report of vehicle volume per day of each road. We use this vehicle volume in our proposed system to calculate the expected bandwidth of each road.

Conflicts of Interest: The authors declare no conflict of interest.

References

1. Zanella, A.; Bui, N.; Castellani, A.; Vangelista, L.; Zorzi, M. Internet of things for smart cities. *IEEE Internet Things J.* **2014**, *1*, 22–32. [CrossRef]
2. Ten of the Biggest IoT Data Generators. Available online: https://www.cbronline.com/internet-of-things/10-of-the-biggest-iot-data-generators-4586937/ (accessed on 5 September 2019).
3. Nyeng, P.; Ostergaard, J. Information and communications systems for control-by-price of distributed energy resources and flexible demand. *IEEE Trans. Smart Grid* **2011**, *2*, 334–341. [CrossRef]
4. He, Y.; Chen, M.; Ge, B.; Guizani, M. On WiFi Offloading in Heterogeneous Networks: Various Incentives and Trade-Off Strategies. *IEEE Commun. Surv. Tutor.* **2016**, *18*, 2345–2385. [CrossRef]
5. Galli, S.; Scaglione, A.; Wang, Z. For the grid and through the grid: The role of power line communications in the smart grid. *Proc. IEEE* **2011**, *99*, 998–1027. [CrossRef]
6. Deconinck, G. An evaluation of two-way communication means for advanced metering in Flanders (Belgium). In *IEEE Instrumentation and Measurement Technology Conference*; IEEE: Victoria, BC, Canada, 2008; pp. 900–905.
7. Kinney, P. Zigbee technology: Wireless control that simply works. In Proceedings of the Communications Design Conference, San Jose, CA, USA, 30 September–2 October 2003; pp. 1–7.
8. Liang, H.; Choi, B.J.; Abdrabou, A.; Zhuang, W.; Shen, X.S. Decentralized economic dispatch in microgrids via heterogeneous wireless networks. *IEEE J. Sel. Areas Commun.* **2012**, *30*, 1061–1074. [CrossRef]
9. Fehske, A.; Fettweis, G.; Malmodin, J.; Biczok, G. The global footprint of mobile communications: The ecological and economic perspective. *IEEE Commun. Mag.* **2011**, *49*, 55–62. [CrossRef]
10. Mobile Industry in New Zealand Performance and Prospects. Available online: https://www.sparknz.co.nz/content/dam/telecomcms/sparknz/content/news/NZIER-Mobile-Industry-in-NZ.pdf (accessed on 5 September 2019).
11. Baron, B.; Spathis, P.; Rivano, H.; de Amorim, M.D. Vehicles as big data carriers: Road map space reduction and efficient data assignment. In Proceedings of the 2014 IEEE 80th Vehicular Technology Conference (VTC2014-Fall), Vancouver, BC, Canada, 14–17 September 2014; pp. 1–5.

12. Naseer, S.; Liu, W.; Sarkar, N.I.; Chong, P.H.J.; Lai, E.; Ma, M.; Prasad, R.V.; Danh, T.C.; Chiaraviglio, L.; Qadir, J.; et al. A Sustainable Marriage of Telcos and Transp in the Era of Big Data: Are We Ready? In Proceedings of the International Conference on Smart Grid Inspired Future Technologies, Auckland, New Zealand, 23–24 April 2018; pp. 210–219.

13. Mayer, C.P.; Waldhorst, O.P. Offloading infrastructure using delay tolerant networks and assurance of delivery. In Proceedings of the 2011 IFIP Wireless Days (WD), Niagara Falls, ON, Canada, 10–11 October 2011; pp. 1–7.

14. Andrae, A.S.; Edler, T. On global electricity usage of communication technology: Trends to 2030. *Challenges* **2015**, *6*, 117–157. [CrossRef]

15. Finlay, A.; Adera, E. *Application of ICTS for Climate Change Adaptation in the Water Sector: Developing Country Experiences an Emerging Research Priorities*; Association for Progressive Communications: Johannesburg, South Africa, 2012.

16. Cho, B.; Gupta, I. Budget-constrained bulk data transfer via internet and shipping networks. In Proceedings of the 8th ACM International Conference on Autonomic Computing, Karlsruhe, Germany, 14 –18 June 2011; pp. 71–80.

17. Munjal, R.; Liu, W.; Li, X.J.; Gutierrez, J.; Furdek, M. Sustainable massive data dissemination by using software defined connectivity approach. In Proccedings of the 2017 27th International Telecommunication Networks and Applications Conference (ITNAC), Melbourne, Australia, 22–24 November 2017; pp. 1–6.

18. Internet Service Provider Survey: 2018. Available online: https://www.stats.govt.nz/information-releases/internet-service-provider-survey-2018 (accessed on 5 September 2019).

19. A Smart City of 1 Million will Generate 180 Million Gigabytes of Data per Day by 2019, Predicts Cisco Study. Available online: https://www.dqindia.com/a-smart-city-of-1-million-will-generate-180-million-gigabytes-of-data-per-day-by-2019-predicts-cisco-study/(accessed on 30 January 2019).

20. Automotive Industry Trends: IoT Connected Smart Cars & Vehicles. Available online: https://www.businessinsider.com/internet-of-things-connected-smart-cars-2016-10 (accessed on 5 September 2019).

21. Transport Volume: Fleet Information. Available online: https://www.transport.govt.nz/resources/tmif/transport-volume/tv004/(accessed on 5 January 2019).

22. Future Demand New Zealand transport and society: Trends and projections. Available online:https://www.transport.govt.nz/assets/Uploads/Our-Work/Documents/10760e4aa4/fd-trends-and-projections.pdf (accessed on 5 September 2019).

23. Sitaram, D.; Manjunath, G. *Moving to the Cloud: Developing Apps in the New World of Cloud Computing*; Elsevier: Amsterdam, Netherlands, 2011.

24. Malik, A.W.; Mahmood, I.; Ahmed, N.; Anwar, Z. Big Data in Motion: A Vehicle-Assisted Urban Computing Framework for Smart Cities. *IEEE Access* **2019**, *7*, 55951–55965.

25. Digital Globe Case Study: "Digital Globe Moves Petabytes of Data Quickly and Securely Using AWS Snowmobile". Available online: https://aws.amazon.com/solutions/case-studies/digitalglobe/ (accessed on 18 September 2019).

26. Soares, V.N.; Farahmand, F.; Rodrigues, J.J. A layered architecture for vehicular delay-tolerant networks. In Proceedings of the 2009 IEEE Symposium on Computers and Communications, Sousse, Tunisia, 18 August 2009; pp. 122–127.

27. Kashihara, S.; Sanadidi, M.; Gerla, M. Mobile, personal data offloading to public transport vehicles. In Proceedings of the 2012 The Sixth International Conference on Mobile Computing and Ubiquitous Networking, Okinawa, Japan, 23–24 May 2012.

28. Hunjet, R.; Fraser, B.; Stevens, T.; Hodges, L.; Mayen, K.; Barca, J.C.; Cochrane, M.; Cannizzaro, R.; Palmer, J.L. Data ferrying with swarming UAS in tactical defence networks. In Proceedings of the 2018 IEEE International Conference on Robotics and Automation (ICRA), Brisbane, Australia, 21–26 May 2018; pp. 6381–6388.

29. Usbeck, K.; Gillen, M.; Loyall, J.; Gronosky, A.; Sterling, J.; Kohler, R.; Newkirk, R.; Canestrare, D. Data ferrying to the tactical edge: A field experiment in exchanging mission plans and intelligence in austere environments. In Proceedings of the 2014 IEEE Military Communications Conference, Washington, DC, USA, 6–8 October 2014; pp. 1311–1317.

30. Baron, B.; Spathis, P.; Rivano, H.; de Amorim, M.D.; Viniotis, Y.; Ammar, M.H. Centrally controlled mass data offloading using vehicular traffic. *IEEE Trans. Netw. Serv. Manag.* **2017**, *14*, 401–415. [CrossRef]

31. Dressler, F.; Handle, P.; Sommer, C. Towards a vehicular cloud-using parked vehicles as a temporary network and storage infrastructure. In Proceedings of the 2014 ACM International Workshop on Wireless and Mobile Technologies for Smart Cities, Philadelphia, PA, USA, 11 August 2014; pp. 11–18.
32. Cho, B.; Gupta, I. New algorithms for planning bulk transfer via internet and shipping networks. In Proceedings of the 30th International Conference on Distributed Computing Systems, Genoa, Italy, 21–25 June 2010; pp. 305–314.
33. Marincic, I.; Foster, I. Energy-efficient data transfer: Bits vs. atoms. In Proceedings of the 2016 24th International Conference on Software, Telecommunications and Computer Networks (SoftCOM), Split, Croatia, 22–24 September 2016; pp. 1–6.
34. Balasubramanian, A.; Mahajan, R.; Venkataramani, A. Augmenting mobile 3G using WiFi. In Proceedings of the 8th International Conference on Mobile systems, Applications, and Services, San Francisco, CA, USA, 15–18 June 2010; pp. 209–222.
35. Hou, X.; Deshpande, P.; Das, S.R. Moving bits from 3G to metro-scale WiFi for vehicular network access: An integrated transport layer solution. In Proceedings of the 2011 19th IEEE International Conference on Network Protocols, Vancouver, BC, Canada, 17–20 October 2011; pp. 353–362.
36. Cheng, N.; Zhou, H.; Lei, L.; Zhang, N.; Zhou, Y.; Shen, X.; Bai, F. Performance analysis of vehicular device-to-device underlay communication. *IEEE Trans. Veh. Technol.* **2016**, *66*, 5409–5421. [CrossRef]
37. Naseer, S.; Liu, W.; Sarkar, N.I.; Chong, P.H.J.; Lai, E.; Prasad, R.V. A sustainable vehicular based energy efficient data dissemination approach. In Proceedings of the 2017 27th International Telecommunication Networks and Applications Conference (ITNAC), Melbourne, Australia, 22–24 November 2017; pp. 1–8.
38. Li, Y.; Jin, D.; Wang, Z.; Zeng, L.; Chen, S. Coding or not: Optimal mobile data offloading in opportunistic vehicular networks. *IEEE Trans. Intell. Transp. Syst.* **2014**, *15*, 318–333. [CrossRef]
39. State Highway Traffic Volumes 1975–2018. Available online: https://www.nzta.govt.nz/resources/state-highway-traffic-volumes/ (accessed on 5 September 2019).
40. McKendry, P. Energy production from biomass (part 1): Overview of biomass. *Bioresour. Technol.* **2002**, *83*, 37–46. [CrossRef]
41. The Power to Know: Statistical Analysis System (SAS) Customer Support. Available online: https://support.sas.com/en/support-home.html (accessed on 5 September 2019).
42. Energy & the Environment: Emissions & Generation Resource Integrated Database (eGRID). Available online: https://www.epa.gov/energy/emissions-generation-resource-integrated-database-egrid (accessed on 5 September 2019).
43. Convert MJ to kwh—Conversion of Measurement Units. Available online: https://www.convertunits.com/from/MJ/to/kwh (accessed on 5 September 2019).

Article

Three Dimensional UAV Positioning for Dynamic UAV-to-Car Communications

Seilendria A. Hadiwardoyo [1,*], **Carlos T. Calafate** [2,*], **Juan-Carlos Cano** [2,*], **Kirill Krinkin** [3], **Dmitry Klionskiy** [3], **Enrique Hernández-Orallo** [2] and **Pietro Manzoni** [2]

[1] Department of Electronics ICT—IDLab, Universiteit Antwerpen—imec, 2000 Antwerp, Belgium
[2] Department of Computer Engineering (DISCA), Universitat Politècnica de València, 46022 Valencia, Spain; ehernandez@disca.upv.es (E.H.-O.); pmanzoni@disca.upv.es (P.M.)
[3] Department of Software Engineering & Computer Applications (MOEVM), St. Petersburg Electrotechnical University "LETI", 197022 St. Petersburg, Russia; kvkrinkin@etu.ru (K.K.); dmklionsky@etu.ru (D.K.)
* Correspondence: seilendria.hadiwardoyo@uantwerpen.be (S.A.H.); calafate@disca.upv.es (C.T.C.); jucano@disca.upv.es (J.-C.C.)

Received: 29 November 2019; Accepted: 29 December 2019; Published: 8 January 2020

Abstract: In areas with limited infrastructure, Unmanned Aerial Vehicles (UAVs) can come in handy as relays for car-to-car communications. Since UAVs are able to fully explore a three-dimensional environment while flying, communications that involve them can be affected by the irregularity of the terrains, that in turn can cause path loss by acting as obstacles. Accounting for this phenomenon, we propose a UAV positioning technique that relies on optimization algorithms to improve the support for vehicular communications. Simulation results show that the best position of the UAV can be timely determined considering the dynamic movement of the cars. Our technique takes into account the current flight altitude, the position of the cars on the ground, and the existing flight restrictions.

Keywords: PSO; genetic algorithm; ITS; UAV; simulation; dynamic positioning; 3D placement; vehicular communications

---

1. Introduction

UAVs can be easily deployed as information relays for emergency scenarios thanks to their flexibility. Since they can be positioned at high altitudes compared to ground infrastructure deployments, UAVs can achieve long-range signal transmissions with better Line-of-Sight (LOS) conditions. Hence, compared to standard ground infrastructure relays, UAVs can offer significant advantages [1]. For instance, UAVs can become mobile infrastructure elements in situations where the existing infrastructures are limited, as in rural areas or even in urban areas when emergency situations take place. Other uses of UAVs include improving Intelligent Transportation Systems (ITS) on disaster assistance operations [2] and performing remote sensing [3], among others.

In terms of supporting vehicular communications, UAVs can relay information for vehicles on the ground when a direct multihop link between them is not available [4]. A UAV can move freely in three-dimensional space without having to follow routes or specific trajectories, a feature that ground vehicles do not have. Also, a UAV can be deployed as a mobile Road Side Unit (RSU) since it flies above high buildings in urban areas that can obstruct the communication between cars on the ground [5]. This can be the solution to providing continuous connectivity among the cars. The presence of UAVs as relays for vehicular communication can assist in forwarding messages when direct car-to-car communication is not possible. Differently from cars, in which the movement is limited to the two dimensional space and to specific roads, UAVs can become an alternative to provide connectivity since they have no space constraints [6].

In three dimensional scenarios that involve UAVs, the diffraction caused by buildings, mountains, or high-level terrains blocking the signal is likely to occur [6]. Placing the UAV at a high altitude can be a solution to avoid the Non-Line-of-Sight (NLOS) conditions caused by high-level terrains such as hills or mountains. Nonetheless, the communication range decreases if the UAV is placed too high, and in addition, there are regulations that restrict the permitted flight height of the UAV. Hence, the position of the UAV has to be determined in the 3D environment, where it can still maintain the signal range towards the receiver by adapting its position to avoid NLOS conditions, while still respecting the maximum flight height allowed, among other flight restrictions. Thus, to account for various dynamic constraints, a solution able to determine the optimal location throughout the time should be found [7].

In this paper, a dynamic positioning technique for UAVs is proposed by implementing optimization algorithms to find the best position of a UAV in a dynamic vehicular communications environment. The position adopted by the UAV at any instant of time will allow it to support the communication between cars on the ground. Such an optimum position is obtained by considering the signal quality received by the cars acting as signal receivers.

We have used a previously developed model [8] to calculate the signal quality. The model considered the 3D environment with terrain irregularities that might obstruct the communications. The goal is having the UAV as a relay placed in its optimum position at each specific time instant so that it can forward information from one car to another when direct car-to-car communications are obstructed by high terrains in the 3D space. In this research work, we have used a Particle Swarm Optimization (PSO) algorithm and a genetic algorithm (GA) that helped us to find the best position of the UAV throughout the experiment. The ideal and desired case is that the UAV can still achieve adequate signal conditions towards every car on the ground [9]. The results show that the positioning technique can find the optimized position by defining its best altitude adequately that it is not hindered by terrain blockages.

The paper is organized as follows: the research works related to our proposal are discussed in the following section. In Section 3, we highlight the problems to be tackled and our contribution in this paper, where we start by a discussion of the problem and the optimization algorithms used. In Section 4, the simulation framework which we have used to evaluate our algorithm is presented, as well as its setup. Afterwards, in Section 5, we present the results and discuss the main findings. Finally, in Section 6, we present the conclusions and discuss future works.

2. Related Works

Several research efforts have been documented involving UAVs as nodes deployed to provide connectivity. Lin et al. [10] studied the deployment of LTE connectivity for UAVs. These authors highlighted some challenges such as LOS propagation in the sky. Other related research work was performed by Van der Bergh et al. [11], where they present an analysis of the impact of LTE-enabled UAVs on an already existing LTE ground network. The authors studied the case of having UAVs as user equipment and base stations. The work of Nguyen et al. [12] highlighted the interference that might occur between terrestrial and aerial-based radio connectivity.

Using simulation, researchers have investigated the methods to optimize the communications between UAVs and cars. One of the research efforts done was analyzing the deployment of lesser amount of UAVs for communications by having an optimal altitude [13]. In [14], the authors analyze the characteristics of communications between drone and vehicle in terms of delay. When supporting connectivity for the groups of cars that are disconnected, drones can link these groups as relays [15]. Nevertheless, in order to get the optimum connectivity between the UAVs and cars, it is needed to find the best location to place the UAVs so that it can transmit adequate signals to the ground vehicles.

Node placement is crucial when it comes to wireless networks where the nodes are located freely in space. To get the best performance in wireless networks, placement strategies can be useful [16]. An algorithm for base station placement to maximize the network capacity was proposed in [17]. Another proposal that instead seeks to maximize the network lifetime is [18].

In car-to-car communications, the placement of nodes that act as relays affects the performance of information dissemination. Optimal placement of RSUs is discussed in [19] with the aim of improving connectivity at intersections. Looking into the number of vehicle reports in the communication range of the RSUs, the proposed scheme can find the best location for these RSUs. Other works [20] proposed a method to determine the RSU deployment so that it is able to maximize the number of vehicles within radio range. Anther placement method was proposed in [21], being able to minimize the average report time from cars to RSUs.

The idea of having UAVs as mobile infrastructure units has been investigated by various researchers. Chiaraviglio et al. [22] investigated the use of small cells on top of the UAVs to cover hotspot areas. Huang et al. [23] proposed a novel coordinated path planning algorithm for multi-UAVs to deal with the trajectory smoothing problem using optimization algorithms. The work in Reina et al. [24] focused on the application of a metaheuristic algorithm to solve multi-objective coverage problems of UAV networks. The issue with battery fuel capacity of the UAV was analyzed by Song et al. [25]. Another idea to overcome the energy issue is proposed by Trotta et al. [26]. In their work, the UAVs that monitor a set of points of interest make use of the public bus network for recharging. This is solved by Mixed Integer Linear Programming techniques, where the formulation identifies the UAVs, the next bus, and the next point of interest.

Specifically talking about UAV placement in 3D environments, proposals such as [27] aim at determining the best place for a UAV to provide connectivity for nodes in indoor buildings affected by a disaster. Nodes, which are the users inside the building, can be covered by a strategically located UAV so that the total transmitted power is kept to a minimum. Another work [9] proposes a method to position a team of UAVs so as to maximize the user coverage ratio in a 5G network. Determining the optimal placement for a drone can be done using a PSO algorithm as well, as discussed in [28], where a coverage area can be maximized while still considering the drone capacity in the scope of public safety and disaster management. Considering the lifetime of drones to maximize the total throughput of the receivers was also discussed in [29] when deploying and positioning a swarm of drones.

A topic covered in [30] was deploying UAVs as aerial base stations in a three-dimensional environment. The proposed work was about deploying UAVs to support cellular networks as it allows extending the coverage area. In another similar work [31], UAVs are deployed to support cellular networks in the presence of dynamic events. When deploying UAVs that support vehicular communications, a UAV position can be determined using an algorithm proposed by [32], so that it offers Quality of Service (QoS) communications to the cars on the ground. In our work, we provide a technique to support the communication of the moving nodes on the ground, in which in this case the nodes are the cars.

In this work, we seek to determine the optimal position of a UAV involved in UAV-to-car communications. As the cars on the ground are moving dynamically, the UAV position should be adjusted throughout the time so that it adapts to the radio links towards the different cars involved. In our work, we used optimization algorithms such as Particle Swarm Optimization (PSO) and Genetic Algorithm (GA) to determine the optimum position of the UAV in terms of achieving the best coverage for the cars on the ground at each specific time. Since the environment is three dimensional, the altitudes of the cars and of the UAV are taken into account at all times. In addition, the irregularities of the terrains are also considered when determining the UAV location and the channel quality towards the different ground vehicles. Hence, the optimization algorithm is used for finding the best position of a UAV in an area where the connectivity is minimum due to the lack of infrastructure. In such scenarios, the connection between cars cannot be maintained since the Line of Sight in the mountainous area is hindered by mountains. Thus, the whole three dimensional space (above ground) is considered as feasible for the UAV to explore with the aim of offering connectivity to the mobile targets on the ground.

3. Optimum UAV Positioning

In this section, we describe the formulation of the positioning problem. In particular, we will analyze how the optimization algorithm solves the positioning problem of nodes. We will also explain in detail the two optimization algorithms used in this research work, which are the Particle Swarm Optimization (PSO) algorithm and a genetic algorithm (GA). In addition, by considering a 3D environment, irregularities of the terrains might affect the signal transmitted; hence, we will also discuss in more detail the path loss model taken into account when proposing the different UAV positioning strategies.

3.1. Problem Formulation

For our study, a rural area scenario is chosen since it has hilly roads and significant terrain irregularities. This way, we can test whether the UAV position in 3D is affected by the terrain levels. Notice that terrain irregularities often cause that, when two cars are communicating, there might be a high chance that there is a hill in between. Thus, the cars experience NLOS conditions in their radio link due to the presence of hills, which make necessary the presence of information relays to maintain the connectivity between these cars. Considering that most rural areas lack enough infrastructure so as to help in information relaying in such cases, alternatives must be sought.

As depicted in Figure 1, cars on the ground might not have the possibility to connect directly with each other, or they can experience very poor link conditions due to the NLOS features of the wireless channel caused by hills. One solution to provide the necessary connectivity among them is to deploy a UAV that can act as a mobile relay. This way, the UAV can forward information from one car to another. In our work, the experiments are made in a simulated environment where the cars follow a specific trajectory on an existing road. Regarding the UAV location, it should change throughout time so as to adjust to the locations of the different cars involved, which are updated every second.

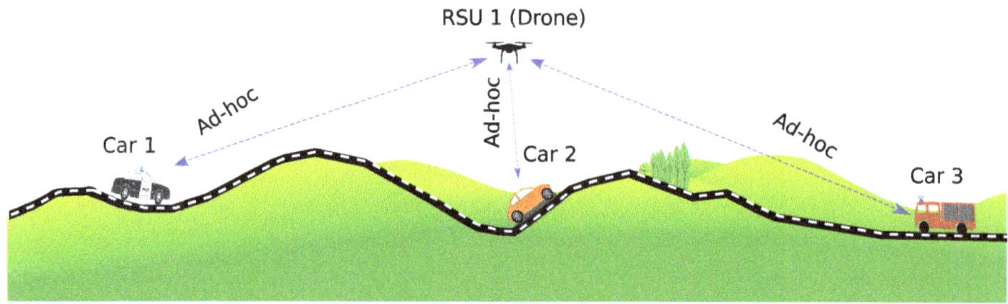

Figure 1. Unmanned Aerial Vehicle (UAV) acting as a Mobile Road Side Unit (RSU) [33].

The position of the UAV is adjusted depending on the signal strength received by the cars on the ground. A new optimal position has to be found at every time step as cars are moving. In particular, the new position should allow the UAV to maintain the best possible communications link with the cars by considering signal levels. As the sender, or the UAV in this case, transmits the signal in a three-dimensional environment to the receivers (the different cars in our case), possible NLOS conditions must be accounted for in the end-to-end link. Hence, the UAV is positioned at a certain altitude in a way that it can achieve good visibility levels towards the current location of a target car.

3.2. Optimization Problem

To determine whether our positioning technique is optimal or not, the metric used is the Received Signal Strength Indicator (RSSI) value measured at the different receivers (cars). So, the optimal UAV location is defined based on the average RSSI value. In particular, the average RSSI should be greater than a minimum threshold of -89 dBm to achieve full coverage connectivity. This value is the main criteria to determine that the placement meets the desired conditions. In detail, we analyze the RSSI at each car when the UAV's position is selected. The RSSI is calculated through a network simulator. Both the position of the UAV and the car are inserted as inputs to the simulator. This repeats when a candidate position of the UAV is to be tested, and when the position of the car changes.

To calculate the RSSI, the path loss model is taken into account depending on the location of the nodes. This way we can determine whether an obstacle might be present and blocking the communication between two nodes. In other words, this condition is called knife-edge diffraction, where the blocking obstacle is the knife. The main objective in our work is to find the best location of the UAV as a sender in a three-dimensional environment in such a way that the signals transmitted by the UAV can be received by the cars on the ground with adequate reception power.

A simple but inefficient solution would be doing an exhaustive search. By doing this, every location in the search space should be explored and tested so that the best definitive location can be found at every time step. Nevertheless, doing an exhaustive search can take a lot of time and consume significant resources. A more sophisticated yet effective solution would be adopting a meta-heuristic optimization algorithm. Using this strategy, optimum values could be obtained without having to test every possibility in the search space. Possibly, random values of candidate locations would be defined at the beginning, but the algorithm would search the other best possible candidates in an iterative manner. In the case of finding the UAV position, to get the most precise optimum position we can do an exhaustive search. However, since the possibilities to explore are too many with respect to the values of longitude, latitude, and altitude as inputs, the search may take forever. Hence, to offer quicker solution in finding a position in 3D area, the optimization techniques can be performed. Thus, based on two different meta-heuristic optimization techniques, two algorithms will be implemented in our positioning technique, which will be covered in more detail in the following subsection.

3.3. Particle Swarm Optimization

An optimization algorithm inspired by the social behavior of animals, such as flocks of birds an schools of fish, is the Particle Swarm Optimization (PSO) [34] algorithm. As part of swarm intelligence, the algorithm improves a candidate solution, or particle, through iterations starting from random solutions, and picking the best experience of each iteration (*BestLoc*), and the best global experience from all the iterations (*GlobalBest*). A particle (*Loc*), or a candidate location, is a possible position for our UAV (*pos*). In each iteration ($t + 1$), the best location of each particle (loc_i) and the best global location are updated. The main parameter affecting the updates is the velocity (V). The velocity in PSO is a distance achieved by a particle, or a candidate following location, from its current or previous location at an iteration. The velocity is affected by the inertial weight (W) when it is varied (in most cases it has a value of 1), the acceleration coefficients ($c1$ and $c2$, which both have the value of 2), and the random numbers ($r1$ and $r2$) uniformly distributed in the interval between 0 and 1. At the beginning, the algorithm has a population of candidate solutions. The equation to calculate the velocity that defines the updates is:

$$V_i(t+1) = W \times V_i(t) + r1 \cdot c1 \cdot (BestLoc_i(t) - Loc_i(t)) + \\ r2 \cdot c2 \cdot (GlobalBest(t) - Loc_i(t)) \tag{1}$$

The following or next location of the particle is obtained by adjusting the velocity to the current or previous location of the particle as:

$$Loc_i(t+1) = Loc_i(t) + V_i(t+1) \tag{2}$$

Algorithm 1 shows the implementation of PSO algorithm. The algorithm begins by having a population with random locations. Hence, a set of particles representing candidate locations makes up our population. The candidate locations, which are obtained randomly, are limited by the variables set as maximum (var_{min}) and minimum (var_{min}). With the initial velocity, which is zero at the beginning and initial locations, the function will find the cost and calculate the initial value, which in our case is the RSSI (*rssi*) value for each particle. The best RSSI of all particles (*Bestrssi*) will be defined as the global best. PSO will then find a new value after having the initial position and RSSI for all the particles. The new value is obtained by calculating the new velocity based on the local best position (*Bestpos*), and the global best position (*GlobalBestpos*). The value of each position will be updated according to the new velocity value, which in turn can return a new *GobalBestPos*. The *Bestpos* is updated if the new position is better. The same happens to the *GlobalBestpos* if its new value after the iteration is better than the one from the previous iteration. By doing more iterations with newly calculated velocities, we can get refine the solution until the best value for the *rssi* is found.

Algorithm 1 Particle Swarm Optimization (PSO) Algorithm.

Input:
 Maximum number of iterations ($MaxIt$).
 Population size ($PopSize$).
 Lower and upper bound variables (var_{min}, var_{max}).
Output:
 Best value of all particles ($GlobalBestrssi$).
 1: **for** i=1:$PopSize$ **do**
 2: $V_i = 0$
 3: $pos_i = rand(var_{min}, var_{max})$
 4: $rssi_i = pathloss(pos_i)$
 5: $Bestpos_i = pos_i$
 6: **if** $Bestrssi_i < GlobalBestrssi$ **then**
 7:   $GlobalBestrssi = Bestrssi_i$
 8: **end if**
 9: **end for**
10: **for** t=1:$MaxIt$ **do**
11: **for** i=1:$PopSize$ **do**
12:   $V_i(t) = W * V_i(t) + r1 * c1 * (Bestpos_i(t) - Loc_i(t)) + r2 * c2 * (GlobalBestpos(t) - pos_i(t))$
13:   $pos_i(t) = pos_i(t) + V_i(t))$
14:   $rssi_i = pathloss(pos_i)$
15:   **if** $rssi_i < Bestrssi_i$ **then**
16:    $Bestpos_i(t) = pos_i(t)$
17:    $Bestrssi_i = rssi_i$
18:    **if** $Bestrssi < GlobalBestrssi$ **then**
19:     $GlobalBestrssi = Bestrssi_i$
20:    **end if**
21:   **end if**
22: **end for**
23: **end for**

3.4. Genetic Algorithm

A Genetic Algorithm (GA) is a non-deterministic optimization method based on genetic theory [35]. GA simulates the evolution of a population of candidate solutions to optimize a problem. The population or candidate solution adapts to the environment over the generations, being these generations renewed through iterations. This iterative process resembles biological behaviors like the crossovers of chromosomes, mutations of genes, and inversions of genes, processes that occur to living organisms over generations. In our work, we use a GA to simulate the evolution of a population of UAV locations adapting to the cost function. The cost function will be the same as the one in PSO: the average RSSI towards the different receivers.

As depicted in Algorithm 2, we first define that the population is a candidate set of optimum locations for the UAV acting as a relay for cars. The candidate set of optimum locations, defined as latitude, longitude, and altitude, is limited by the lower and upper bound variables (var_{min}, var_{max}). At first, the populations are generated randomly, but having specific genes or characteristics. The genes, in this case are the latitude, longitude, and altitude associated with the UAV's location. From those characteristics, we define the fitness, which in this case is the RSSI. Afterwards, we build a new generation. This is done by selecting the parents or the chromosomes in the current generation. The parents are chosen by randomly selecting two sets of chromosomes. After selecting the chromosomes, the genes inside the chromosomes are crossed over to create new chromosomes. This is when the crossover process occurs, consisting of combining the genes, or, in the case of this work, the location parameters (latitude, longitude, and altitude). With the new chromosomes, we then do a mutation. The mutation is performed to maintain the genetic diversity, or, in this case, to increase the number of candidate locations. With the mutated chromosomes, we determine the fitness, which in this case is the RSSI. After getting a new generation, this process is repeated according to the limit of the generations. With more and more generations produced, better RSSI values can be obtained.

Algorithm 2 Genetic Algorithm (GA)

Input:
    Maximum number of generations ($MaxGen$).
    Population size ($PopSize$).
    Lower and upper bound variables (var_{min}, var_{max}).
Output:
    Best value of all chromosomes ($GlobalBest_{rssi}$).
 1: **for** i=1:$PopSize$ **do**
 2:      $pos_i = rand(var_{min}, var_{max})$
 3:      $rssi_i = pathloss(pos_i)$
 4:      $Bestpos_i = pos_i$
 5: **end for**
 6: **for** t=1:$MaxGen$ **do**
 7:    **for** i=1:$PopSize$ **do**
 8:        $SelectParents$
 9:        $Crossover$
10:        $Mutation$
11:        $pos_i = pos_{i+1}$
12:        $Bestpos_i = pos_i$
13:        $Bestrssi_i = rssi_i$
14:        $GlobalBestrssi = Bestrssi_i$
15:    **end for**
16: **end for**

3.5. Path Loss Model

The RSSI value can be obtained through the path loss model. To derive the path model used in our work, we relied on our previous work [8]. With this model, the signal loss can be calculated by considering terrain features that act as obstacles. By considering the elevation information retrieved from a Digital Elevation Model (DEM), we can determine when the terrain affects the LOS between a sender and a receiver. A knife-edge is detected when there is a blocking terrain, as depicted in Figure 2. Through multiple knife-edge diffraction effects, the actual end-to-end loss is obtained. On our previous work [8], we opted for the Bullington model [36] as it offers a good trade-off between performance and computational costs.

According to the information obtained from the DEM, the path loss or signal attenuation can be calculated. By taking into account the elevation level of the terrain, we can get the height of the knife through the difference of altitude between the UAV and the car. Notice that, in rural areas, signal interferences from external sources are minimal, meaning that the terrain becomes the main factor affecting signal quality.

A DEM provides real-world terrain data that, for the purpose of our current work, can provide information about the elevation of the terrains with respect to the sea level. Since the terrains have different levels of elevation in the area, we can easily find whether the terrain is hilly, mountainous, or flat. This elevation information is obtained by indicating the latitude and the longitude of each selected location.

A LOS segment connecting the sender and the receiver can help us to spot the knife or the blocking terrains. The height of this segment and the terrain elevation in the locations along the segment are compared to determine whether the LOS is blocked by the terrain or not. An obstacle is present whenever the terrain level or the elevation is higher than the LOS segment height. This way, the signal loss can be calculated as the diffraction effect occurs. The signal attenuation is obtained by accounting for the height of the obstacle, the wavelength, and the distance between the obstacle and the sender and receiver terminals. In this case, the Fresnel-Kirchoff diffraction parameter will also define the signal attenuation. In our simulation framework, this path loss model is incorporated. This will be explained in more detail in Section 4.

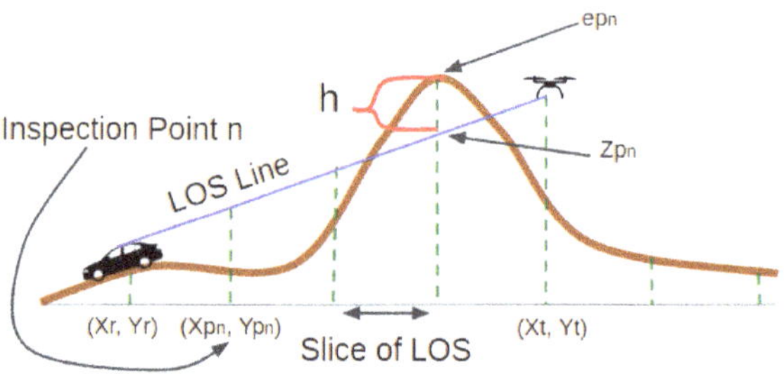

Figure 2. Detecting the Hills as Obstacles [8].

4. Overview of Simulation Tools Used

In this section, we provide the details on how our proposed solutions are implemented in the simulation tools adopted. A testing framework was developed to get the best position of the UAV in a simulated environment. In addition, the simulated environment setup is also covered in this section.

4.1. Testing Framework

An application that runs the simulation was developed as a testing framework, which will implement both the PSO and the GA algorithms to determine the optimum placement of the UAV. This framework integrates OMNeT++ [37] as a network simulation tool, SUMO [38] as a vehicular mobility simulation tool to characterize the movement of the cars on the ground, as well as Veins [39] as a vehicular network simulation tool. Our application determines the location by having the average value of the signal received by the cars (RSSI) as the algorithm's cost function value. This is obtained by executing the simulation tools and extracting the results.

Figure 3 shows the architecture of our testing framework. The algorithm was implemented in a separate developed application as a framework. The cost function value or the best fitness of the algorithm was obtained by running the simulation tools combined (OMNeT++, SUMO, and Veins). The variables that affect the RSSI values obtained are the receivers' locations as returned by SUMO. To detect the signal blockages or knife-edge effects that affect the signal strength, a path loss model in [8] was implemented.

By default, the parameters are selected when running the PSO and GA. For the PSO, the number of particles was set to 50, and the number of iterations was also set to 50, whereas in GA both the number of populations and the number of generations were set to 50. The exploration space for both algorithms is limited by the map space in the simulation. Both PSO and GA algorithms will have 50 candidate locations at first, before iterating or going through generations. By considering the cars' locations determined by SUMO, Veins can determine the best RSSI value by testing every possible location.

In this particular case, the position of the UAV dynamically changes depending on the mobility of the cars on the ground. Default values of the commonly used parameters are chosen by using either PSO or GA. As a side note, these parameters can affect the optimality of the position. Having more iterations on the algorithm used can effectively result in a more optimum position. However, we should not neglect the fact that, in real deployments, more iteration times might result in having the cars' position to change, making the results less effective. Since we are working with simulation time, and not real time, for practical purposes we consider the calculation involved to be instantaneous. Hence, the main goal is to determine the best-case performance achieved through the sequence of positions determined by PSO and GA.

The performance of finding the best UAV position depending on the cars' position on the ground differs from time to time, and subject to the positions of the cars on the ground. Our main focus on this work is making a comparison of both algorithms while having similar parameters. Due to this fact, we did not highlight the matter of selecting the best parameters for algorithmic convergence. As a side note, making these algorithms converge can be a challenge since both PSO and GA are heuristics where no exact configuration or parameter choice guarantees obtaining the optimal result in a search space. Our maximum recorded time to get the optimum solution was with 320 particles, and the minimum recorded time was with 16 particles. Concerning GA, the maximum recorded time in getting the solution was with 760 generations, and the minimum recorded time was with 32 generations. The primary goal of this work is uniformly using the commonly used parameters in both PSO and GA, and then compare which one is more accurate and efficient.

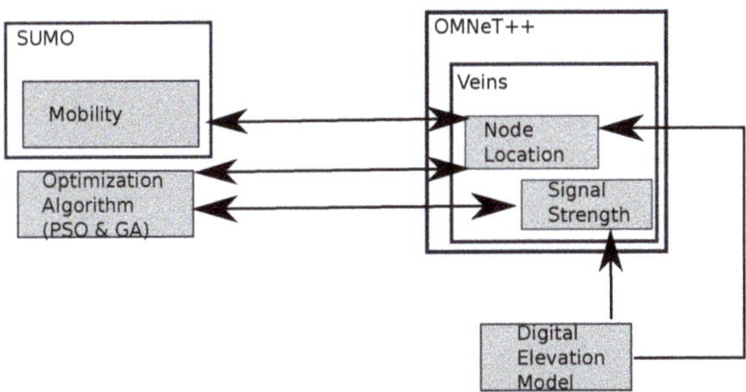

Figure 3. Testing Framework Architecture.

4.2. Simulation Setup

We have defined a scenario to test the optimization algorithms. Specifically, we have chosen a mountainous rural area in Pont de Suert, Catalonia, Spain, near the Pyrenees Mountains. A real map of Pont the Suert was imported from Open Street Map (OSM) [40] to make the simulation more real. To complement the environment imported into the simulation, we have also considered the elevation information, which allowed us to have a complete characterization of the levels of the terrain in the chosen area. In particular, we have obtained it from the SRTM DEM [41]. One of the reasons why we have chosen this particular location is because, in this area, irregular terrains exist that are prone to hinder communications. The area imported has a size of 5000 × 5000 m in Cartesian coordinates. With this in mind, the exploration space for the optimization algorithm is defined. As for the altitude, it is limited to 120 m above the ground, which is the maximum flight height permitted in Spain.

In the scenario, we have placed three cars that have their trajectories defined according to Figure 4. In theory, the UAV that acts as a relay can offer the connectivity between two nodes or cars. However, we have considered three cars in the scenario to show that the UAV can simultaneously act as a data relay for more than two nodes. However, the number of cars served by the UAV should be taken into consideration carefully, since a growing number of cars will complicate the UAV positioning strategy.

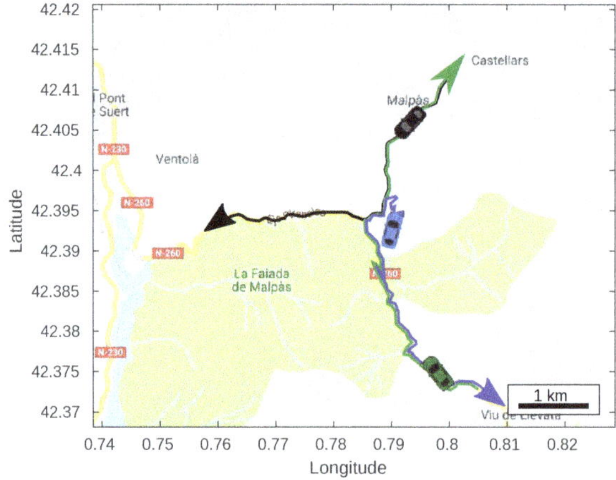

Figure 4. Trajectories for ground vehicles in our experiments.

As for simulating the UAV, which is of the hexacopter type, in the scenario, we assume that it has no limitations in terms of energy. The UAV is deployed on-demand at a specific time to act as a mobile relay. In addition, we have a single UAV in the scenario since our main focus is to find the optimal position for this UAV, and not on how multiple UAVs could be deployed to provide coverage, an alternative that is outside the scope of this work. It is assumed that the UAV carries a small embedded system having some computational power, performing better than e.g., a Raspberry Pi, in addition to an IEEE 802.11p wireless interface for communication with the vehicles, and a 4G interface for remote monitoring tasks. Overall, the payload weight remains below than 1 kg.

The position of the UAV is calculated by considering the current location of each car every second. The car positions are obtained by the UAV from their beacons. This way, we can assess whether the UAV offers optimal coverage when communicating with the cars. The SUMO traffic simulator generates the cars' movements, and it simulates the scenario for 280 s. This limited time is due to the fact that, after that mentioned time limit, the cars will no longer be in the exploration space boundaries, in which the cars will be too far to each other that the communication will surely be out of range.

The UAV generates UDP packets in the scenario and broadcasts them to the cars on the ground. The mode of communication is ad-hoc in this case. The packets that are transmitted by the UAV are Basic Safety Messages (BSMs). Each second, the UAV sends 10 packets. In this scenario, an 802.11p connection is considered, with broadcast communications taking place. Table 1 details the parameters that are set in our simulation experiments.

Table 1. Simulation parameters.

Parameter	Value
Transmission Power	200 mW
Antenna	5 dBi
Packet Size	1.4 kB
Message Type	BSM
Packet Sending Rate	10 Hz

5. Results

As explained in the previous section, our proposed solution to find optimal UAV positions was developed using simulation tools. After executing the test framework that runs the simulation, we have obtained the optimum sequence of UAV positions, as well as other useful information, such as the RSSI, time, and the position of the other nodes.

By using two different optimization algorithms, we have defined the optimal positions and, in turn, can derive the trajectory. Other results include the impact on altitude, received signal strength, total path length, and speed.

5.1. Uav Positions

We have obtained the results concerning the best locations of the UAV through the simulation experiments. The location information consists of latitude, longitude, and altitude. In addition, the value for the RSSI is also obtained for each UAV location at every second of the simulation. Thus, the locations of the UAV in a real map can be plotted using this information, allowing us to draw the trajectory of the UAV throughout the simulation time.

The locations of the UAV at specific times are represented in Figure 5 as a result from either the PSO or the GA. At $t = 0$ s in Figure 5a, both the positions obtained using PSO and GA are located in the center of the map. Specifically, such location is at a central position with respect to the three cars on a 2D map. At $t = 90$ s (Figure 5b), the location of the UAV obtained using PSO is near of the two cars. At this time it is not near to the other car because this car is located at a lower altitude, and hence still within LOS. On the other hand, with the GA, the UAV location is a bit more towards the other car that is far away. However, this does not make much difference as this is the time where all the cars cross with each other, and so the expected RSSI result is high. At the time where the cars move away, as depicted in Figure 5c, the positions obtained using PSO and GA are not too close from each other. In fact, the location obtained using GA is quite far away from the car moving west. On the other hand, the location obtained using PSO is closer to the center. At $t = 270$ s, as presented in Figure 5d, the position of the UAV obtained using either PSO or GA is readjusted near to the center. This time, the cars are spreading around, and hence the UAV assumes a central position for both PSO and GA strategies.

5.2. UAV Trajectory Based on Positions

The locations points obtained throughout the simulation can be sequenced with respect to time in order to get the trajectory of the UAV. This allows us to observe how the UAV should be adjusting its position in order to maintain the best connectivity towards all the cars on the ground, as shown in Figure 5. We have plotted the trajectory for both optimization algorithms in a two-dimensional and three-dimensional environment in Figure 6. These trajectories represent the most ideal path to be followed by the UAV to maintain the connection throughout the time.

The trajectory represents the optimum positions of the UAV, which change through every second, as a response to the movement of cars. Notice that the UAV should have a dynamic behavior, similarly to the mobility of the cars on the ground. As the UAV movement is exploring the 3D space, a trajectory was drawn as well, as depicted in Figure 6, where we can see the altitude variations for both sequences of UAV best positions corresponding to PSO and GA, respectively.

5.3. Impact on Flight Height and Altitude

As the terrain is irregular, or hilly in the scenario, we have also obtained the altitude of the ground nodes along the time so as to have their detailed elevation information. The altitude changes as the roads have different elevation levels as they are hilly. Cars are moving at an altitude between 850 and 1100 m above sea level, as depicted in Figure 7, whereas the UAV flies at altitudes between about 970 to 1300 m above sea level if using PSO, and up to 1400 m above sea level if using GA.

Looking into Figure 7, we can observe that, at a midpoint of the simulation time, at about $t = 120$ s, the UAV altitude, as a result from both PSO and GA algorithms, is lower than the one for car 3. The UAV flies lower since the cars are located near to each other. Since the distance between ground nodes is small, the UAV does not need to fly high to maintain connectivity. However, at the end of the simulation, since the cars' locations are far from each other, the UAV has to fly higher. So, even though the cars' altitudes are not greater at the endpoint compared to the one at the midpoint, the UAV still has to increase its altitude since it attempts to maintain line-of-sight conditions towards all the cars. Nevertheless, notice that greater altitudes also cause the distance towards all three cars to increase. Thus, the UAV avoids flying too high to prevent losing signal quality due to higher distances, while simultaneously avoiding NLOS conditions.

The maximum altitude achieved by the UAV is at about $t = 230$ s. It is reasonable since, at that particular time, more hills are present that obstruct the signal transmission. The altitude of car 1 is below 900 m, and this might as well result in the UAV having to fly higher to avoid NLOS towards it.

As for comparison between PSO and GA, we can observe in Figure 8 that the trend is more or less similar, with its altitude decreasing starting from about $t = 40$ s, and increasing at about $t = 130$ s. Both algorithms result in a maximum altitude value at about $t = 230$ s as well. However, the difference can be spotted in terms of consistency. With the GA approach, the curve trend is more dynamic than for the PSO approach. An example can be seen at about $t = 170$ s. When the PSO indicates a more stable altitude change, GA shows a drastic change from an altitude of about 1250 m to 1050 m. Overall, flying height results shows that the GA trend is not as stable as the PSO trend. The heights produced from the experiment with GA have more varieties and sudden changes. An example can be seen between $t = 50$ s and $t = 100$ s. Within this time range, the changes are drastic for the height of the UAV, dropping from 97 m to 28 m, and then rising up again to 79 m. On the other hand, the height only rises from 28 m to 55 m with the PSO approach. In this case, if using GA, the UAV needs more effort in flying as in terms of height and altitude, it tends to be higher than when using PSO.

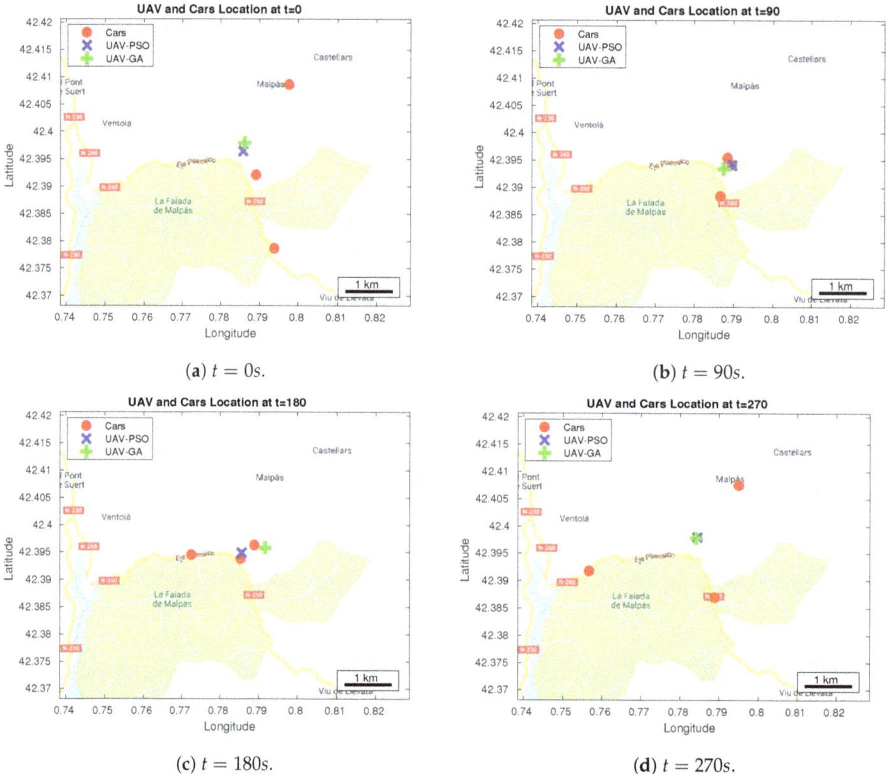

(**a**) $t = 0s$.

(**b**) $t = 90s$.

(**c**) $t = 180s$.

(**d**) $t = 270s$.

Figure 5. UAV and cars' locations obtained at different timings by using both algorithms.

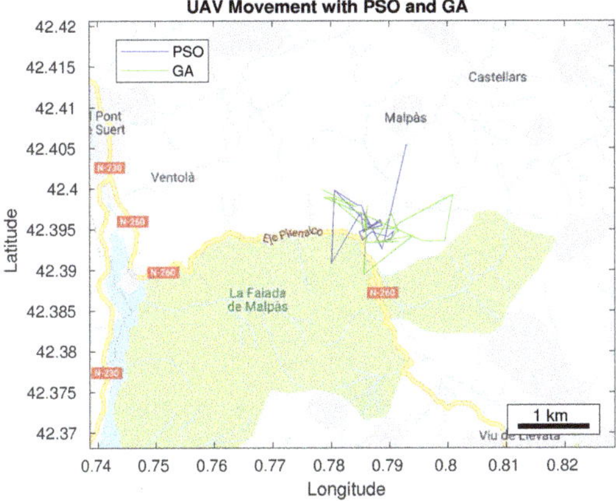

(**a**) 2D mobility patterns.

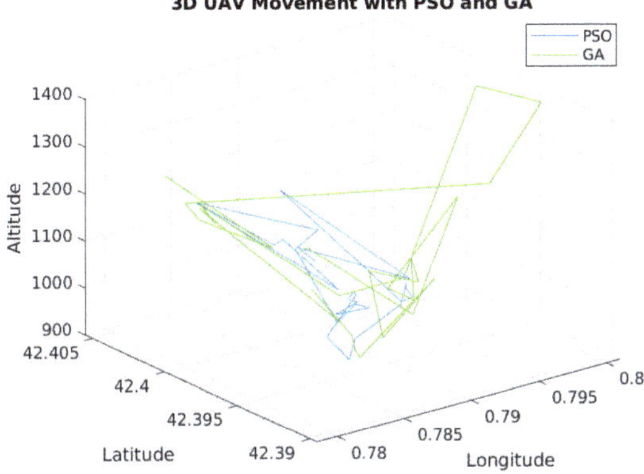

(**b**) 3D mobility patterns.

Figure 6. UAV trajectories generated by the PSO and GA algorithms.

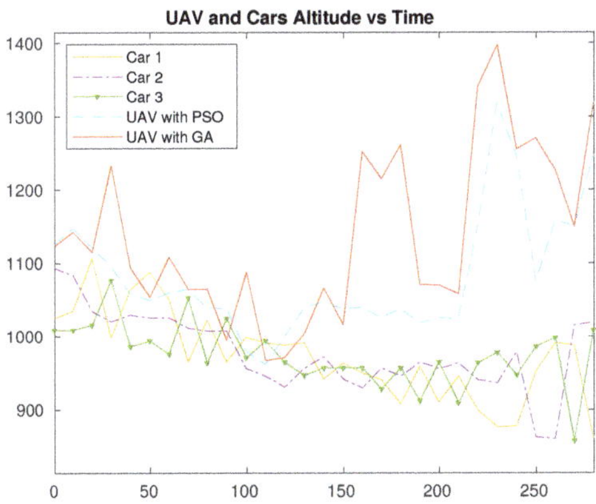

Figure 7. Altitude variations throughout time for the UAV and ground vehicles.

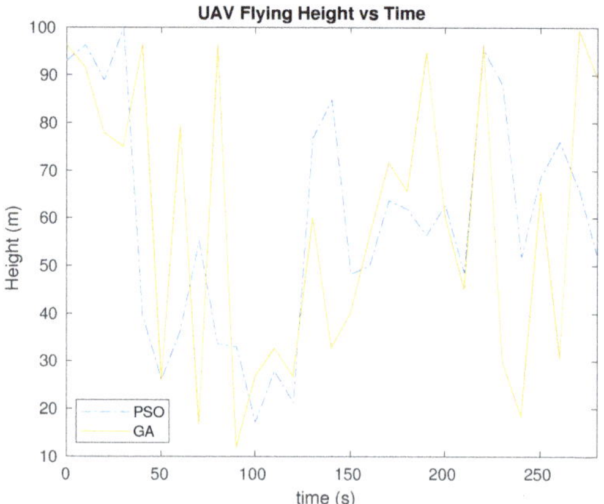

Figure 8. Flying height variations throughout time for the UAV.

5.4. Impact on Received Signal Strength

The average values of the received signal quality on the cars throughout the simulation time are represented in Figure 9 for the PSO and GA approaches. According to the figure, the lowest average values ($\leq$85 dBm) are achieved at the beginning and end of the simulation for both GA and PSO. This is due to the fact that, at those times, the cars are spread in the area and are far from each other. Hence, the UAV is placed in-between the cars, but, since the cars are distant, the UAV had to fly higher to achieve LOS towards all three cars by adjusting its height in order to avoid blockages from the high-level terrains. This will, in turn, weaken the signal strength.

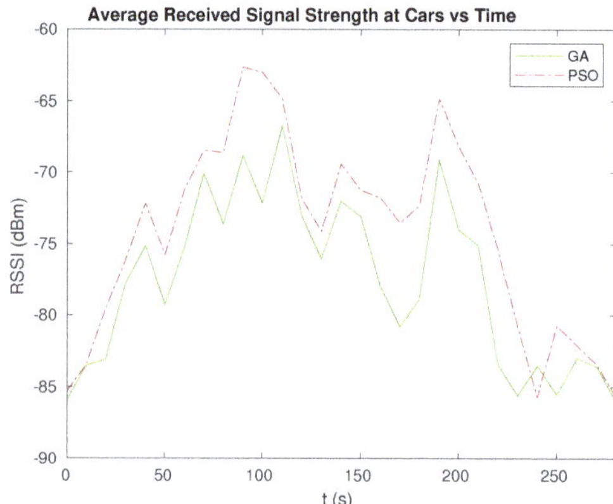

Figure 9. Average Received Signal Strength Indicator (RSSI) values at the receivers throughout time.

The best signal recorded in the simulation for PSO is at about $t = 70$ s, where the RSSI is around -63 dBm. On the other hand, for GA, the best RSSI recorded is -78 dBm when the simulation time is $t = 130$ s. At these times, the cars are located near to each other. Since the distance is smaller, the UAV does not have to fly high to achieve good transmission conditions.

Figure 10 further evidences how these values are distributed, highlighting the differences between PSO and GA. From all the average RSSI values gathered, PSO shows the better results, being the majority of its values between -79 dBm and -69 dBm. On the other hand, the majority of the results for GA are between -83 dBm and -73 dBm. The mean for both algorithms also evidences the differences found: for PSO the mean result is -73 dBm, whereas for GA the mean result is -77 dBm. Overall, we clearly find that PSO is more efficient at achieving better RSSI values than GA.

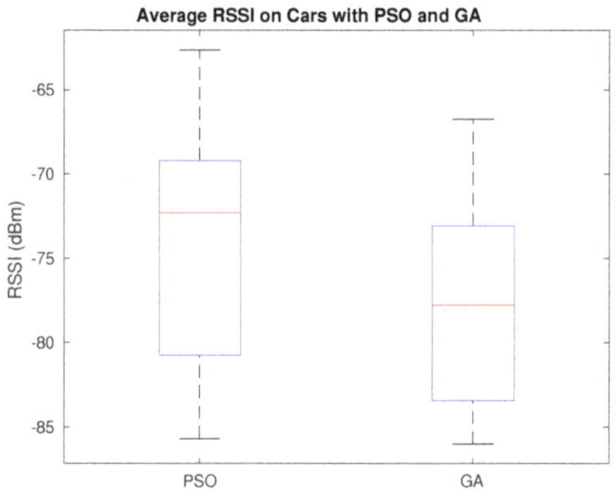

Figure 10. Average RSSI values at the receivers.

5.5. Impact on Total Path Length

Through the sequence of locations defining the UAV trajectory, we are able to find the total length of the UAV's path for each algorithm; these results are presented in Table 2. When using PSO as the positioning technique, the total path length is about 9.306 kilometers. On the other hand, the total path length for GA is longer, accounting to about 11.224 kilometers. The total path length using GA is longer due to the fact that not only the optimum locations obtained are sparse and quite far from each other in terms of latitude and longitude, but also, if we look at altitude variations throughout the time (see Figure 7), the position in terms of altitude changes drastically. Through these results, we can observe that the locations obtained using PSO introduce less burden to the UAV in terms of travelling distance.

Table 2. Path length and average flight speed for each algorithm.

Optimization Algorithm	Total Path Length (m)	Speed (km/h)
PSO	9305.90	119.65
GA	11223.69	144.30

5.6. Impact on Speed

Aside from total path length, we have also calculated how fast the UAV has to fly from one point to another for every second in the simulation. Since this kind of UAV is expected to be used for mission-specific purposes, the UAV is allowed to move at maximum speed at all times if necessary. The speed maximum values closely match the average ones as the scenario dynamics prevent it from remaining still. According to Table 2, the average speed needed for the UAV to passes through every optimum location is more than 100 km/h for both PSO and GA. To be more precise, when using PSO, the average speed needed is 119.65 km/h, which is less than the speed needed if using GA, which reaches 144.3 km/h. Again, when comparing these two optimization algorithms in terms of the average speed needed for the UAV to pass through all the optimum locations, PSO introduces lower requirements, being thus the option of choice.

6. Conclusions

The issue of placing UAVs to support car-to-car communications considering the restrictions of three-dimensional environments was investigated in this paper. In particular, we analyzed how different optimization algorithms can be used to find the best and optimum placement for a UAV providing support for car communications on the ground. Two types of optimization algorithms were included in our proposed placement technique: Particle Swarm Optimization, or PSO, and Genetic Algorithm, or GA. The exact sequence of UAV locations that are able to offer the best signal levels towards moving cars throughout time can be determined using PSO and GA. The quality of the signals received by the cars is the optimization parameter used to designate the location of the UAV sending the signals. By simulation, the signal quality can be calculated considering a path loss model that also counts on the elevation information for the area tested. The simulation tool can calculate the signal attenuation due to terrain blockages present based on elevation data. The positioning technique thus optimizes the position of the UAV defining its best altitude so that it can avoid terrain blockages. Based on our findings, the PSO can offer more optimized results than the GA in terms of efficiency.

To extend the work, in the future we will propose a method to find a more realistic trajectory of a UAV which takes into consideration the same parameters as in our positioning technique. The idea can be proposing a mobility model for a UAV that is aware of the dynamic movement of the cars on the ground but introducing more realistic speed requirements for the UAV.

Author Contributions: Conceptualization, S.A.H. and C.T.C.; Formal analysis, S.A.H.; Funding acquisition, C.T.C. and J.-C.C.; Investigation, S.A.H.; Methodology, S.A.H., C.T.C. and K.K.; Data Curation, S.A.H.; Visualization, S.A.H.; Project administration, C.T.C.; Resources, J.-C.C., D.K. and P.M.; Software, S.A.H.; Supervision, C.T.C., J.-C.C., E.H.-O. and K.K.; Validation, S.A.H., C.T.C., J.-C.C., E.H.-O., D.K., K.K. and P.M.; Writing—original draft, S.A.H.; Writing—review & editing, S.A.H., C.T.C., J.-C.C., E.H.-O., D.K., K.K. and P.M. All authors have read and agreed to the published version of the manuscript.

Funding: This work was partially supported by the "Ministerio de Ciencia, Innovación y Universidades, Programa Estatal de Investigación, Desarrollo e Innovación Orientada a los Retos de la Sociedad, Proyectos I+D+I 2018", Spain, under Grant RTI2018-096384-B-I00, and grant BES-2015-075988, Ayudas para contratos predoctorales 2015.

Conflicts of Interest: The authors declare no conflict of interest.

References

1. Gupta, L.; Jain, R.; Vaszkun, G. Survey of important issues in UAV communication networks. *IEEE Commun. Surv. Tutor.* **2016**, *18*, 1123–1152. [CrossRef]
2. Yanmaz, E.; Quaritsch, M.; Yahyanejad, S.; Rinner, B.; Hellwagner, H.; Bettstetter, C. Communication and coordination for drone networks. In *Ad Hoc Networks*; Springer: Cham, Switzerland, 2017; pp. 79–91.
3. Daniel, K.; Dusza, B.; Lewandowski, A.; Wietfeld, C. AirShield: A system-of-systems MUAV remote sensing architecture for disaster response. In Proceedings of the 2009 3rd Annual IEEE Systems Conference, Vancouver, BC, Canada, 23–26 March 2009; pp. 196–200.
4. Zhou, Y.; Cheng, N.; Lu, N.; Shen, X.S. Multi-UAV-aided networks: Aerial-ground cooperative vehicular networking architecture. *IEEE Veh. Technol. Mag.* **2015**, *10*, 36–44. [CrossRef]
5. Oubbati, O.S.; Lakas, A.; Zhou, F.; Güneş, M.; Lagraa, N.; Yagoubi, M.B. Intelligent UAV-assisted routing protocol for urban VANETs. *Comput. Commun.* **2017**, *107*, 93–111. [CrossRef]
6. Hadiwardoyo, S.A.; Hernández-Orallo, E.; Calafate, C.T.; Cano, J.C.; Manzoni, P. Evaluating UAV-to-Car Communications Performance: Testbed Experiments. In Proceedings of the 32-nd IEEE International Conference on Advanced Information Networking and Applications (AINA-2018), Krakow, Poland, 16–18 May 2018.
7. Chen, Y.; Feng, W.; Zheng, G. Optimum placement of UAV as relays. *IEEE Commun. Lett.* **2017**, *22*, 248–251. [CrossRef]
8. Hadiwardoyo, S.A.; Calafate, C.T.; Cano, J.C.; Ji, Y.; Hernández-Orallo, E.; Manzoni, P. Evaluating UAV-to-Car Communications Performance: From Testbed to Simulation Experiments. In Proceedings of the 2019 16th IEEE Annual Consumer Communications & Networking Conference (CCNC), Las Vegas, NV, USA, 11–14 January 2019; pp. 1–6.

9. Shi, W.; Li, J.; Xu, W.; Zhou, H.; Zhang, N.; Zhang, S.; Shen, X. Multiple Drone-Cell Deployment Analyses and Optimization in Drone Assisted Radio Access Networks. *IEEE Access* **2018**, *6*, 12518–12529. [CrossRef]

10. Lin, X.; Yajnanarayana, V.; Muruganathan, S.D.; Gao, S.; Asplund, H.; Maattanen, H.L.; Bergstrom, M.; Euler, S.; Wang, Y.P.E. The sky is not the limit: LTE for unmanned aerial vehicles. *IEEE Commun. Mag.* **2018**, *56*, 204–210. [CrossRef]

11. Van der Bergh, B.; Chiumento, A.; Pollin, S. LTE in the sky: Trading off propagation benefits with interference costs for aerial nodes. *IEEE Commun. Mag.* **2016**, *54*, 44–50. [CrossRef]

12. Nguyen, H.C.; Amorim, R.; Wigard, J.; Kovács, I.Z.; Sørensen, T.B.; Mogensen, P.E. How to ensure reliable connectivity for aerial vehicles over cellular networks. *IEEE Access* **2018**, *6*, 12304–12317. [CrossRef]

13. Jia, S.; Zhang, L. Modelling unmanned aerial vehicles base station in ground-to-air cooperative networks. *IET Commun.* **2017**, *11*, 1187–1194. [CrossRef]

14. Seliem, H.; Ahmed, M.H.; Shahidi, R.; Shehata, M.S. Delay analysis for drone-based Vehicular Ad-hoc Networks. In Proceedings of the 2017 IEEE 28th Annual International Symposium on Personal, Indoor, and Mobile Radio Communications (PIMRC), Montreal, QC, Canada, 8–13 October 2017; pp. 1–7.

15. Shilin, P.; Kirichek, R.; Paramonov, A.; Koucheryavy, A. Connectivity of VANET segments using UAVs. In *International Conference on Next Generation Wired/Wireless Networking*; Springer: Cham, Switzerland, 2016; pp. 492–500.

16. Younis, M.; Akkaya, K. Strategies and techniques for node placement in wireless sensor networks: A survey. *Ad Hoc Netw.* **2008**, *6*, 621–655. [CrossRef]

17. Shi, Y.; Hou, Y.T.; Efrat, A. Algorithm design for base station placement problems in sensor networks. In *Proceedings of the 3rd International Conference on Quality of Service in Heterogeneous Wired/Wireless Networks*; ACM: New York, NY, USA, 2006; p. 13.

18. Efrat, A.; Har-Peled, S.; Mitchell, J.S. Approximation algorithms for two optimal location problems in sensor networks. In Proceedings of the 2nd International Conference on Broadband Networks, Boston, MA, USA, 7 October 2005; pp. 714–723.

19. Lee, J.; Kim, C.M. A roadside unit placement scheme for vehicular telematics networks. In *Advances in Computer Science and Information Technology*; Springer: Berlin/Heidelberg, Germany, 2010; pp. 196–202.

20. Trullols, O.; Fiore, M.; Casetti, C.; Chiasserini, C.F.; Ordinas, J.B. Planning roadside infrastructure for information dissemination in intelligent transportation systems. *Comput. Commun.* **2010**, *33*, 432–442. [CrossRef]

21. Aslam, B.; Amjad, F.; Zou, C.C. Optimal roadside units placement in urban areas for vehicular networks. In Proceedings of the 2012 IEEE Symposium on Computers and Communications (ISCC), Cappadocia, Turkey, 1–4 July 2012; pp. 423–429.

22. Chiaraviglio, L.; D'Andreagiovanni, F.; Choo, K.K.R.; Cuomo, F.; Colonnese, S. Joint Optimization of Area Throughput and Grid-connected Microgeneration in UAV-based Mobile Networks. *IEEE Access* **2019**, *7*, 69545–69558. [CrossRef]

23. Huang, L.; Qu, H.; Ji, P.; Liu, X.; Fan, Z. A novel coordinated path planning method using k-degree smoothing for multi-UAVs. *Appl. Soft Comput.* **2016**, *48*, 182–192. [CrossRef]

24. Reina, D.; Tawfik, H.; Toral, S. Multi-subpopulation evolutionary algorithms for coverage deployment of UAV-networks. *Ad Hoc Netw.* **2018**, *68*, 16–32. [CrossRef]

25. Song, B.D.; Kim, J.; Kim, J.; Park, H.; Morrison, J.R.; Shim, D.H. Persistent UAV service: An improved scheduling formulation and prototypes of system components. *J. Intell. Robot. Syst.* **2014**, *74*, 221–232. [CrossRef]

26. Trotta, A.; D'Andreagiovanni, F.; Di Felice, M.; Natalizio, E.; Chowdhury, K.R. When UAVs ride a bus: Towards energy-efficient city-scale video surveillance. In Proceedings of the IEEE INFOCOM 2018-IEEE conference on computer Communications, Honolulu, HI, USA, 16–19 April 2018; pp. 1043–1051.

27. Shakhatreh, H.; Khreishah, A.; Alsarhan, A.; Khalil, I.; Sawalmeh, A.; Othman, N.S. Efficient 3d placement of a uav using particle swarm optimization. In Proceedings of the 2017 8th International Conference on Information and Communication Systems (ICICS), Irbid, Jordan, 4–6 April 2017; pp. 258–263.

28. He, X.; Yu, W.; Xu, H.; Lin, J.; Yang, X.; Lu, C.; Fu, X. Towards 3D Deployment of UAV Base Stations in Uneven Terrain. In Proceedings of the 2018 27th International Conference on Computer Communication and Networks (ICCCN), Hangzhou, China, 30 July–2 August 2018; pp. 1–9.

29. Chou, S.F.; Yu, Y.J.; Pang, A.C.; Lin, T.A. Energy-Aware 3D Aerial Small-Cell Deployment over Next Generation Cellular Networks. In Proceedings of the 2019 IEEE 88th Vehicular Technology Conference (VTC Spring), Porto, Portugal, 3–6 June 2018.

30. Alzenad, M.; El-Keyi, A.; Yanikomeroglu, H. 3-D placement of an unmanned aerial vehicle base station for maximum coverage of users with different QoS requirements. *IEEE Wirel. Commun. Lett.* **2018**, *7*, 38–41. [CrossRef]

31. Arribas, E.; Mancuso, V.; Cholvi, V. Fair Cellular Throughput Optimization with the Aid of Coordinated Drones. In Proceedings of the 2019 Mission-Oriented Wireless Sensor, UAV and Robot Networking (MiSARN), Paris, France, 29 April–2 May 2019.

32. Pourbaba, P.; Manosha, K.S.; Ali, S.; Rajatheva, N. Full-Duplex UAV Relay Positioning for Vehicular Communications with Underlay V2V Links. In Proceedings of the 2019 IEEE 88th Vehicular Technology Conference (VTC-Spring), Kuala Lumpur, Malaysia, 28 April–1 May 2019.

33. Hadiwardoyo, S.A.; Hernández-Orallo, E.; Calafate, C.T.; Cano, J.C.; Manzoni, P. Experimental characterization of UAV-to-car communications. *Comput. Netw.* **2018**, *136*, 105–118. [CrossRef]

34. Kennedy, J. Particle swarm optimization. In *Encyclopedia of Machine Learning*; Springer, Boston, MA, USA, 2010; pp. 760–766.

35. Holland, J. *Adaptation in Natural and Artificial Systems*; MIT Press: Cambridge, MA, USA, 1975.

36. Bullington, K. Radio propagation at frequencies above 30 megacycles. *Proc. IRE* **1947**, *35*, 1122–1136. [CrossRef]

37. Varga, A.; Hornig, R. An overview of the OMNeT++ simulation environment. In *Proceedings of the 1st International Conference on Simulation Tools and Techniques for Communications, Networks and Systems & Workshops*; ICST (Institute for Computer Sciences, Social-Informatics and Telecommunications Engineering): Brussels, Belgium, 2008; p. 60.

38. Behrisch, M.; Bieker, L.; Erdmann, J.; Krajzewicz, D. Sumo–simulation of urban mobility. In Proceedings of the Third International Conference on Advances in System Simulation (SIMUL 2011), Barcelona, Spain, 23–29 October 2011; Volume 42.

39. Sommer, C.; German, R.; Dressler, F. Bidirectionally coupled network and road traffic simulation for improved IVC analysis. *IEEE Trans. Mob. Comput.* **2011**, *10*, 3–15. [CrossRef]

40. Haklay, M.; Weber, P. Openstreetmap: User-generated street maps. *IEEE Pervasive Comput.* **2008**, *7*, 12–18. [CrossRef]

41. Farr, T.G.; Rosen, P.A.; Caro, E.; Crippen, R.; Duren, R.; Hensley, S.; Kobrick, M.; Paller, M.; Rodriguez, E.; Roth, L.; et al. The shuttle radar topography mission. *Rev. Geophys.* **2007**, *45*. [CrossRef]

Article

Caching Transient Contents in Vehicular Named Data Networking: A Performance Analysis

Marica Amadeo [1,*]**, Claudia Campolo** [1]**, Giuseppe Ruggeri** [1]**, Gianmarco Lia** [1] **and Antonella Molinaro** [1,2]

[1]   DIIES Department, University Mediterranea of Reggio Calabria, Via Graziella, Loc. Feo di Vito,
89100 Reggio Calabria, Italy; claudia.campolo@unirc.it (C.C.); giuseppe.ruggeri@unirc.it (G.R.);
gianmarco.lia@unirc.it (G.L.); antonella.molinaro@unirc.it (A.M.)

[2]   Laboratoire des Signaux et Systémes (L2S), CentraleSupélec, Université Paris-Saclay,
91190 Gif-sur-Yvette, France

*   Correspondence: marica.amadeo@unirc.it; Tel.: +39-0965-1693276

Received: 28 February 2020; Accepted: 1 April 2020; Published: 2 April 2020

Abstract: Named Data Networking (NDN) is a promising communication paradigm for the challenging vehicular ad hoc environment. In particular, the built-in pervasive caching capability was shown to be essential for effective data delivery in presence of short-lived and intermittent connectivity. Existing studies have however not considered the fact that multiple vehicular contents can be transient, i.e., they expire after a certain time period since they were generated, the so-called *FreshnessPeriod* in NDN. In this paper, we study the effects of caching transient contents in Vehicular NDN and present a simple yet effective freshness-driven caching decision strategy that vehicles can implement autonomously. Performance evaluation in ndnSIM shows that the *FreshnessPeriod* is a crucial parameter that deeply influences the cache hit ratio and, consequently, the data dissemination performance.

Keywords: Caching; Named Data Networking; Information Centric Networking; Vehicular Ad Hoc Networks

---

1. Introduction

Recent advancements in the fields of sensing, computing, communication, and networking technologies are contributing in making vehicles multi-faceted elements (equipped with cameras, sensors, radars, storage, processing, and positioning capabilities) of smart and connected cities.

Vehicular on-board units (OBUs) are able to interact among each other through vehicle-to-vehicle (V2V) communications, and with nearby road-side units (RSUs), traffic lights, pedestrians, and edge nodes through vehicle-to-infrastructure (V2I) communications. Thanks to vehicle-to-everything (V2X) connectivity, interactions with remote entities, such as cloud servers and Internet facilities, are also available. V2X connectivity would overall increase the driving and traveling experience, by enabling a rich set of applications, ranging from safety and traffic efficiency to infotainment and, more recently, cooperative and automated driving. Data exchanged by such revolutionary vehicular application will exhibit different features (e.g., size, lifetime, generation frequency, dissemination scope, popularity) and requirements (e.g., latency, throughput, reliability).

Named Data Networking (NDN) has natural advantages to greatly overcome the challenges of vehicular ad-hoc networks (VANETs), such as rapidly changing topology, harsh propagation environments and short-lived and intermittent connectivity, thanks to its name-based routing and native in-network caching capabilities [1]. Moreover, being focused on *what* content to retrieve, instead of *where* the content is located, NDN well matches vehicular applications where typically, *(i)* communicating entities are interested in retrieving content (e.g., road congestion information, weather conditions) regardless of the identity of the node(s) producing it and *(ii)* the requested contents have a spatial and/or temporal scope.

NDN implements a naïve caching strategy that lets nodes cache *all* the received contents. However, indiscriminate caching may waste network resources and reduce the cache efficiency and it is poorly suited for contents that exhibit a limited validity, which are frequently exchanged in VANET. Examples of transients contents requested by vehicular applications are, for instance, those related to parking lots availability, road congestion, maps of the surroundings [2,3]. If the content lifetime is not properly conveyed in packets and accounted for in the caching decision, stale contents risk to be propagated by affecting the behaviour of applications relying on them. Such an effect is particularly exacerbated in VANETs, due to the broadcast nature of the wireless medium that facilitates sharing of data, also of useless data, if they expired. However, literature on caching transient contents in NDN is almost unexplored in VANETs and still in its early stage for Internet of Things (IoT) contents [4–6] and wired networks [7,8]. To fill this gap, the following contributions are provided in this paper:

- We showcase the benefits of tracking the content lifetime in NDN packets to prevent stale information from becoming disseminated in the vehicular network.
- We propose a simple but effective NDN-compliant caching strategy that accounts for the content lifetime, not only for replacement purposes but also for the caching decision.
- We perform a comprehensive simulation campaign in ndnSIM [9], the official ns-3-based simulator of the NDN community, to study the impact of the content lifetime in the caching decision when considering two distinct vehicular scenarios, urban and highway, and varying traffic load and content popularity settings.

The remainder of this paper is organized as follows: Section 2 introduces the NDN paradigm. An overview of vehicular applications and connectivity options is provided in Section 3. Section 4 discusses the status quo on Vehicular NDN (V-NDN), with special focus on caching strategies in the literature. Section 5 motivates our study, by also providing early results, and presents the proposal. More comprehensive results are reported in Section 6. Section 7 concludes the paper with hints at future work.

2. NDN Basics

The NDN architecture was originally conceived for named content dissemination in the Internet [10], but today it is considered to be an enabling networking technology in different application domains, such as Wireless Ad Hoc Networks [11], IoT [12], and Edge/Fog Computing [13–15]. NDN is based on two packet types that carry hierarchical content names: the Interest, used by consumers for requesting contents, and the Data, used for carrying the content.

Data packets are originated by a producer/source node, which also signs them to allow per-packet authentication and integrity checks. Any node receiving Data packets can cache them to satisfy further requests. In the following, we refer to as content provider, or simply *provider*, any node in the network that acts as producer or cacher.

Each NDN node maintains three data structures: *(i) Content Store (CS)*, used for caching incoming Data packets, *(ii) Pending Interest Table (PIT)*, used for recording Interests that were not yet satisfied, and *(iii) Forwarding Information Base (FIB)*, used to forward the Interests.

As shown in Figure 1, at the Interest reception, each node first looks for a matching in the CS. In case of failure, it checks in the PIT for the same pending request. If a matching is found, the Interest is discarded. Otherwise, it looks in the FIB for an outgoing interface (or multiple ones) over which sending the request. Data packets follow the PIT entries back to the consumer(s); they can be cached by on-path nodes according to their storage space. The NDN reference caching implementation is Cache Everything Everywhere (CEE), where nodes cache indiscriminately every incoming Data packet. CEE is usually coupled with the Least Recently Used (LRU) replacement policy.

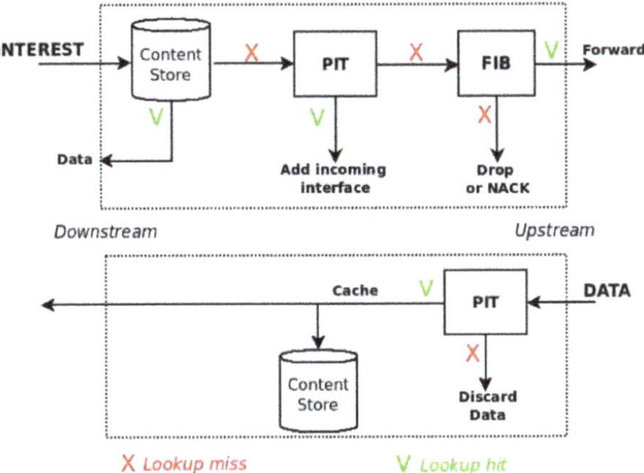

Figure 1. Forwarding Process in NDN.

Compared to traditional caching systems, such as web caching or content delivery networks, NDN caching shows some distinct features: it is performed on a per-packet basis and at line speed, during the forwarding process. Therefore, cache decision policies that require complex calculations or multiple interactions between network nodes are not affordable, since they would slow down the content delivery [16].

3. Vehicular Applications and Connectivity Options

A plethora of heterogeneous applications will be supported in the vehicular landscape, targeting different use cases, ranging from autonomous driving to traffic efficiency and infotainment. Vehicular applications exhibit different delivery demands, e.g., in terms of latency, throughput and reliability [17] and typically exchange data with spatial and temporal relevance. For instance, road traffic information (e.g., mean speed in a given road segment) is locally relevant: data collected in one area will be requested in the same area. The time validity of such data spans a few seconds or minutes; whereas the time validity of other types of data, such as the fees of charging stations for electric vehicles and the flyers of points-of-interest in a road area, may span several hours [18].

Vehicular applications rely on the exchange of data among vehicles, between vehicles and roadside infrastructure and nearby sensors, pedestrians and remote server facilities, enabled through V2X connectivity. The V2X term covers, among others, both V2V and V2I communications, as shown in Figure 2.

Although more than 20 years passed since a dedicated spectrum at 5.9 GHz was allocated to vehicular communications, the decision about the V2X radio access technology is still under debate and revolving between two mainstream technologies, i.e., IEEE 802.11 and Cellular-V2X (C-V2X).

IEEE 802.11 initially captured the interest of the research community, due to operation simplicity and native support for V2V communications in an ad hoc manner. The IEEE 802.11p amendment, now superseded and part of the IEEE 802.11 standard [19], was conceived as an enhancement of the IEEE 802.11a specification, with physical and medium access control (MAC) layers' settings and procedures properly adjusted to support outdoor communications under high speed mobility. At the MAC layer, 802.11p relies on the Carrier Sense Multiple Access with Collision Avoidance (CSMA/CA) protocol. A node wishing to transmit senses the medium to detect if it is busy. If it is the case, a mechanism based on random backoff is performed to reduce the probability of collisions, which, however, cannot be prevented. More recently, the interest for 802.11-based V2X connectivity revamped thanks to the creation of a new IEEE task group, now preparing the IEEE 802.11bd amendment.

The group aims to investigate evolved physical-layer technologies that enhance the .11p coverage and throughput [20].

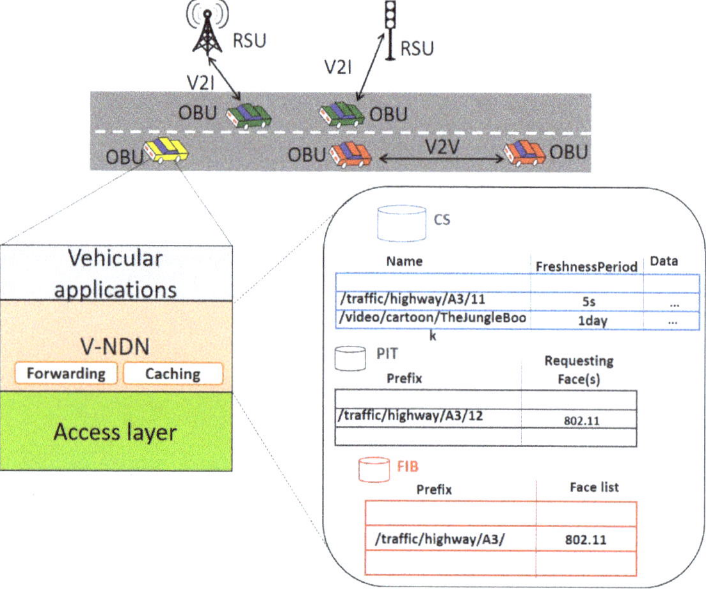

Figure 2. Vehicular communications and V-NDN reference architecture.

4. V-NDN: Design Concepts and Caching Strategies

4.1. V-NDN

Originally designed in [21], V-NDN extends the NDN model to accommodate the distinctive and challenging features of VANETs, namely ad hoc intermittent connectivity and mobility, to fit the spatio-temporal validity of contents. A reference V-NDN architecture is shown in Figure 2. As with the vanilla NDN implementation, each V-NDN node maintains CS, PIT and FIB tables but, to take full advantage from the broadcast nature of the channel and maximize the possibility of content sharing, major modifications are introduced in the forwarding and caching process.

Forwarding. Due to the high dynamicity of vehicular topologies, V-NDN does not implement a proactive routing protocol to build the FIBs. Instead, it assumes that Interest and Data packets are always broadcasted over the wireless medium, and it designs a reactive distance-based forwarding scheme that limits packet collision and redundancy and speeds up the data retrieval. More specifically, when sending an Interest, each node S includes its Global Positioning System (GPS) coordinates. Each receiving node R calculates its distance from S and sets a random *Defer Time* that is inversely proportional to such distance. The smaller the distance the larger the time a node waits before transmitting; therefore, the farthest node from the sender has higher transmission priority. This speeds up the Interest dissemination. If, during the *Defer Time*, R overhears the same packet broadcasted by another node in the same area, then it can cancel its own transmission. Re-broadcasted Interests act also as an implicit acknowledgment for the sender S. In case no rebroadcasting is overheard, then S will retransmit the packet up to 7 times before giving up. To avoid the unrestrained dissemination of Interests, a maximum hop count limit is set to 5 [21].

Caching. Unlike other wireless terminals, such as smartphones or sensors, vehicles do not have strict energy or memory constraints. Therefore, V-NDN nodes can, in principle, cache all Data overheard over the wireless channel, even if they do not have a matching PIT entry. We call this strategy

Cache Everything in the Air (CEA). By implementing CEA, vehicles can serve as data mules between disconnected areas and enable opportunistic delivery services. In practice, however, this strategy may lead to inefficient performance due to the high cache redundancy (which is even higher than CEE) and it is not convenient in areas with a high density of vehicles [22]. This is why other caching strategies were proposed in the literature, as discussed in the following section.

4.2. Cooperative vs. Autonomous Caching

In wired networks, cooperative schemes usually lead to better performance than autonomous ones, in terms of low cache redundancy and reduced retrieval latency, at the expense of a potentially high signalling overhead [23]. Things however change in vehicular environments where, due to node mobility and unstable connectivity, it is difficult to exchange consistent information about the status of the network and the CSs of vehicles and take decisions at line speed. Traditional approaches for ad hoc networks (outside the NDN context) such as the one in [24], which takes decisions based on information density estimated during an inference phase, seem not affordable in NDN, since they would introduce a slowdown in the forwarding fabric [16,25].

To cope with the dynamicity of vehicular topologies, some studies considered mobility-aware caching strategies that can be applied in the presence of a full or partial RSU infrastructure. In [26], a proactive caching policy is proposed that takes into account the content popularity and the vehicle mobility prediction. The latter information is used to prefetch the contents at the RSUs the vehicles will be connected to during their journey, thus their requests will be satisfied with lower latency. However, the strategy requires the collection of prior mobility data over which applying offline a mobility prediction algorithm. In [27], a scheme called Cooperative Caching based on Mobility Prediction (CCMP) is designed where urban areas are divided into hot regions based on users' mobility patterns and a prediction model is applied to compute the probabilities that vehicles re-visit the hot regions. Nodes with higher chances of staying in hot regions and for longer times are chosen to cache contents. In [28], mobility prediction is used to create clusters of vehicles. The most suitable vehicles (e.g., the ones with the best channel quality) are selected as cluster heads and act as cachers: they receive contents from the closest RSUs and cache them to serve requests from other vehicles. A shortcoming of such approach is the lack of fairness, since only few nodes handle caching operations.

When the mobility prediction is not available, implementing cooperative caching schemes may be disadvantageous. For instance, the work in [22] shows that a notable caching scheme with implicit coordination, Leave Copy Down (LCD) [29], has results comparable to a Never Cache policy, in which vehicles do not cache packets. In the rest of the paper, we focus on a general scenario where vehicles mobility patterns are a priori unknown and communications are mainly based on short-lived V2V interactions. Autonomous caching schemes better suit this type of situation: they are lightweight solutions that do not require additional knowledge of the network topology and do not incur in further signalling.

4.3. Autonomous Schemes for Non-Transient Contents

CEE and CEA are the simplest autonomous caching schemes available in the literature. To limit their intrinsic cache redundancy effects while maintaining a simple decision criterion, random schemes with a static caching probability were devised [30], where nodes cache Data packets with a pre-defined probability p, with $0 \leq p \leq 1$. If $p = 0$ the node never caches packets, while if $p = 1$ the scheme behaves like CEE. The most common value used in NDN implementations is $p = 0.5$, which limits the cache redundancy without underusing the available storage space [31]. The caching probability can be also computed dynamically at each node based on the perceived information about the network status and the content demands in the neighbourhood. In this context, decision strategies largely vary depending on the communication type, namely V2I or V2V.

In [32], an autonomous probabilistic caching scheme, for RSUs only, is deployed, with the targets of minimizing the average number of hops to retrieve the requested content and maximizing the cache hit ratio. The caching probability is computed according to the content popularity evaluated adaptively to the distinct request patterns. However, the strategy is deployed in an infrastructured network of RSUs that retrieve contents from a remote server and it does not consider caching at vehicles. Conversely, in [33], the focus is on V2V communications and the caching probability is computed by vehicles according to three parameters: *(i)* the content popularity, inferred from the received Interest packets, *(ii)* the vehicle's topological centrality, and *(iii)* the relative movement of the receiver and the sender. Performance evaluation shows that the strategy outperforms CEE and probabilistic caching with $p = 0.5$. In [34], a cache probability utility function is defined that takes into account the content popularity and two new defined parameters, the *moving similarity* and the *content similarity*. The moving similarity indicates if a content is requested by vehicles moving on the same route of the potential cacher. The higher the moving similarity the longer the connectivity between vehicles, and therefore the higher the caching probability. Vice versa, the content similarity indicates if drivers/passengers request similar contents. The higher the content similarity the higher the probability that a vehicle will cache a received Data packet.

5. Caching Transient Contents in V-NDN

5.1. Contributions of This Paper

The above-mentioned caching schemes do not consider that many contents exchanged in a vehicular environment have a limited time validity, which can vary from a few seconds to a couple of minutes. Intelligent driving assistance systems, parking lots availability, high-definition maps are just a few examples of services that rely on contents that may change with time due to the variation of the driving environment [2,3]. They entail fresh content retrieval and also short latency. Such a transient feature may largely influence the performance of a caching system and cannot be ignored in the caching decision. Therefore, in the following sections, we aim at exploring the impact of transient contents in the caching systems of V-NDN nodes. In particular:

- We detail how the vanilla NDN forwarding fabric deals with transient data, by emphasizing the potential weaknesses, and report the few related literature studies in the field.
- We perform a first basic simulation campaign to quantify the effects of caching transient contents in V-NDN by studying crucial metrics such as the cache hit ratio and the cache inconsistency, which is due to the wrong awareness about the content lifetime supported by the vanilla NDN implementation.
- We design a new autonomous strategy, named Freshness-Driven Caching (FDC), which addresses the cache inconsistency issues and takes caching decisions based on the content lifetime. The rationale behind our proposal is pretty intuitive: caching contents that are ready to expire (possibly at the expenses of contents with a larger lifetime) does not efficiently use the storage space. Indeed, regardless of the request pattern, short-lasting contents will be quickly removed from the CS. Therefore, FDC aims at caching long-lasting contents with a higher probability.
- Performance of FDC is evaluated in two different mobility scenarios, urban and highway, and compared against two NDN benchmark schemes, CEE and random caching.

5.2. Cache Inconsistency in Existing Solutions

In mobile wireless and broadcast environments such as VANETs, vehicles act like data mule that collect packets and move them across distances thus offering a valuable dissemination service. Transient data, however, require an ad hoc caching strategy that is aware of their lifetimes. Storing stale data can lead to cache inconsistency, i.e., distinct cachers can have inconsistent copies of the same content that result in multiple adverse effects in the real life. For instance, if a vehicle looks for an empty space by transmitting an Interest in its neighborhood, it may uselessly reach a wrong (busy) space, by wasting both fuel and time, if stale data are received as a reply.

In vanilla NDN, the transient feature of a content is expressed in term of a *FreshnessPeriod* (in the following shortened as FP), a field in the Data packet header indicating the lifetime of the content. The parameter is application-specific and it is set by the original producer. If the lifetime is not expired, the content can be considered still *fresh*. Otherwise, the original source may have produced new content. Tracking the freshness is, therefore, crucial in presence of transient data not to incur in cache inconsistency effects. A policy that honors content freshness (we refer to as "FP-Aware" policy) is implemented in vanilla NDN and applied in conjunction with a standard replacement policy such as LRU. Basically, when caching a received Data packet, the NDN node sets a timer equal to the FP value; when this latter expires, the content is removed.

Literature on caching transient data in NDN is still at its infancy and almost unexplored in vehicular environments. Some proposals were devised in IoT sensor networks, with the main target of reducing the data retrieval latency and limit the energy consumption [4–6]. Other works considered caching transient contents in wired networks segments [7,8].

With focus on V-NDN, a Multi-Metric Cache Replacement (M^2CRP) scheme is presented in [25]. There, content popularity, freshness, and distance between the content producer and the cacher are used to select the packet that must be evicted from the CS. M^2CRP is coupled with the CEE policy: when a new Data packet is received, it is always cached and, if the CS is full, then an existing cached item must be replaced. Popularity, freshness and distance metrics are used to compute a score for every cached item; the one with the minimum score is selected as the candidate for eviction. A similar replacement strategy, but implemented in RSUs only, is deployed in [35]. Works [25,35], however, do not consider the effect of freshness in the caching decision process. By using CEE, all the contents are indiscriminately cached regardless of their freshness period. On the one hand, CEE does not create content diversity within the vehicles' neighbourhood: the CSs of vehicles in the same area are filled with the same information, thus resulting in an inefficient use of the distributed storage space. On the other hand, caching contents with a long lifetime could be more convenient than caching contents with a shorter lifetime, since these latter must be evicted more frequently from the CS.

We also observe that the FP information is static, i.e., it is not decreased by caching nodes when answering requests. Indeed, NDN Data packets are immutable [36]: if some information in the packet changes, the producer must generate a new packet and sign it. Under such conditions, cache inconsistency can still occur. As an example, we consider a scenario where an RSU monitors the average speed on a certain road and produces new Data packets named */RoadY/avgSpeed* every 60 s. An NDN vehicle A, implementing CEE+LRU+FP-Aware policy, requests a Data packet at time $t = 0$ s and it is allowed to cache it for 60 s. At time $t = 50$ s a vehicle B asks for the same content and receives it from A. According to the FP, the packet could be stored in the CS of node B for 60 s, but the residual lifetime of the packet is actually 10 s. If, at $t = 70$ s, a vehicle C asks for the */RoadY/avgSpeed* Data packet and receives it from B, it will actually receive a stale information.

5.3. Quantifying Cache Inconsistency Effects

To quantify the cache inconsistency in an urban V-NDN environment, we performed a preliminary simulation campaign with ndnSIM [37]. We consider a first case, where nodes implement the legacy CEE+LRU scheme without applying the FP-Aware policy, and a second one where, instead, the FP-Aware policy is implemented.

The simulation scenario is a Manhattan Grid of size 1 km^2 with 2 lanes per direction, where 100 vehicles move at speeds ranging between 20 and 40 km/h, according to the Simulation of Urban MObility (SUMO) model [38]. One RSU acting as the original producer of transient contents is deployed in the middle of the topology. Vehicles and RSU interact through the broadcast transmissions of Interest/Data packets, according to the V-NDN forwarding strategy in [21]. IEEE 802.11 is considered to be the access layer technology.

We consider a catalog of 10,000 transient contents, each one composed of 100 1kbyte-long Data packets. As with [16], we assume that nodes have the same storage space, which summed up accounts for 1% of the overall content catalog size. We also assume that 20 vehicles are selected as consumers, and the content request pattern follows a Zipf distribution [39], which is commonly used to model content popularity in the current Internet, NDN networks and VANETs [33,40,41].

Given a catalog of content items, the Zipf distribution assumes that the access probability of the ith, $1 \leq i \leq m$, content item is represented as:

$$P(i, \alpha, m) = \frac{1/i^{\alpha}}{\sum_{z=1}^{m}(1/z^{\alpha})} \tag{1}$$

where the α exponent, which is typically denoted as *skewness parameter*, characterizes the distribution popularity. The higher the value of α, the higher the number of requests concentrated on a few (popular) contents.

In this simulation, we consider a skewness parameter α equal to 1 or 2. Content requests start asynchronously: the time between two consecutive Interest transmissions for the first Data packet by different consumers is exponentially distributed with rate $\lambda = 0.3 request/sec$.

Two distinct metrics are reported in this preliminary evaluation stage:

- the *cache hit ratio*, computed as the ratio, in percentage, between the received Interests satisfied by the local CS and the total number of received Interests;
- the *cache inconsistency*, computed as the ratio, in percentage, between the received Data packets that were expired and the total number of received Data packets.

Table 1 reports the results averaged over 15 runs, in presence of CEE+LRU, when considering a first case where all the Data packets have the *FP* set to 20s, and a second case where the parameter is set to 10s. It can be observed that, reasonably, the cache hit ratio largely increases when parameter α passes from 1 to 2, since a larger number of requests are issued for the same popular contents. Not surprisingly, the lower the FP the higher the cache inconsistency, which can reach even 62.06% when $\alpha = 2$ and FP = 10s. This means that more than half of the cached Data packets are actually expired, but they are not removed from the CS, since the caching system is not able to recognize it, and only the LRU replacement policy applies. It is also worth noticing that the higher is α, the higher is the cache inconsistency, since the dissemination of stale cached packets is higher over the shared broadcast medium.

Table 1. Cache hit ratio and inconsistency metrics in presence of CEE+LRU policy, when varying the Zipf skewness α and the FP parameter.

	Hit Ratio [%]	Inconsistency (FP = 20s)	Inconsistency (FP = 10s)
$\alpha = 1$	16.5%	19.42%	25.96%
$\alpha = 2$	38.2%	50.65%	62.06 %

Table 2 shows the cache hit ratio and inconsistency metrics, when instead considering the CEE+LRU+FP-Aware policy in the same scenario. Compared to Table 1, it can be observed that reported values are considerably lower. Indeed, thanks to the FP-Aware policy, the nodes can cache contents for a time equal to their FP and, when this latter expires, the packets are removed from the CS. As a result, compared to the previous case, the cache hit ratio is lower and the cache inconsistency reduces, although a not negligible percentage, in the range 3–9%, is still present.

We can conclude that the vanilla caching system in V-NDN is not able to guarantee cache consistency in presence of transient contents. This motivates our proposal in the next Section.

Table 2. Cache hit ratio and inconsistency metrics in presence of CEE+LRU+FP-Aware policy, when varying the Zipf skewness parameter α and the freshness period.

	Hit Ratio [%]	Inconsistency (FP = 20s)	Inconsistency (FP = 10s)
$\alpha = 1$	9.12%	3.81%	5.46%
$\alpha = 2$	20.27%	7.01%	9.52 %

5.4. Freshness-Driven Caching (FDC) Strategy

In this Section, we present a simple and fully distributed freshness-driven caching strategy that V-NDN nodes can apply without exchanging any additional control message. FDC is designed with two main targets in mind: to avoid cache inconsistency effects and to privilege caching contents with a longer residual lifetime.

To overcome the cache inconsistency of the FP-Aware policy, FDC requires that information about the generation time (i.e., a timestamp) is added in the Data packet by the producer. The timestamp can be included as an additional METAINFO field of the packet header, after the FP information, see Figure 3.

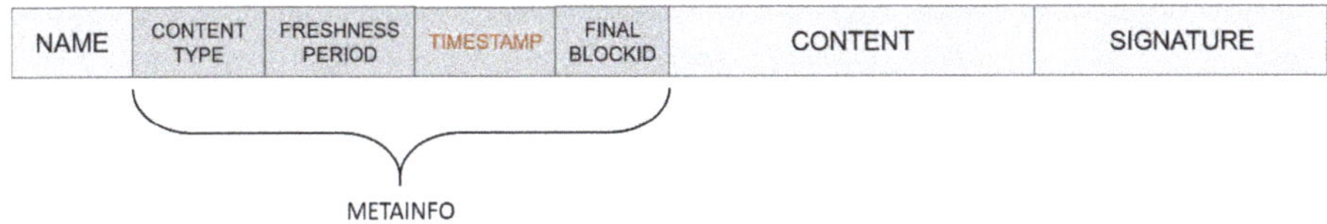

Figure 3. New structure of the NDN Data packet.

Instead of using the FP information for setting the time a content can remain in the CS, the caching system must consider the residual freshness period (RFP), defined as:

$$RFP = FP + timestamp - currentTime \tag{2}$$

where *currentTime* is the instant the vehicle is receiving the Data packet. A proper computation of the *RFP* parameter is ensured by the fact that all vehicles maintain strict synchronization with the Coordinated Universal Time (UTC) that can be acquired from the Global Navigation Support System (GNSS) [42].

When caching the Data packet, the node sets a timeout equal to *RFP*. When the timeout expires, the content is erased from the CS and, therefore, cache inconsistency is avoided. In FDC, RFP-based eviction is integrated with a traditional replacement policy, such as LRU. Therefore, in principle, Data packets could be erased also before the *RFP* timeout expires.

FDC also implements a probabilistic caching decision strategy based on the *RFP* value: the target is to cache with higher probability the Data packets with a longer residual lifetime. The distinction between long- or short-lasting packets is done by setting a dynamic threshold value, Th_{RFP}, obtained as the exponential weighted moving average (EWMA) of the RFP values carried by the received Data packets, regardless of their senders.

More specifically, when a Data packet, i, traverses a node, it is cached with probability $P_c(i)$, which is computed as:

$$P_c(i) = \begin{cases} 1 & if \quad RFP_i \geq Th_{RFP_i} \\ \dfrac{RFP_i}{Th_{RFP_i}} & otherwise \end{cases} \tag{3}$$

where:
- Th_{RFP_i} is the current value of the threshold, as available at the reception of packet i;
- RFP_i is the RFP value computed starting from the fields carried by packet i.

After the caching decision is taken, regardless of its outcome, the node updates the threshold as follows:

$$Th_{RFP_{i+1}} = (1 - \beta)Th_{RFP_i} + \beta \cdot RFP_i \tag{4}$$

where parameter $\beta \in (0, 1)$ is set to 0.125 to avoid large fluctuations in the estimation and give more relevance to the historical values in front of the instantaneous ones, as commonly agreed in multiple works in the literature, e.g., [43,44].

At the reception of a subsequent Data packet, $i + 1$, the novel value $Th_{RFP_{i+1}}$ will be used for the caching decision.

6. Performance Evaluation

To assess the performance of FDC, we performed a simulation campaign in two distinct vehicular scenarios: the same urban topology described previously, and a highway topology, which consists of a 2 km-long highway road segment, where 100 vehicles move at a maximum speed of 90 km/h. In both scenarios, we assume that a RSU in the middle of the topology acts as original producers of transient contents. The *FreshnessPeriod* of Data packets varies uniformly in the range [5–100] s to match a realistic and heterogeneous data traffic pattern.

CEE and Random Caching (RC) with probability $p = 0.5$ are considered to be benchmark schemes. They were selected as the most representative baseline solutions in the literature for V-NDN. As with FDC, they have the virtue of simplicity and incur no overhead, being completely autonomous. This is a crucial feature in the vehicular domain. For the sake of a fair comparison, all the schemes implement LRU coupled with RFP-based replacement. By doing so, cache inconsistency is always null. The proposal as well as the benchmark schemes were implemented in ndnSIM [37].

The main simulation settings are reported in Table 3.

Table 3. Main simulation settings.

Parameter	Value
Content catalog size	10,000 contents
Content size	100 Data packets
Data packet size	1000 bytes
Content Popularity	Zipf distributed with $\alpha \in [1 - 2.5]$
Propagation	Nakagami fading
Scenario	Urban topology (Manhattan Grid of size 1 km^2) Highway topology (2 km-long highway road segment)
Number of vehicles	100
Number of consumers	20–50
Number of producers	1 RSU

In addition to the cache hit ratio, the following metrics are considered:

- *Content retrieval delay*, computed as the average time for retrieving a content.
- *Number of hops*, computed as the average number of hops travelled by the Interest packets for retrieving the content.
- *Number of Data packets* as the total number of Data packets broadcasted by vehicles in the simulation. It, therefore, includes also re-transmitted and redundant packets.

Results are averaged over ten independent runs and reported with 95% confidence intervals.

6.1. Urban Scenario

The first set of results in Figure 4 focuses on the urban scenario, when varying the Zipf skewness parameter, α, in the range 1, 1.5, 2, 2.5 to model different content popularity distributions.

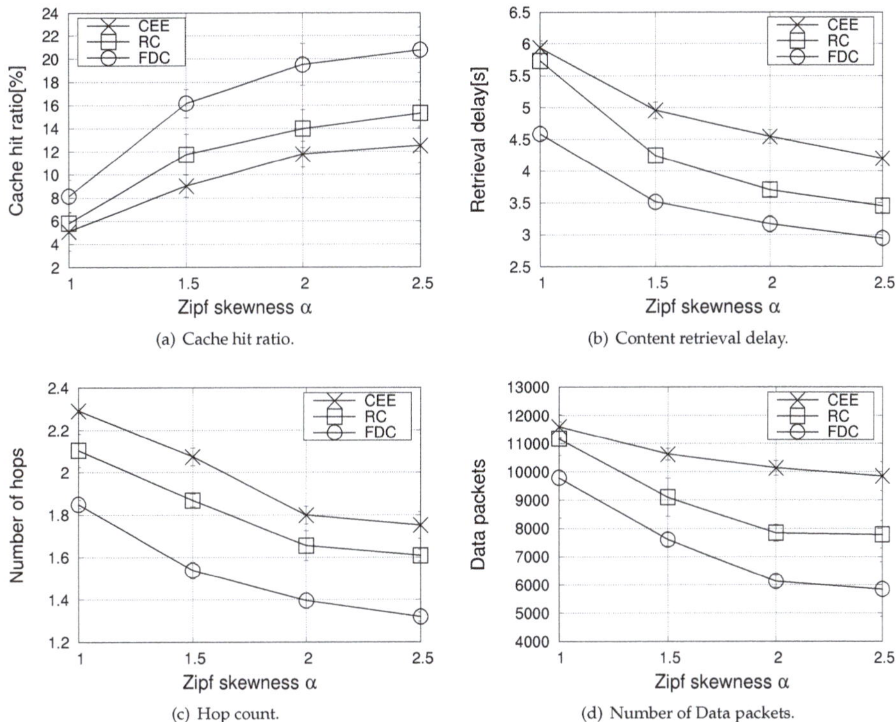

Figure 4. Performance metrics in the urban scenario, when varying the Zipf skewness parameter α (number of consumers equal to 20).

Figure 4a reports the cache hit ratio (the metric equally applies for the different lifetimes of contents). Not surprisingly, the CEE strategy exhibits the poorest performance. Due to the broadcast nature of the wireless medium, in fact, many neighbouring vehicles are likely to receive the same Data packets at the same time instant and they all cache the same information, with a neat penalty in terms of content diversity. Better performance is obtained with the RC strategy which, by introducing the probabilistic caching decision, increases the content diversity in the network and facilitates cache hits.

In FDC, a probabilistic decision is also foreseen, but it prioritizes caching of contents with a longer residual lifetime, hence it increases the cache hit ratio, and, consequently, reduces the content retrieval delay, as shown in Figure 4b. The cache hit ratio metric increases for all the compared schemes when the Zipf skewness parameter increases. Indeed, as α increases, requests from multiple consumers concentrate on a few contents from the catalog and the more popular contents are kept in the cache. This increases the chance for a request to be satisfied and a lower number of hops for the content retrieval is experienced, see Figure 4c.

Figure 4d shows that, as a further benefit, the proposal allows reducing the number of Data packets which are exchanged into the network compared to the benchmark schemes. The largest load is experienced by the CEE scheme and it can be observed that the latter one is the less sensitive strategy to the Zipf skewness parameter. This happens because, by caching all incoming contents indiscriminately, CEE unavoidably results in a high redundancy. When multiple nearby vehicles receive a broadcast content request and have a match in the CS, they all try to answer with the Data packet. Although V-NDN implements collision avoidance techniques, based on defer times and overhearing, hidden terminal phenomena cannot be completely avoided and multiple redundant packet are transmitted.

Figure 5 reports the cache hit ratio and retrieval delay metrics when varying the number of consumers from 20 to 50, for the Zipf skewness parameter α set equal to 1.5. It can be observed that performance improves as the number of consumers increases from 20 to 40. In this case, the value of α ensures that many requests by multiple consumers concentrate on the more popular contents, thus, ensuring a high cache hit ratio and a low retrieval delay. Moreover, content sharing is facilitated due to the broadcast nature of the wireless medium. The supremacy of the proposed solution compared to the benchmark schemes is confirmed also under such settings.

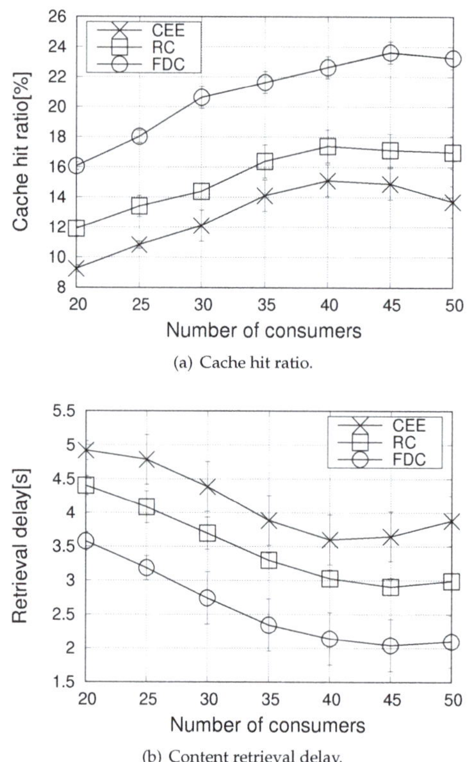

(a) Cache hit ratio.

(b) Content retrieval delay.

Figure 5. Performance metrics in the urban scenario, when varying the number of consumers ($\alpha = 1.5$).

Notwithstanding, no more improvements are experienced when the number of consumers reaches the value of 40. Indeed, with a higher number of consumers also the number of distinct contents increases: more contents are requested that are less popular, which implies a higher traffic congestion in the network and a lower cache hit ratio.

6.2. Highway Scenario

The second set of results, in Figure 6, focuses on the highway scenario, when varying the Zipf skewness parameter α in the range 1, 1.5, 2, 2.5.

The same trends can be observed already seen for the urban scenario. First, the proposed solution, FDC, outperforms the benchmark schemes, under all settings. Second, performance gets better for all schemes as the α parameter increases. The main difference regarding the previous scenario is that all solutions exhibit slightly worse performance. For instance, in Figure 6b, it can be seen that the lowest retrieval delay is 6.1s for FDC when $\alpha = 2.5$, while it was about 3s in the urban scenario in the same settings. This has to be ascribed to the topology and the higher vehicle speed which make contact times among vehicles shorter, hence also reducing the caching events. As a result, the number of hops increases as well as the delay in retrieving contents which entails the exchange of more Data packets.

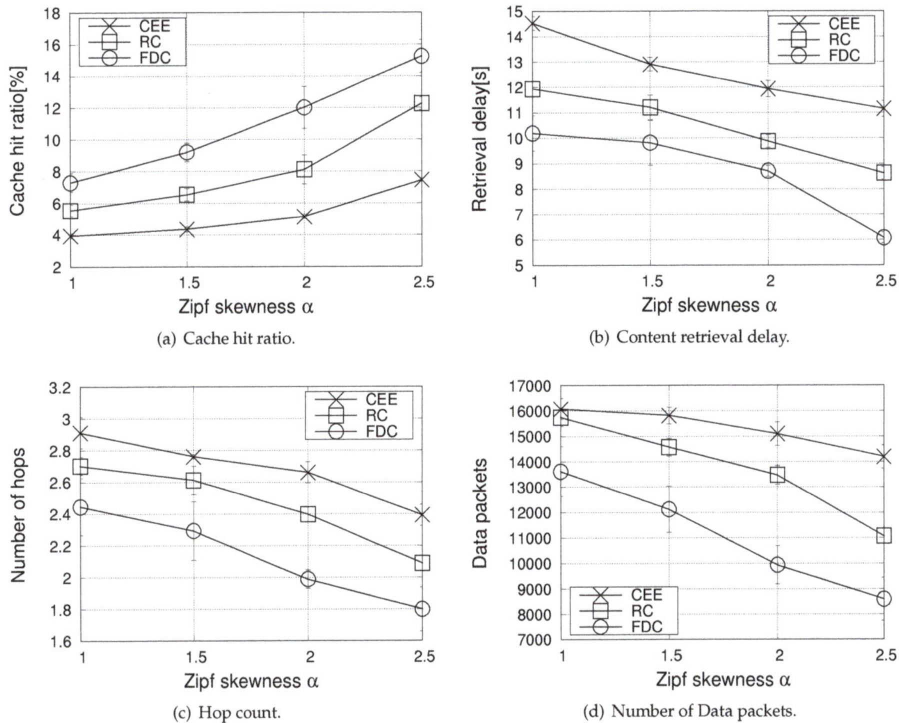

(a) Cache hit ratio.

(b) Content retrieval delay.

(c) Hop count.

(d) Number of Data packets.

Figure 6. Performance metrics in the highway scenario, when varying the Zipf skewness parameter α (number of consumers equal to 20).

As shown in Figure 7, the increasing number of consumers affects the achieved performance, similarly to the urban scenario. However, the trend of the experienced improvements in terms of both cache hit ratio and retrieval delay gets steeper as the number of consumers increases, compared to the urban scenario. Indeed, less congestion is experienced in this topology, due to the higher volatile nature of connectivity and the smaller size of the one-hop neighborhood per vehicle.

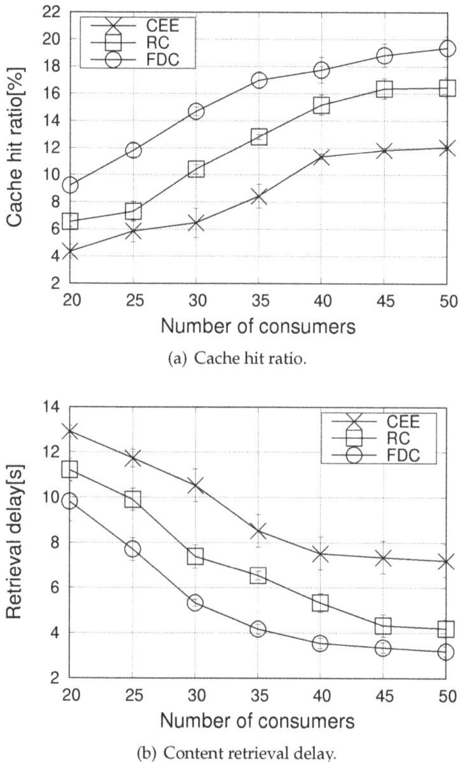

(a) Cache hit ratio.

(b) Content retrieval delay.

Figure 7. Performance metrics in the highway scenario, when varying the number of consumers ($\alpha = 1.5$).

7. Conclusions

In this paper we investigated the issues related to the caching of transient contents in V-NDN. We conceived a novel autonomous caching strategy, FDC, in which the caching decision is taken according to the content lifetime. The solution is meant to be as compliant as possible with the legacy NDN caching routines and not to add additional signaling overhead, which could uselessly overwhelm highly dynamic vehicular links. The addition of a single field, i.e., the timestamp, to the NDN Data packet is foreseen to allow nodes infer the actual residual lifetime of contents.

Simulation results conducted under realistic mobility and data pattern settings confirm the supremacy of the proposal against two representative benchmark solutions in terms of valuable metrics, i.e., content retrieval latency, cache hit ratio, number of hops, and exchanged Data packets.

FDC can be integrated in traditional caching approaches in order to let them deal with transient content. As future work, we plan to investigate this aspect and to design a more sophisticated caching strategy, e.g., relying on content popularity and topological information of cachers, besides content freshness.

Author Contributions: Conceptualization and methodology, M.A., C.C., G.R., G.L. and A.M.; software, M.A.; validation, M.A.; investigation, M.A.; writing–original draft preparation, M.A., C.C. and G.L.; writing–review and editing, M.A., C.C., G.R., A.M. and G.L.; supervision, A.M. and G.R. All authors have read and agreed to the published version of the manuscript.

Funding: This research received no external funding.

Conflicts of Interest: The authors declare no conflict of interest.

References

1. Khelifi, H.; Luo, S.; Nour, B.; Moungla, H.; Faheem, Y.; Hussain, R.; Ksentini, A. Named Data Networking in Vehicular Ad Hoc Networks: State-of-the-Art and Challenges. *IEEE Commun. Surv. Tutor.* **2019**, *22*, 320–351. [CrossRef]
2. Zhang, S.; Luo, H.; Li, J.; Shi, W.; Shen, X.S. Hierarchical Soft Slicing to Meet Multi-Dimensional QoS Demand in Cache-Enabled Vehicular Networks. *IEEE Trans. Wirel. Commun.* **2020**, *19*, 2150–2162. [CrossRef]
3. Zhang, J.; Letaief, K.B. Mobile Edge Intelligence and Computing for the Internet of Vehicles. *Proc. IEEE* **2019**, *108*, 246–261. [CrossRef]
4. Hail, M.A.; Amadeo, M.; Molinaro, A.; Fischer, S. Caching in Named Data Networking for the Wireless Internet of Things. In Proceedings of the IEEE International Conference on Recent Advances in Internet of Things (RIoT), Singapore, 7–9 April 2015; pp. 1–6.
5. Hahm, O.; Baccelli, E.; Schmidt, T.C.; Wählisch, M.; Adjih, C.; Massoulié, L. Low-power Internet of Things with NDN & Cooperative Caching. In Proceedings of the ACM Conference on Information-Centric Networking (ICN), Berlin, Germany, 26–28 September 2017; pp. 98–108.
6. Hail, M.A.M.; Amadeo, M.; Molinaro, A.; Fischer, S. On the Performance of Caching and Forwarding in Information-Centric Networking for the IoT. In Proceedings of the International Conference on Wired/Wireless Internet Communication (WWIC), Malaga, Spain, 25–27 May 2015; pp. 313–326.
7. Vural, S.; Wang, N.; Navaratnam, P.; Tafazolli, R. Caching Transient Data in Internet Content Routers. *IEEE/ACM Trans. Netw.* **2016**, *25*, 1048–1061. [CrossRef]
8. Meddeb, M.; Dhraief, A.; Belghith, A.; Monteil, T.; Drira, K.; AlAhmadi, S. Cache Freshness in Named Data Networking for the Internet of Things. *Comput. J.* **2018**, *61*, 1496–1511. [CrossRef]
9. Mastorakis, S.; Afanasyev, A.; Zhang, L. On the Evolution of ndnSIM: An Open-Source Simulator for NDN Experimentation. *ACM SIGCOMM Comput. Commun. Rev.* **2017**, *47*, 19–33. [CrossRef]
10. Zhang, L.; Afanasyev, A.; Burke, J.; Jacobson, V.; Claffy, K.; Crowley, P.; Papadopoulos, C.; Wang, L.; Zhang, B. Named Data Networking. *ACM SIGCOMM Comput. Commun. Rev.* **2014**, *44*, 66–73. [CrossRef]
11. Liu, X.; Li, Z.; Yang, P.; Dong, Y. Information-Centric Mobile Ad Hoc Networks and Content Routing: A Survey. *Ad Hoc Netw.* **2017**, *58*, 255–268. [CrossRef]
12. Baccelli, E.; Mehlis, C.; Hahm, O.; Schmidt, T.C.; Wählisch, M. Information Centric Networking in the IoT: Experiments with NDN in the Wild. In Proceedings of the ACM Conference on Information-Centric Networking (ICN), Paris, France, 24–26 September 2014, pp. 77–86.
13. Mtibaa, A.; Tourani, R.; Misra, S.; Burke, J.; Zhang, L. Towards Edge Computing over Named Data Networking. In Proceedings of the IEEE International Conference on Edge Computing (EDGE), San Francisco, CA, USA, 2–7 July 2018, pp. 117–120.
14. Amadeo, M.; Campolo, C.; Molinaro, A. NDNe: Enhancing Named Data Networking to Support Cloudification at the Edge. *IEEE Commun. Lett.* **2016**, *20*, 2264–2267. [CrossRef]
15. Amadeo, M.; Ruggeri, G.; Campolo, C.; Molinaro, A. IoT Services Allocation at the Edge via Named Data Networking: From Optimal Bounds to Practical Design. *IEEE Trans. Netw. Serv. Manag.* **2019**, *16*, 661–674. [CrossRef]
16. Rossi, D.; Rossini, G. Caching Performance of Content Centric Networks Under Multi-Path Routing (and More). *Relatório técnico, Telecom ParisTech* **2011**, 1–6. Available online: https://pdfs.semanticscholar.org/8fcc/e9e4865a950723f93bb97b5d5aa7e793037a.pdf (accessed on 2 April 2020).
17. MacHardy, Z.; Khan, A.; Obana, K.; Iwashina, S. V2X Access Technologies: Regulation, Research, and Remaining Challenges. *IEEE Comm. Surv. Tutor.* **2018**, *20*, 1858–1877. [CrossRef]
18. Amadeo, M.; Campolo, C.; Molinaro, A. Priority-based Content Delivery in the Internet of Vehicles through Named Data Networking. *J. Sens. Actuator Netw.* **2016**, *5*, 17. [CrossRef]
19. *IEEE Std. 802.11-2012: IEEE Standard for Information Technology - Part 11: Wireless LAN Medium Access Control (MAC) and Physical Layer (PHY) Specifications*; IEEE: Piscataway, NJ, USA, 2012. [CrossRef]
20. Naik, G.; Choudhury, B.; Park, J.M. IEEE 802.11bd & 5G NR V2X: Evolution of Radio Access Technologies for V2X Communications. *IEEE Access* **2019**, *7*, 70169–70184.
21. Grassi, G.; Pesavento, D.; Pau, G.; Vuyyuru, R.; Wakikawa, R.; Zhang, L. VANET via Named Data Networking. In Proceedings of the IEEE Conference on Computer Communications Workshops (INFOCOM WKSHPS), Toronto, ON, Canada, 27 April–2 May 2014; pp. 410–415.

22. Modesto, F.M.; Boukerche, A. An Analysis of Caching in Information-Centric Vehicular Networks. In Proceedings of the IEEE International Conference on Communications (ICC), Paris, France, 21–25 May 2017; pp. 1–6.

23. Zhang, G.; Li, Y.; Lin, T. Caching in Information Centric Networking: A Survey. *Comput. Netw.* **2013**, *57*, 3128–3141. [CrossRef]

24. Fiore, M.; Casetti, C.; Chiasserini, C.F. Caching strategies based on information density estimation in wireless ad hoc networks. *IEEE Trans. Veh. Technol.* **2011**, *60*, 2194–2208. [CrossRef]

25. Ostrovskaya, S.; Surnin, O.; Hussain, R.; Bouk, S.H.; Lee, J.; Mehran, N.; Ahmed, S.H.; Benslimane, A. Towards Multi-metric Cache Replacement Policies in Vehicular Named Data Networks. In Proceedings of the IEEE 29th Annual International Symposium on Personal, Indoor and Mobile Radio Communications (PIMRC), Bologna, Italy, 9–12 September 2018, pp. 1–7.

26. Abani, N.; Braun, T.; Gerla, M. Proactive Caching with Mobility Prediction under Uncertainty in Information-Centric Networks. In Proceedings of the ACM Conference on Information-Centric Networking (ICN), Berlin, Germany, 26–28 September 2017, pp. 88–97.

27. Yao, L.; Chen, A.; Deng, J.; Wang, J.; Wu, G. A cooperative caching scheme based on mobility prediction in vehicular content centric networks. *IEEE Trans. Veh. Technol.* **2017**, *67*, 5435–5444. [CrossRef]

28. Huang, W.; Song, T.; Yang, Y.; Zhang, Y. Cluster-based Cooperative Caching with Mobility Prediction in Vehicular Named Data Networking. *IEEE Access* **2019**, *7*, 23442–23458. [CrossRef]

29. Laoutaris, N.; Che, H.; Stavrakakis, I. The LCD Interconnection of LRU Caches and its Analysis. *Perform. Eval.* **2006**, *63*, 609–634. [CrossRef]

30. Tarnoi, S.; Suksomboon, K.; Kumwilaisak, W.; Ji, Y. Performance of Probabilistic Caching and Cache Replacement Policies for Content-Centric Networks. In Proceedings of the IEEE Conference on Local Computer Networks (LCN), Edmonton, AB, Canada, 8–11 September 2014; pp. 99–106.

31. Pfender, J.; Valera, A.; Seah, W.K. Performance Comparison of Caching Strategies for Information-Centric IoT. In Proceedings of the ACM Conference on Information-Centric Networking (ICN), Boston, MA, USA, 21–23 September 2018; pp. 43–53.

32. Ma, L.; Dong, X.; Xu, Z.; Wu, Y.; Shen, L.; Xing, S. Distributed Probabilistic Caching with Content-location Awareness in VNDNs. In Proceedings of the 21st ACM International Conference on Modeling, Analysis and Simulation of Wireless and Mobile Systems, Montreal, QC, Canada, 28 October–2 November, 2018; pp. 311–314.

33. Deng, G.; Wang, L.; Li, F.; Li, R. Distributed Probabilistic Caching Strategy in VANETs through Named Data Networking. In Proceedings of the IEEE Conference on Computer Communications Workshops (INFOCOM WKSHPS), San Francisco, CA, USA, 10–14 April 2016, pp. 314–319.

34. Zhao, W.; Qin, Y.; Gao, D.; Foh, C.H.; Chao, H.C. An Efficient Cache Strategy in Information Centric Networking Vehicle-to-Vehicle Scenario. *IEEE Access* **2017**, *5*, 12657–12667. [CrossRef]

35. Wei, Z.; Pan, J.; Wang, K.; Shi, L.; Lyu, Z.; Feng, L. Data Forwarding and Caching Strategy for RSU Aided V-NDN. In Proceedings of the International Conference on Wireless Algorithms, Systems, and Applications, Honolulu, HI, USA, 24–26 June 2019; pp. 605–612.

36. Zhang, Z.; Yu, Y.; Zhang, H.; Newberry, E.; Mastorakis, S.; Li, Y.; Afanasyev, A.; Zhang, L. An Overview of Security Support in Named Data Networking. *IEEE Commun. Mag.* **2018**, *56*, 62–68.

37. Mastorakis, S.; Afanasyev, A.; Moiseenko, I.; Zhang, L. ndnSIM 2.0: A new version of the NDN simulator for NS-3. NDN, Technical Report NDN-0028. 2015. Available online: https://www.researchgate.net/profile/Spyridon_Mastorakis/publication/281652451_ndnSIM_20_A_new_version_of_the_NDN_simulator_for_NS-3/links/5b196020a6fdcca67b63660d/ndnSIM-20-A-new-version-of-the-NDN-simulator-for-NS-3.pdf (accessed on 2 April 2020).

38. Behrisch, M.; Bieker, L.; Erdmann, J.; Krajzewicz, D. SUMO–simulation of urban mobility: An overview. In Proceedings of the Third International Conference on Advances in System Simulation (SIMUL), Barcelona, Spain, 23–29 October 2011.

39. Breslau, L.; Cao, P.; Fan, L.; Phillips, G.; Shenker, S. Web caching and Zipf-like distributions: Evidence and implications. In Proceedings of the IEEE Conference of Computer Communications (INFOCOM), New York, NY, USA, 21–25 March 1999; pp. 126–134.

40. Zhang, Y.; Zhao, J.; Cao, G. Roadcast: A popularity aware content sharing scheme in vanets. *ACM SIGMOBILE Mob. Comput. Commun. Rev.* **2010**, *13*, 1–14. [CrossRef]

41. Grassi, G.; Pesavento, D.; Pau, G.; Zhang, L.; Fdida, S. Navigo: Interest Forwarding by Geolocations in Vehicular Named Data Networking. In Proceedings of the IEEE 16th International Symposium on A World of Wireless, Mobile and Multimedia Networks (WoWMoM), Boston, MA, USA, 14–17 June 2015; pp. 1–10.
42. Hasan, K.F.; Wang, C.; Feng, Y.; Tian, Y.C. Time synchronization in vehicular ad-hoc networks: A survey on theory and practice. *Veh. Commun.* **2018**, *14*, 39–51. [CrossRef]
43. Tseng, F.H.; Hsueh, J.H.; Tseng, C.W.; Yang, Y.T.; Chao, H.C.; Chou, L.D. Congestion Prediction with Big Data for Real-time Highway Traffic. *IEEE Access* **2018**, *6*, 57311–57323. [CrossRef]
44. Carrascal, V.; Diaz, G.; Zavala, A.; Aguilar, M. Dynamic Cross-layer Framework to Provide QoS for Video Streaming Services over Ad Hoc Networks. In Proceedings of the International Conference on Heterogeneous Networking for Quality, Reliability, Security and Robustness (QShine), Hong Kong, China, 28–31 July 2008; pp. 1–7.

Article

DCS: Distributed Caching Strategy at the Edge of Vehicular Sensor Networks in Information-Centric Networking

Yahui Meng [1,†], Muhammad Ali Naeem [1,†], Rashid Ali [2], Yousaf Bin Zikria [3,*] and Sung Won Kim [3,*]

[1] School of Science, Guangdong University of Petrochemical Technology, Maoming 525000, China; mengyahui@gdupt.edu.cn (Y.M.); malinaeem7@gmail.com (M.A.N.)
[2] School of Intelligent Mechatronics Engineering, Sejong University, Seoul 05006, Korea; rashidali@sejong.ac.kr
[3] Department of Information and Communication Engineering, Yeungnam University, Daegu 38541, Korea
[*] Correspondence: yousafbinzikria@ynu.ac.kr (Y.B.Z.); swon@yu.ac.kr (S.W.K.)
[†] Equal Authorship: Y.M. and M.A.N. contributed equally.

Received: 3 September 2019; Accepted: 9 October 2019; Published: 11 October 2019

Abstract: Information dissemination in current Vehicular Sensor Networks (VSN) depends on the physical location in which similar data is transmitted multiple times across the network. This data replication has led to several problems, among which resource consumption (memory), stretch, and communication latency due to the lake of data availability are the most crucial. Information-Centric Networking (ICN) provides an enhanced version of the internet that is capable of resolving such issues efficiently. ICN is the new internet paradigm that supports innovative communication systems with location-independent data dissemination. The emergence of ICN with VSNs can handle the massive amount of data generated from heterogeneous mobile sensors in surrounding smart environments. The ICN paradigm offers an in-network cache, which is the most effective means to reduce the number of complications of the receiver-driven content retrieval process. However, due to the non-linearity of the Quality-of-Experience (QoE) in VSN systems, efficient content management within the context of ICN is needed. For this purpose, this paper implements a new distributed caching strategy (DCS) at the edge of the network in VSN environments to reduce the number of overall data dissemination problems. The proposed DCS mechanism is studied comparatively against existing caching strategies to check its performance in terms of memory consumption, path stretch ratio, cache hit ratio, and content eviction ratio. Extensive simulation results have shown that the proposed strategy outperforms these benchmark caching strategies.

Keywords: information-centric networking (ICN); client-cache (CC); video on demand (VoD); vehicular sensor network (VSN)

---

1. Introduction

The increasing demands for novel applications due to advancements in technology have led to increased interest in finding a means by which to deliver popular data contents to remote physical locations such as in Vehicular Sensor Networks (VSNs), mainly for Vehicular Ad Hoc Networks (VANET) [1]. In VSNs, vehicles are equipped with diverse onboard units (sensors) for the communication of information. The exponentially-increasing usage of the internet has posed problems for current VSNs due to its need for diverse facilities such as the dissemination of immense amounts of data from heterogeneous consumers along with periodic connectivity in harsh signal propagation, spare roadside conditions, and high levels of mobility [2,3]. It is difficult to provide these facilities to vehicular networks using the IP-based protocols of the present, host-centric connectivity network paradigm [2,3]. However, the Information-Centric Internet (ICN) has proposed emerging technologies to provide novel applications to fulfill future internet requirements. In recent years, the ICN has received significant interest from the research community because of its rapid growth and flexible nature vis-a-vis data communication services. It delivers a unique computing environment in which the router turns into a server. As a result, these servers can modify, understand, and measure the surrounding environment for data dissemination [4]. The immense growth of today's internet traffic requires high-quality communication services because of network congestion, which has been increasing exponentially [5]. Therefore, the internet is currently facing several problems related to network traffic. For example, the internet requires extra content retrieval latency along with high bandwidth consumption during data dissemination.

Moreover, the usage of resources and energy have also increased. Connected devices are resource-constrained, and connected devices have a significant impact upon communication in everyday life. The basic concept behind ICN technology is that all objects have to operate through processing, identifying, and caching abilities to communicate within a diverse environment and achieve good data dissemination performance [6]. The reason is that the current internet supports an outdated, location-based paradigm in which all devices need to connect through IP addresses that indicates their location. As a result, the IP-based internet is facing several issues, such as communication latency, data searching overhead due to high network congestion, and the dissemination of identical contents many times from remote servers [7,8]. Moreover, the IP-based internet architecture is insufficient to achieve better results in data communication through a large number of devices because location-based communication needs a high amount of energy; this is a fundamental limitation of internet architecture. New research and big data technology will deliver an enormous amount of data that will be challenging for the current, IP-based internet architecture [8,9].

ICN-based projects were designed combining several modules, such as caching, naming, forwarding, mobility, and security. However, caching is the first module that pays full attention to determining the ICN from the IP-based internet architecture. It delivers many benefits during data dissemination such as short stretch, and provides fast data delivery services [10]. ICN focuses on data delivery without location dependency. Thus, this approach makes the ICN architecture beneficial for the internet environment. ICN does not require IP addresses for data dissemination between sources and consumers; rather, it uses unique names to send and retrieve data contents [11]. The cache is the most significant feature of the ICN; it is used to store the transmitted contents near the desired consumers. In vehicular networks, vehicles obtain their required contents from neighboring vehicles in short time periods with small stretch. Therefore, there is no need to forward the coming interests to remote providers, and a large number of user requests can be served locally.

In the ICN, consumers send their interests directly to the network, and the whole system is responsible for sending the corresponding data to the appropriate consumer. A copy of the disseminated content is cached at different locations between consumers and providers, according to the selected caching strategy. This makes it possible to store the contents in a location which is geographically close to the consumers [12]; therefore, it can reduce latency by caching contents near consumers, because subsequent interests will be satisfied with the cached content. The purpose of the implementation of the in-network cache is to enhance data transmission services and reduce the high amount of network traffic that causes link congestion and increases bandwidth consumption [13]. Moreover, in-network caching can reduce energy and resource consumption because, in the ICN, subsequent interests do not require traversing towards remote servers [14]. ICN caching is divided into two categories, i.e., off-path and on-path caching. In off-path caching a particular entity named Name Resolution System (NRS) is used to broadcast the published contents' names with their locations. Initially, all the consumer's interests are transmitted to the NRS, and the NSR forwards these interests to the appropriate data sources, as shown in Figure 1 (Off-Path Caching). In on-path caching, consumers are directly sending their Interests to the network, and the network directly sends back the corresponding contents to the consumer, as illustrated in Figure 1 (On-Path caching). Therefore, it can reduce the communication and computation overhead in data dissemination [15].

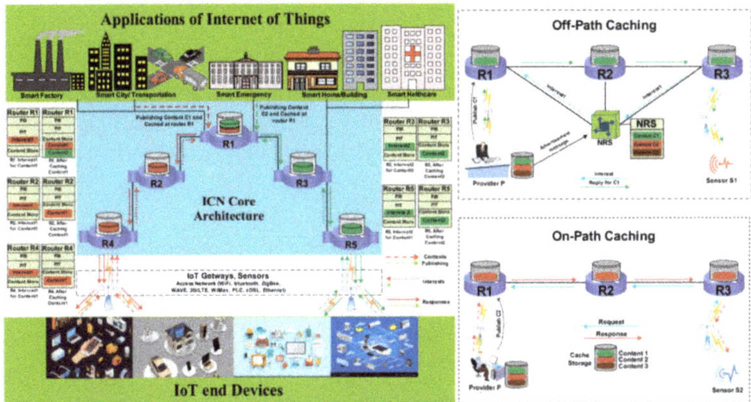

Figure 1. Caching architecture, On-path and Off-path caching.

In contrast to the ICN cache strategy, there are three exceptional features to take into account when applying ICN cache to VSNs. First, in view of their protection and selfishness, drivers of vehicles may play a tentative role in terms of obeying the guidelines of a cache-sharing strategy [2]. Furthermore, vehicles' frequent and dynamic topology changes increase the unpredictability of the cache strategy [16]. In addition, vehicles have weak computational and storage resources compared to conventional network base stations (such as access points) and routers, and the cache redundancy of the strategy ought to be diminished [17].

Most of the work done by researchers in this domain has not taken into account and explored the characteristics of VSNs. A vehicle-to-infrastructure scenario cache policy in VSNs is proposed in [18]. The authors proposed an Integer Linear Programming (ILP) definition of the issue of optimally appropriating the contents in the VSN while thinking about the accessible storage limit and connection ability to expand the likelihood that a vehicle will be able to retrieve the desired content. However, due to weak wireless links and mobility, vehicles cannot directly access servers or access points (APs). Therefore, a VSN cache strategy is needed at the edge of the network. For this purpose, this paper implements a new distributed caching strategy (DCS) at the edge of the network in VSN environments to reduce the number of data dissemination problems. The proposed DCS mechanism is studied comparatively against existing caching strategies to check the performance in terms of memory consumption, path stretch ratio, cache hit ratio, and content eviction ratio.

Section 2 provides an overview of related studies. Section 3 defines the problems that still exist in associated studies. In Section 4, the proposed model is explained. In Section 5, the performance evaluation of related and proposed research is done using a simulation platform. In Section 6, the paper is concluded. Finally, Section 7 presents some future directions for Vehicular Sensor Networks.

2. Related Study

ICN is an emerging environment in which devices have the ability to respond to their surroundings with the help of caching [19]. Data dissemination is the most fundamental phenomenon of all internet architectures, in which the current IP addresses-based internet is supported by the old version of the architecture for data transmission between remote locations. Therefore, data is distributed when a consumer's interest is received [20]. The reason for this is that the IP-based internet architecture supports location-based data dissemination that produces serious issues for future communication processes due to the exponential increase in the amount of data traffic. At the same time, ICN delivers location-independent data dissemination and offers lots of benefits in terms of improving the overall data communication process [21]. Therefore, ICN can reduce the critical issues of the IP-based architecture, and can fulfill future internet requirements.

2.1. Client-Cache (CC)

In Client-Cache Strategy (CC), the validity of cached contents is observed. The concept of CC is derived from central-based caching, in which the content is cached at routers that are linked to more routers [22]. The aim of CC is to increase the validity of a given content. The validity is measured according to the lifespan of the cached content at intermediate routers and from the publisher. The content is selected as valid if its lifespan of at the publisher is higher than its lifespan which has been cached at an intermediate router.

In Figure 2 (Client-Cache scenario), various interests from Consumers A and B are sent to retrieve the Content C1. Primarily, the lifespan of Content C1, Content C2, and Content C3 are shown by VC6, VC4, and VC5, respectively, in Figure 2. In CC, the lifespan of the content is taken as VC, which shows the validity of the content. Therefore, the lifespans of contents C1 and C2 are higher at the publisher than at router R5. This indicates that contents C1 and C2 should be cached at router R5; thus, C1 will be cached at router R5, as shown in Figure 2.

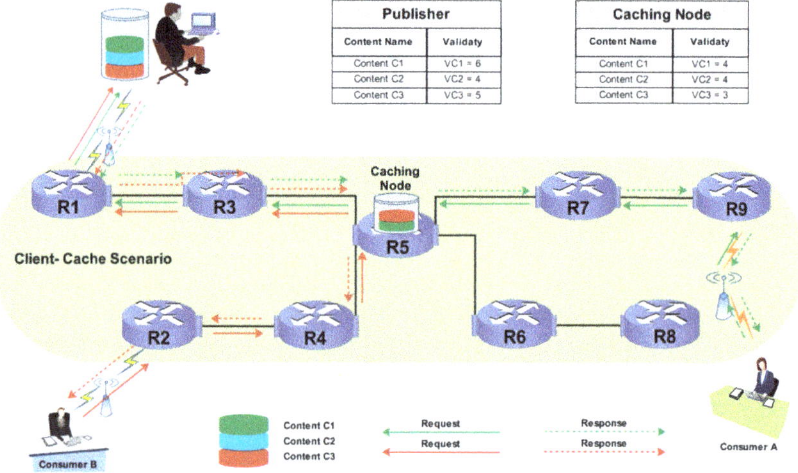

Figure 2. Client-Cache.

2.2. Flexible Popularity-Based Caching Strategy (FlexPop)

The FlexPop caching strategy compiles two mechanisms to complete its content caching procedure [23]. Primarily, it performs a content caching procedure to cache transmitted content alongside the data routing path. It executes a second content eviction procedure if the disseminated content does not identify the free cache space for accommodation at the intermediate routers. FlexPop requires the maintenance of a popularity table that helps to count the number of interests at each router for all content names. On the basis of the received interests, the popularity of a given piece of content is calculated in the PT using the content counter and popularity tag. Initially, the content is stored in the PT to calculate its popularity. If the content within the PT indicates that its popularity is equal or greater than the threshold, it forwards it to the comparison table (CT). The CT is responsible for maintaining information about the popular content. It compares the popularity of the new content with the popularities of the previous popular content; if the new content demonstrates more significant demand than the other content, it is labeled as popular, and the CT is shared with the neighboring routers. When the popularity of that content reaches a threshold, the content is forwarded to the router that has the maximum number of outgoing interfaces to be cached. If the cache of the router having the maximum outgoing interfaces is overflowing, the content is recommended for caching at the router that is associated with the second-highest number of outgoing interfaces.

Figure 3 illustrates the content caching procedure in FlexPop. Initially, two contents, C2 and C3, are cached at router R5. Router R5 is associated with the maximum outgoing interfaces, and only two pieces of content can reside in its cache owing to its limited capacity. Three interests from consumers A and B are sent to router R2 to retrieve content C2. In response to the received interests, the router R2 becomes the provider and sends content C1 to consumers A and B. At the same time, the popularity of content C1 is measured on the basis of the received interests for content C1. According to FlexPop, C1 gains the highest popularity, as shown by the CT in Figure 3; therefore, it is labeled "popular" and recommended for caching at the router with the maximum number of outgoing interfaces (i.e., router tR5). However, there is no free space at router R5 for caching content C1; therefore, it will be cached at the router having the second-highest number of outgoing interfaces. Thus, C1 will be cached at routers R4 and R6.

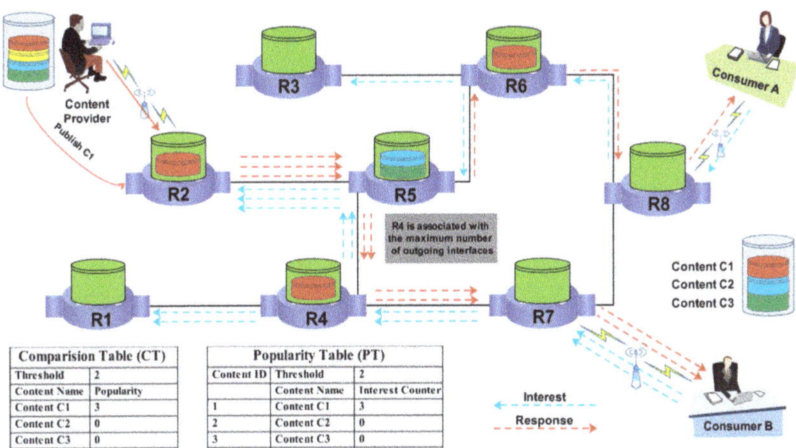

Comparision Table (CT)	
Threshold	2
Content Name	Popularity
Content C1	3
Content C2	0
Content C3	0

Popularity Table (PT)		
Content ID	Threshold	2
	Content Name	Interest Counter
1	Content C1	3
2	Content C2	0
3	Content C3	0

Figure 3. Flexible Popularity-based Caching Strategy.

2.3. Centrality-Based Caching Strategy (CCS)

This content caching mechanism requires two approaches. First, it determines the betweenness centrality node by calculating the links associated with each node. Second, it decides how to cache the transmitted content along the data routing path [24]. In this caching mechanism, the interesting content is forwarded to the node that has the maximum number of outgoing interfaces or the maximum number of paths associated with it. If a node is associated with a high number of data routing paths, it has more opportunities to cache the disseminated content [25]. Figure 4 illustrates the content caching mechanism using centrality-based caching in which Consumers A, B, and C are associated with routers R4, R7, and R9, respectively. These consumers sent three interests to retrieve content C1, as that content is already published in the network by the content provider (P). As the interests for content C1 reach router R3, the required content is obtained. Therefore, router R3 acts as a provider and transmits content C1 to the interested consumers (i.e., A, B, and C). During the transmission of the content, each router calculates the number of data routing paths associated with it. According to the caching nature of the CCS, router R6 is selected as the betweenness centrality router because it has the highest number of paths associated with it along the data delivery path between the provider and the consumers. Hence, content C1 will be cached at R5.

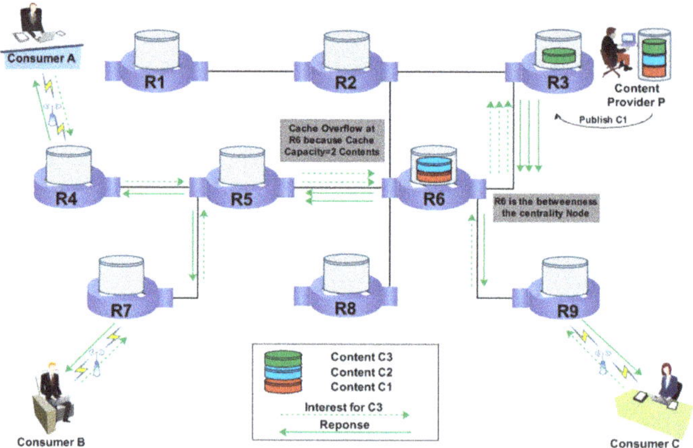

Figure 4. Centrality-Based Caching Strategy.

3. The Problem Description

ICN provides centrality-based caching strategies in which the transmitted content is cached at a betweenness centrality location to fulfill the requirements of subsequent interests [26]. However, these caching strategies have been facing some critical issues due to the limited capacity of cache storage at the betweenness centrality location.

The CC tries to improve content validity, but also introduced some problems such as content eviction ratio and the stretch ratio between the consumer and the provider, because the content's legality must be measured at all the routers, which takes time. According to CC, the interesting content will be cached at a betweenness centrality router that increases the number of significant issues which occur due to caching the transmitted content only at one router, such as memory consumption. Moreover, it increases the path length due to the high content eviction rate between the consumer and the provider. The reason for this is that all the interests need to be forwarded to the primary publisher due to the limited cache capacity at the betweenness centrality location. Another issue of CC is that if a large number of interests are received for Content C, and the validity of C is the same at the betweenness centrality node and the server, then according to the CC, Content C will not cache at the betweenness centrality router even, it is deemed to be popular. Therefore, all the interests for popular content will be forwarded to the main server that maximizes the stretch, and the cache hit ratio will automatically be decreased. The amount of cache storage is limited, and it is difficult to accommodate all the content at the betweenness centrality router. Therefore, certain problems arise in CCS that demonstrate the increased congestion which can occur at the centrality position, leading to a high number of evictions within short intervals of time. The reason for this is that if the cache of the betweenness centrality position becomes full, all the interest for content must to be forwarded to the remote provider. In addition, it does not care about the content popularities, which increases the caching of contents with lower popularities. Thus, the overall cache hit ratio decreases because several interests have to be accomplished from remote providers owing to the large accommodation of less popular content. Hence, the overall caching performance is decreased [27].

FlexPop was developed to solve important problems such as high memory consumption, high evictions, and stretch. However, it increases content homogeneity through multiple replications of the same content. Consequently, it retains the process of content evictions and higher resource utilization. Moreover, there is no criterion by which to choose popular content according to time consumption. We assumed a case where three interests were generated for content C1 in 5 s, and two for content C2 in 1 s. According to FlexPop, C1 will be the most popular because no time distinction is included for the selection of popular content. Consequently, the most recently-used content will remain unpopular, which causes a low cache hit ratio that affects the efficiency of the content dissemination and increases the content eviction ratio. Moreover, in FlexPop, two tables, PT and CT, must be computed for each piece of content and to identify popular content, which increases the searching overhead during the selection of popular content, because several attempts must be made to calculate the popularity.

Consequently, this increases the source (cache) utilization. The cache size is limited compared to the giant volume of data being communicated. Owing to the enormous number of replications of similar content, the hit ratio cannot retain its beneficial level to strengthen the caching performance. Another concern is the procedure of changing the cache location based on popular content, which increases the number of eviction-caching operations caused while searching for an empty cache space and for content that has to be replaced.

- How could the content memory consumption be minimized with an improved cache hit ratio?
- How could we enhance the caching mechanism by selecting the centrality position by reducing the stretch ratio?

To answer these questions, a new ICN-based caching strategy is proposed that has the ability to reduce memory consumption with a high cache hit ratio and short stretch for subsequent interests. In addition, it has the ability to minimize content eviction operations.

4. Proposed Distributed Caching Strategy (DCS)

In previous studies, it was observed that the ideal structure of the network could affect the overall performance of the system. Cache management is an optimal feature of content centrism, and many researchers have focused on diverse methods of managing disseminated content over networks. Recently, several content caching mechanisms have been developed to increase the efficiency of in-network caching by distributing the transmitted content according to the diverse nature of caching approaches. However, in existing caching mechanisms, several problems related to multiple replications of homogeneous content persist, thereby increasing memory wastage. Content caching mechanisms must implement the optimal objectives to actualize the basic concept of the NDN cache and overcome issues in the data dissemination process which are faced by the aforementioned caching mechanisms [28]. Consequently, in this study, a new, flexible mechanism for content caching has been designed to improve the overall caching performance [29]. The distributed caching strategy works on the popularities of contents. Popularity-based caching strategies are more efficient in terms of improving content dissemination, because these strategies only cache the popular content that can fulfill the requirements of large numbers of consumers, as compared to offensive content. Therefore, the level of popularity of a given piece of content has a significant influence on the caching performance. Mostly, consumers are interested in downloading popular content, and it is a substantial undertaking to cache popular content at the central position. The reason for this is that most incoming interests will be forwarded through the central location. Therefore, if a popular piece of content is cached at the central location, the communication distance will be decreased because all the interests traversing a central position will be accomplished there. Moreover, the central position may also be used to reduce the overall bandwidth consumption. Thus, in this strategy, it becomes more important to cache popular content at centrality positions. This caching strategy is divided into three sections, as shown below:

Case 1

The selection of popular content in this strategy is made by taking the sum of the received interests for a specific content name. In the DCS caching strategy, each node is associated with a distinctive statistic table in which information about content name, interest count, and a threshold value is stored. Whenever user interest for particular content occurs, the interest count for a specific piece of content name is incremented with the number of received interests to calculate the popularity of that content. The threshold is a value that is specified to measure the popularity of the content. As a result, if the content receives a number of interests which is equal to the threshold value, it is recommended for classification as "popular". In earlier popularity-based caching strategies, the threshold is used as statically defined by the strategy algorithm, as described in MPC. However, DCS represents the dynamic threshold to calculate the popularity of a given piece of content. According to DCS, the threshold will be equal to half the total number of received interests for all the contents at a router. Algorithm 1 illustrates the mechanism of selecting popular content. According to the proposed algorithm, if the number of received interests for a particular piece of content is greater than half the total number of interests for all the pieces of content, that content is recommended for classification as "popular"; otherwise, it is ignored. Figure 5 illustrates the mechanism for measuring content popularity. Suppose that 14 searches are generated for Contents C1, C2, C3, and C4, as shown in Figure 5a. According to DCS, Content C4 recommended for classification as "popular" because it has surpassed the threshold value as shown in Figure 5b. Hence, Content C4 is recommended for caching at the intermediate routers along the data delivery path between the user and the provider. Therefore, the first caching operation for popular content will be performed at the closeness centrality router, and secondly, a copy of these contents will also be cached at the edge nodes.

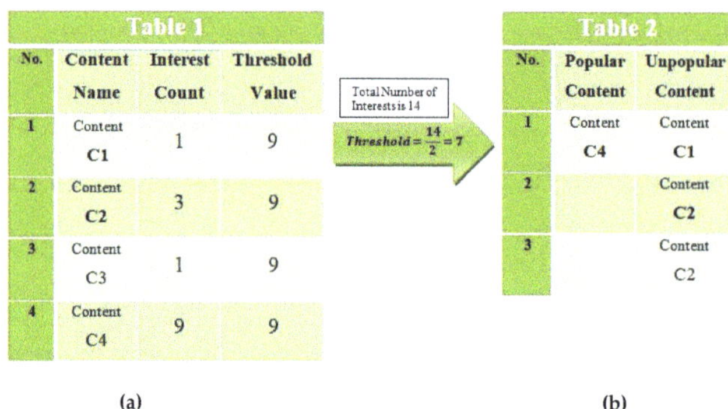

(a)   (b)

Figure 5. Selection of Popular Content.

Algorithm 1: Selection of Popular Content

```
1 def Get_Popular_Content()
2   interestRequests=[]
3   interestCount={}
4   popularContents={}
5   unpopularContents={}
6   for interest in interestRequests:
7     if interestCount.__contains__(interest):
8       interestCount[interest]+=1
9     else
10 interestCount.Add(interest,1)
11   threshold=len(interestCount)/2
12   for interest in interestCount.keys():
13     if interestCount[interest] > threshold:
14       popularContents.Add(interest, interestCount[interest])
15     else
16       unpopularContents.Add(interest, interestCount[interest])
17 return popularContents, unpopularContents
```

Case 2

Popular content cannot be cached in the same way as has been already implemented. The selected contents will be cached in chunk form to reduce the usage of memory and congestion. The reason for this is that the betweenness centrality router is associated with a large number of other routers, which increases the congestion in data dissemination because all interests and contents need to be forwarded through the betweenness centrality router. Therefore, the centrality router has fewer chances to accommodate all popular content at the same time. Thus, the new model increases the ability to cache the maximum quantity of popular content. In DCS, when a content is selected as popular, it is recommended for caching at the closeness centrality router in chunks, as shown in Figure 6 (Distributed Caching Strategy).

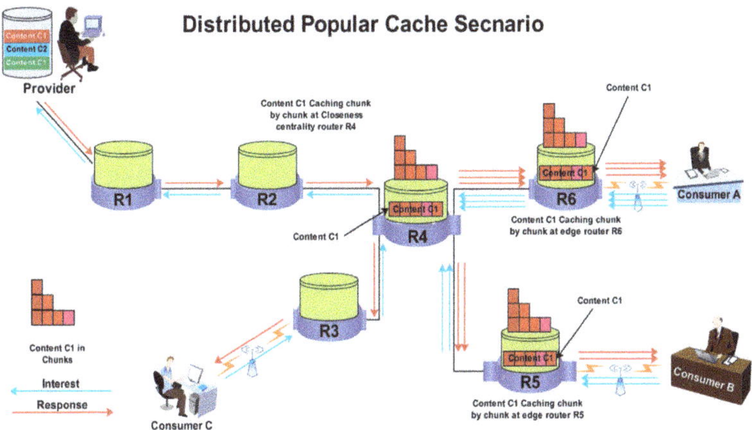

Figure 6. Distributed Caching Strategy.

Moreover, in terms of chunks, the cache will be used efficiently because its availability will increase to accommodate more content in chunk form. When content is deemed to be popular, it cannot be forwarded to the centrality router until it receives more interest; in response to first interest, only one chunk will be delivered to the closeness centrality and edge router. For subsequent reactions, fragments are multiplied to be forwarded for caching at the closeness centrality router and the edge router. This process will remain functional until the content transfers in its entirety to the centrality router. In this way, if the content was popular, but after being popular, it does not receive any interest, then it will not be cached at the centrality router, and the cache of the centrality router will remain unallocated to accommodate subsequent, more popular contents. In this way, DCS resolves the problem of centrality position and uses the cache in an inefficient manner.

Case 3

If a piece of content is deemed to be popular, it will also be forwarded to the edge router at the same time for caching at the closeness centrality router, as shown in Figure 6 (Distributed Caching Strategy). In this way, the path stretch between the consumer and the provider will be reduced for subsequent interests.

Moreover, this will minimize the content retrieval latency for subsequent interests, and reduce the path link congestion by caching the content at edge routers. Therefore, all the following interests will be satisfied with edge routers. If the content is not found at the edge router, then the interest will be satisfied from the closeness centrality router. Moreover, the closeness centrality router is selected for the caching of popular content, because most intents will be accomplished from the centrality router, thereby saving bandwidth consumption with a short stretch path.

For content eviction, the Least Recent (LRU) policy is used to make room for incoming content. The present study proposes a new, ICN-based caching strategy to improve content retrieval latency by reducing the path length between consumers and the provider. Moreover, it reduces the communication path, and network congestion enhances the bandwidth consumption within the limited cache capacity of the network routers.

Figure 6 illustrates the content the caching mechanism in DCS. In the given scenario, Consumers A and B send multiple interests to retrieve Content C1 from the provider. After a while, content C1 becomes popular, because it has received the maximum number of interests that are required to make content popular. Therefore, content C1 is forwarded for caching at closeness centrality router R5. Moreover, the popular content also caches at edge routers R5 and R6. Hence, subsequent interests from Consumers A and B will be satisfied with edge routers R5 and R6. Consequently, Consumer C can download Content C1 from the closeness centrality router.

5. Performance Evaluation

For the evaluation of the proposed caching strategy, a simulation platform is used, in which the SocialCCNsim simulator is selected to evaluate the caching performance. The SocialCCNsim [30] simulator was designed to measure caching performance because, in this simulator, all the network routers are associated with cache storage. Cesar Bernardini [31] developed SocialCCNSim based on SONETOR [32], which is a set of utilities that generates synthetic social network traces. These social network traces represent the interactions of users in a social network or a regular client-server fashion. Any caching strategy can be implemented in SocialCCNSim because it was developed especially for ICN-based caching strategies. Two ISP-level topologies were selected to perform a fair evaluation, i.e., Abilene and GEANT. In the final stage, the DCS evaluation was done using simulations, where the chosen parameters were cache size, catalog size, network topology, Zipf probability model, and simulation time. In our simulations, the Zipf probability distribution is used as the popularity model with the α parameter varying between 0.88 and 1.2; the cache size (which specifies the available space in every node for temporally storing content objects) ranges from 100 to 1000 elements (1 GB to 10 GB); and the catalog (which represents the total number of contents in a network) is 107 lements. The performance of the proposed caching strategy is evaluated in terms of memory consumption and the stretch ratio [31].

Moreover, performance is also comparatively evaluated in terms of network contention to measure the cache hit ratio. The proposed caching strategy is compared to ICN centrality-based caching strategies in which FlexPop, CC, and CCS are included. Moreover, categories of contents (User-Generated Content and Video on Demand) are selected with different cache sizes, such as 1 GB to 10 GB. The x-axis of simulation graphs is divided into ten equal parts, in which each part shows the capacities of the cache storage (e.g., from 1 GB to 10 GB). Accordingly, 100 elements show 1 GB and 1000 items 10 GB of cache size. Table 1 shows the simulation parameters. The proposed strategy is evaluated in terms of checking the performance of the most applicable metrics, i.e., memory consumption, path stretch ratio, and cache hit ratio [33].

Table 1. Simulation Parameters.

Parameter	Description
Simulation time	24 h
Topologies	Abilene and GEANT
Content Size	10 MB each
Catalog Size	10^7 elements
Cache Size	100 to 1000 elements
α Parameter	0.88 and 1.2
Content Categories	UGC and VoD
Simulator	SocialCCNSim
Social Network Topology	Facebook
Traffic Source	SONETOR
Metrics	Memory Consumption, Cache Hit Ratio, Stretch Ratio, and Content Eviction Ratio.

5.1. Memory Consumption

Memory consumption shows the amount of transmitted content that can be cached while downloading the data path for a particular time interval [34]. Consumers can download the contents from multiple routers. In ICN, memory consumption can be clarified as the term of capacity, which shows the volume used by interest and data contents. It can be calculated using the following equation:

$$Memory\ Consumption = \frac{U_m}{T_m} \times 100 \tag{1}$$

where U_m shows the memory that is utilized by the cached content and T_m presents the total memory (cache storage) of the router along the data delivery path.

The DCS performs better than CCS, CC, and FlexPop in terms of memory consumption because it provides the ability of chunk level caching of content, thereby decreasing the usage of memory and congestion in path links. Moreover, it delivers the most popular content near consumers, reducing data traffic and allowing contents to move freely across the network. FlexPop and CC deliver poor performance in terms of memory consumption because of their caching of popular content only at a centrality router, a process that increases the traffic congestion within the limited cache capacity. The CCS caches all the content at the betweenness centrality position without considering the content's popularity, thereby maximizing memory consumption. Figures 7 and 8 show the simulation results on memory consumption using two different topologies (Abilene and GEANT). From these figures, it can be seen that the proposed DCS caching strategy performs much better than FlexPop, CC, and CCS. Thus, we can conclude that DCS is better at enhancing the overall performance of ICN caching in terms of achieving efficient memory consumption.

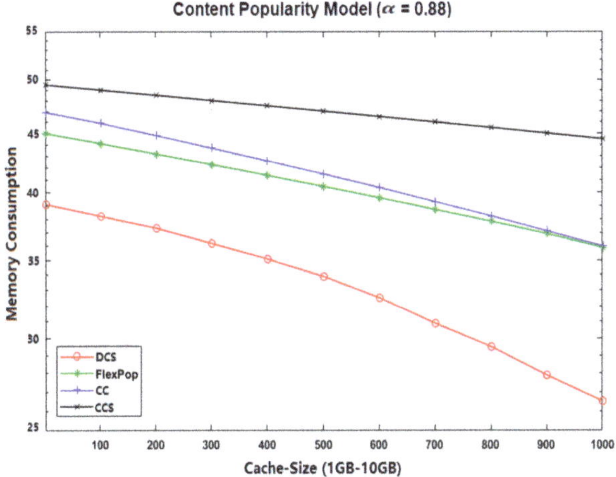

Figure 7. Memory Consumption on Abilene topology with UGC.

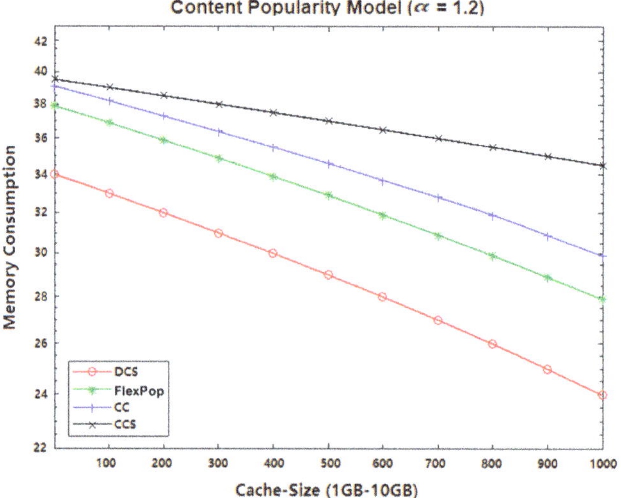

Figure 8. Memory Consumption on GEANT Topology with VoD.

5.2. Stretch

The distance travelled by an interest for a publisher (content provider) is considered as stretch [35,36]. It can be measured using the following equation:

$$Stretch = \frac{\sum_{i=1}^{I} Hop - traveled}{\sum_{i=1}^{I} Total - Hop} \tag{2}$$

where $\sum_{i=1}^{I} Hop - traveled$ represents the number of hops traveled by an interest from the end-user to the content provider. $\sum_{i=1}^{|I|} Total - Hop$ shows the total number of hops from the user to the content provider, and I represent the total number of received interests for a given piece of content.

As the cache capacity is small compared to the disseminating content, less content can be accommodated within the centrality routers. Besides, CCS caches all the content without taking their popularity into account; thus, the most popular contents have fewer chances to be cached at the betweenness centrality position due to the unavailability of a popularity module. Hence, overall performance is reduced in terms of a stretch, because all the interests for the most popular contents need to be forwarded to the remote provider, thereby increasing the path length between the consumer and the provider.

The path length is increased for each interest and response. At the same time, the CC and FlexPop cache provide the ability to accommodate popular contents at intermediate locations for a specific time, that can decrease the path length between consumers and providers. The reason for this is that most interests are satisfied with the centrality positions. However, these strategies provide the ability to store popular contents, but due to the limited capacity of the cache at the betweenness centrality router, CC and FlexPop cannot achieve better results in terms of stretch, because both strategies are used to cache less popular contents due to their small thresholds. On the other hand, DCS caches content in a chunk format, increasing the possibility of accommodating more contents.

Therefore, most incoming interests are satisfied with the centrality location. Moreover, the DCS achieves better results in terms of reducing the path stretch because it provides the ability to store content near consumers. Furthermore, it spreads out the cache ability to store chunk level caching of popular content that is used to increase the space available for new popular content. Moreover, DCS caches popular content at edge routers, thereby reducing the path stretch between consumers and providers; therefore, the proposed caching strategy delivers much better results in terms of reducing the overall stretch ratio. From Figures 9 and 10, results indicating that DCS performs better than CCS, CC, and FlexPop are clearly shown.

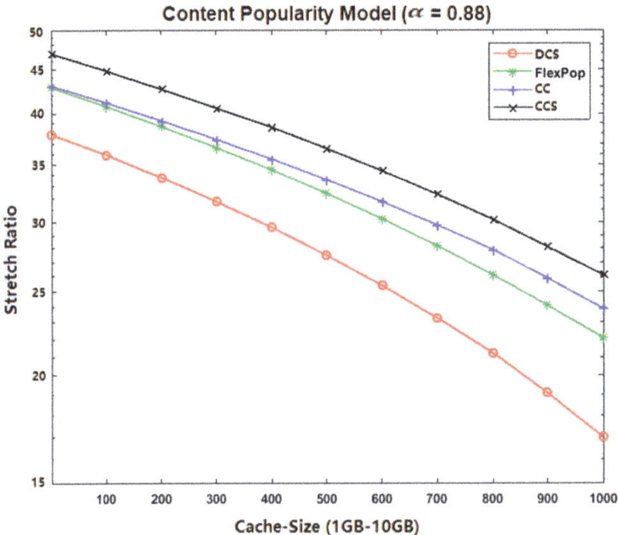

Figure 9. Stretch Ratio on Abilene Topology with UGC.

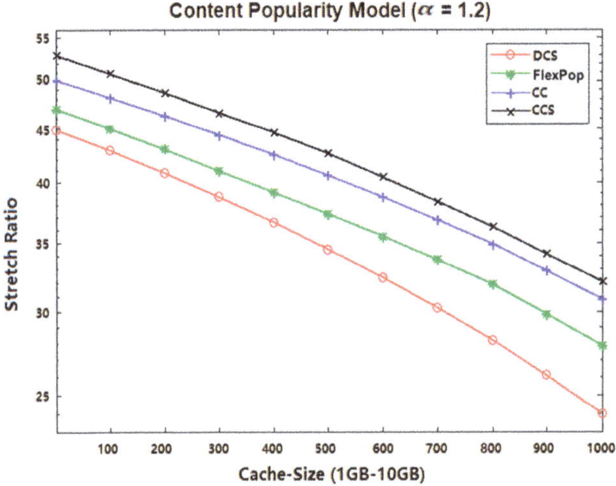

Figure 10. Stretch Ratio on GEANT Topology with VoD.

5.3. Cache Hit Ratio

Cache Hit Ratio refers to the quantity of the current content hits as interests are sent [37–39] by the consumer to the provider. It can be measured as using the following equation:

$$Cache\ Hit\ Ratio = \frac{\sum_{n=1}^{N} hit_i}{\sum_{n=1}^{N} (hit_i + miss_i)} \tag{3}$$

Figures 11 and 12 show the effects of the cache hit ratio on the Abilene and GEANT topologies using different content popularity models. Among the given figures, the DCS caching strategy performed better in terms of a cache hit ratio with both content topologies, because DCS tries to improve the cache allocation of popular contents. Moreover, DCS caches the most popular content at the edge router and closeness centrality routers. Therefore, subsequent interests are satisfied from edge routers, rather than from the remote router.

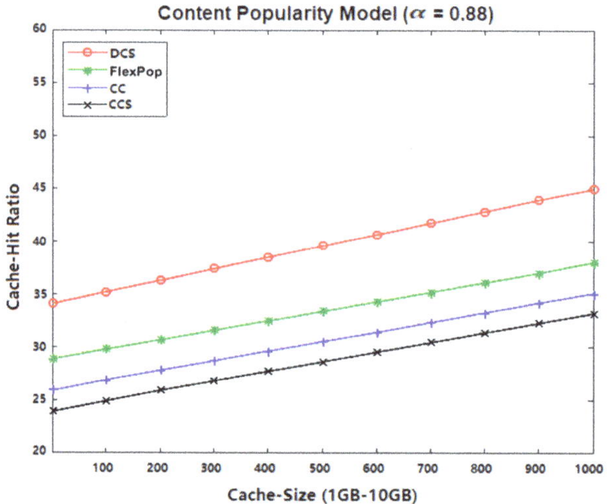

Figure 11. Cache Hit Ratio on Abilene Topology with UGC.

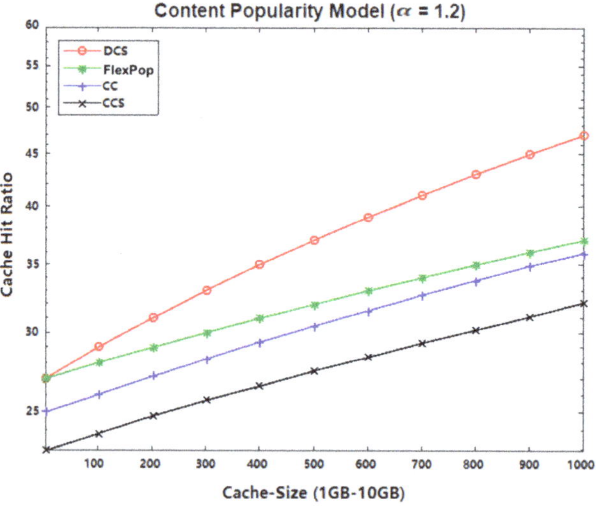

Figure 12. Cache Hit Ratio on GEANT Topology with VoD.

If an interest cannot be served by the edge router, it is satisfied with the closeness centrality router. Meanwhile, the CCS approach does not define any criteria by which to handle popular content when the cache of the centrality router is full. Therefore, all interests needed to be forwarded to the main data source (or remote router), which increases the path length and decreases the cache hit ratio. In comparison to the CCS approach, the CC and FlexPop approaches performed better. However, both strategies produce a low hit ratio, because fewer contents are accommodated between the centrality routers. On the other hand, DCS caches the content in chunks to increase the availability of storage space at the centrality router. Consequently, we conclude that the proposed DCS strategy performed much better by caching content close to consumers at the network edge.

5.4. Eviction Ratio

Content eviction is also one of the significant metrics by which to measure the performance of the caching-based ICN architecture. It can be defined as when the cache of a network node becomes saturated and there is a need to delete some content to accommodate the newly-arriving content. It can be calculated using the following equation:

$$Eviction\ Ratio = \frac{evicted\ content}{total\ content} \tag{4}$$

The last number of content evictions disturbs the network throughput and reduces the cache hit and stretch ratios. The reason for this is that all the incoming interests must be forwarded to the distant source to download the appropriate content due to an excessive number of evictions of popular content. Figures 13 and 14 illustrate the outcomes generated by comparisons of centrality-based caching strategies. In the given figure, we can see that the CCS shows a high content eviction ratio, because CCS generally caches all the contents without considering their popularity, and thus, all arriving interests must be forwarded to the remote provider.

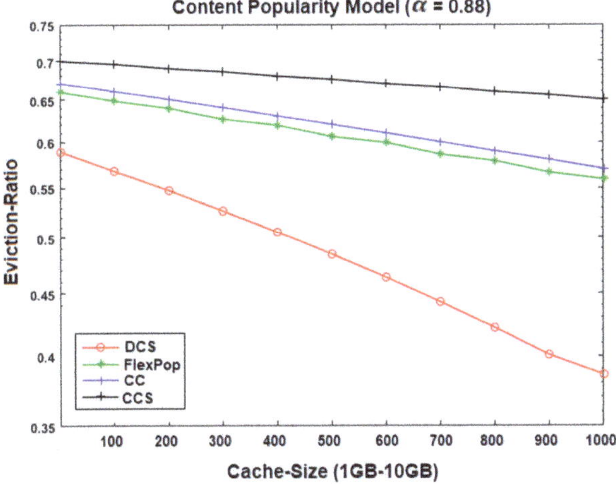

Figure 13. Content Eviction Ratio on Abilene Topology with UGC.

CC and FlexPop seem to show better performance in terms of the content eviction ratio, because both strategies are used to cache popular content at centrality routers. However, due to small and static thresholds, these caching strategies cache the least popular contents as well, causing a high number of content evictions. On the other hand, the proposed DCS caching strategy performed better in terms of reducing content eviction ratio as compared to CC, CCS, and FlexPop caching strategy. The reason is that the DCS distributes and caches the content in chunks format that increases the overall cache storage to accommodate the new contents. Besides, it uses to cache on the most popular content at centrality routers that increase the availability of free cache to provide popular content. Moreover, DCS caches the least popular content at the edge routers, and therefore, the subsequent interests are accomplished from the nearest routers. Thus, DCS minimizes the content eviction ratio by caching the least popular content at edge routers and the most popular content at centrality routers.

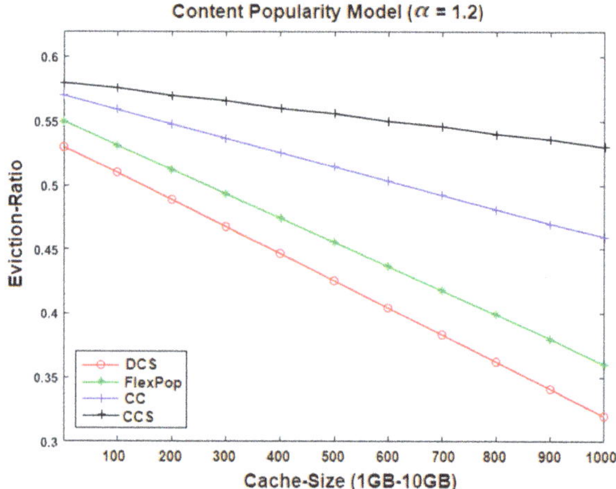

Figure 14. Content Eviction Ratio on GEANT Topology with VoD.

6. Conclusions

The new search and big data technology will deliver a massive amount of data that will be difficult to handle by using the current IP-based internet architecture. The reason is that the existing internet architecture supports the addresses based data communication which will be insufficient to fulfill the future requirements related to location-based data transmission. Similarly, the information dissemination in the current VSNs also depend on physical location in which similar data is transmitted several times across the network. This data replication has led to several problems in which resource consumption (memory), stretch, and communication latency due to the lake of data availability, are the most crucial issues. ICN provides an enhanced version of the internet that can provide the ability to resolve such issues efficiently. ICN is a new internet paradigm that supports innovative communication systems with location-independent data dissemination. ICN with VSN can handle the massive amount of data generated from heterogeneous mobile sensors in surrounding smart environments. Therefore, new ICN paradigms are emerging as a new technology to enhance communication processes for VSNs. Moreover, it can reduce the number of difficulties in the current internet paradigm; it provides edge routers in a VSN that can store the disseminated content for a specific time, while taking the required memory consumption, stretch ratio, and hit ratio into account. To improve the performance of content dissemination in an ICN-based cache of vehicles, a new caching strategy is proposed to provide less memory consumption, a low stretch ratio, a low content eviction ratio, and a high cache hit ratio by caching the most desired content close to consumers.

7. Future Directions

The requirements for enhancing the VNS infrastructure are rapidly expanding, because content generation and dissemination require more volume than the currently network capacities. Consumers are interested in data-needed contents, rather than data source locations. The reason for this is that the existing internet architecture supports location-based content routing, which increases the amount of network traffic; similar contents are transferred multiple times to satisfy consumers' needs. This redundant content routing process generates several problems, e.g., congestion, high bandwidth usage, and resource consumption (power and energy). Consequently, these critical problems have to be resolved by using an efficient, scalable, and reliable (secure) architecture for the internet [40,41]. The VNS is a new promising architecture that integrates several technologies and communication developments for the mobile internet. It provides several benefits, using identification and tracking technologies for wireless networks.

The most significant feature of the ICN is a cache that is used to store popular contents in order to serve user requests. In vehicular networks, vehicles can obtain their required contents from neighboring vehicles in short time with a small stretch [42]. Therefore, there is no need to forward incoming interests to remote providers. A large number of interests are generated for the same content from several vehicles, and vehicles are unable to retrieve the required content directly from the base station within partial coverage situations [43]. In this situation, the proposed caching strategy will significantly decrease the burden on the original provider, and will provide efficient data dissemination services [44]. Moreover, it offers distributed intelligence for smart objects (vehicles) [43]. VNS technology delivers benefits to mobile, interconnected nodes (vehicles), such as informatics, telecommunication, social science, and electronics. However, VSN still faces several complications, owing in no small part to the amount of data that is produced from heterogeneous devices (vehicles). Numerous diverse sensors are required in VSN, thereby increasing power and resource consumption [2]. Furthermore, VSN devices transmit a tremendous amount of content that is difficult to manage using the current IP-based internet architecture. In these situations, DCS introduces an enhanced scheme for data transmission across the internet, and it can overcome the current challenges of the IP-based internet and VSN [1].

The vast number of smart devices generates a significant amount of content that can be managed efficiently by the implementation of the DCS caching strategy. DCS provides content to network nodes, and all the nodes can store the disseminated contents during their transmission near the consumers at the intermediate nodes. Consequently, they can fulfill the requirements of subsequent interests in a shorter period compared to the retrieval of content from remote content providers. Moreover, the DCS caching strategy can reduce the power and resource consumption by caching content near users in chunk form. Thus, if a source node in the VSN is unreachable, consumers can still retrieve their desired content from any other caching node. The integration of DCS within the VSN can increase the reliability of the VSN architecture by deploying content near end users [45].

Author Contributions: Y.M. and M.A.N. formulated the problem statement and proposed the solution; M.A.N. and R.A. structured the comparative study and related work to evaluate the proposed mechanism; Y.B.Z. and S.W.K. supervised and guided throughout the project completion.

Acknowledgments: This research was supported in part by the Brain Korea 21 Plus Program (No. 22A20130012814) funded by the National Research Foundation of Korea (NRF), in part by the MSIT(Ministry of Science and ICT), Korea, under the ITRC(Information Technology Research Center) support program(IITP-2019-2016-0-00313) supervised by the IITP(Institute for Information & communications Technology Planning & Evaluation), and in part by Basic Science Research Program through the National Research Foundation of Korea (NRF) funded by the Ministry of Education (2018R1D1A1A09082266).

Conflicts of Interest: The authors declare no conflict of interest.

References

1. Zhao, W.; Qin, Y.; Gao, D.; Foh, G.H.; Chao, H.-C. An efficient cache strategy in information centric networking vehicle-to-vehicle scenario. *IEEE Access* **2017**, *5*, 12657–12667. [CrossRef]
2. Yan, Z.; Zeadally, S.; Park, Y.-J. A novel vehicular information network architecture based on named data networking (NDN). *IEEE Int. Things J.* **2014**, *1*, 525–532. [CrossRef]
3. Grewe, D.; Wagner, M.; Frey, H. PeRCeIVE: Proactive caching in ICN-based VANETs. In Proceedings of the 2016 IEEE Vehicular Networking Conference (VNC), Columbus, OH, USA, 8–10 December 2016; pp. 1–8.
4. Banerjee, A.; Chen, X.; Erman, J.; Gopalakrishnan, V.; Lee, S.; Merwe, J.V.D. MOCA: A lightweight mobile cloud offloading architecture. In Proceedings of the Eighth ACM International Workshop on Mobility in the Evolving Internet Architecture, Miami, FL, USA, 4 October 2013; pp. 92–101.
5. Ascigil, O.; Sourlas, V.; Psaras, I.; Pavlou, G. A native content discovery mechanism for the information-centric networks. In Proceedings of the 4th ACM Conference on Information-Centric Networking, Berlin, Germany, 26–28 September 2017; pp. 145–155.
6. Cisco, C. Cisco Visual Networking Index: Global Mobile Data Traffic Forecast Update, 2016–2021; Cisco (White Paper). Available online: https://www.cisco.com/c/en/us/solutions/collateral/service-provider/visual-networking-index-vni/white-paper-c11-738429.html (accessed on 18 February 2019).

7. Conti, M. Computer communications: Present status and future challenges. *Comput. Commun.* **2014**, *37*, 1–4. [CrossRef]

8. Zezulka, F.; Marcon, P.; Vesely, I.; Sajdl, O. Industry 4.0—An Introduction in the phenomenon. *IFAC-PapersOnLine* **2016**, *49*, 8–12. [CrossRef]

9. Lee, J.; Bagheri, B.; Kao, H.-A. A cyber-physical systems architecture for industry 4.0-based manufacturing systems. *Manuf. Lett.* **2015**, *3*, 18–23. [CrossRef]

10. Ahlgren, B.; Dannewitz, C.; Imbrenda, C.; Kutscher, D.; Ohlman, B. A survey of information-centric networking. *IEEE Commun. Mag.* **2012**, *50*, 26–36. [CrossRef]

11. Zhang, M.; Luo, H.; Zhang, H. A survey of caching mechanisms in information-centric networking. *IEEE Commun. Surv. Tutor.* **2015**, *17*, 1473–1499. [CrossRef]

12. Xu, Y.; Li, Y.; Ci, S.; Lin, T.; Chen, F. Distributed caching via rewarding: An incentive caching model for icn. In Proceedings of the GLOBECOM 2017–2017 IEEE Global Communications Conference, Singapore, 4–8 December 2017; pp. 1–6.

13. Naeem, M.A.; Nor, S.A. A survey of content placement strategies for content-centric networking. In Proceedings of the AIP Conference Proceedings, Kedah, Malaysia, 11–13 April 2016; pp. 1–6.

14. Alberti, A.M.; Casaroli, M.A.F.; Singh, D.; da Rosa Righi, R. Naming and name resolution in the future internet: Introducing the NovaGenesis approach. *Future Gener. Comput. Syst.* **2017**, *67*, 163–179. [CrossRef]

15. Araldo, A.; Rossi, D.; Martignon, F. Cost-aware caching: Caching more (costly items) for less (ISPs operational expenditures). *IEEE Trans. Parallel Distrib. Syst.* **2016**, *27*, 1316–1330. [CrossRef]

16. Bian, C.; Zhao, T.; Li, X.; Yan, W. Boosting named data networking for efficient packet forwarding in urban VANET scenarios. In Proceedings of the 21st IEEE International Workshop on Local and Metropolitan Area Networks, Beijing, China, 22–24 April 2015; pp. 1316–1330.

17. Grassi, G.; Pesavento, D.; Pau, G.; Vuyyuru, R.; Wakikawa, R.; Zhang, L. VANET via named data networking. In Proceedings of the 2014 IEEE conference on computer communications workshops (INFOCOM WKSHPS), Toronto, Canada, 27 April–2 May 2014; pp. 17–30.

18. Mauri, G.; Gerla, M.; Bruno, F.; Cesana, M.; Verticale, G. Optimal Content Prefetching in NDN Vehicle-to-Infrastructure Scenario. *IEEE Trans. Veh. Technol.* **2017**, *66*, 2513–2525. [CrossRef]

19. Gubbi, J.; Gerla, M.; Bruno, F.; Cesana, M.; Verticale, G. Internet of Things (IoT): A vision, architectural elements, and future directions. *Future Gener. Comput. Syst.* **2013**, *29*, 1645–1660. [CrossRef]

20. Zhang, Z.; Cho, M.C.Y.; Wang, C.; Hsu, C.; Chen, C.; Shieh, S. IoT security: Ongoing challenges and research opportunities. In Proceedings of the IEEE 7th International Conference on Service-Oriented Computing and Applications (SOCA), Matsue, Japan, 17–19 November 2014; pp. 230–234.

21. Zhang, L.; Afanasyev, A.; Burke, J.; Jacobson, V.; Claffy, K.; Crowley, P.; Papadopoulos, C.; Wang, L.; Zhang, B. Named Data Networking. *SIGCOMM Comput. Commun. Rev.* **2014**, *44*, 66–73. [CrossRef]

22. Meddeb, M.; Dhraief, A.; Belghith, A.; Monteil, T.; Drira, K. How to cache in ICN-based IoT environments? In Proceedings of the 2017 IEEE/ACS 14th International Conference on Computer Systems and Applications (AICCSA), Hammamet, Tunisia, 30 October–3 November 2017; pp. 1117–1124.

23. Din, I.U. Flexpop: A Popularity-Based Caching Strategy for Multimedia Applications in Information-Centric Networking. Ph.D. Thesis, Universiti Utara Malaysia, Kedah, Malaysia, 2016.

24. Yan, H.; Gao, D.; Su, W.; Foh, C.H.; Zhang, H.; Vasilakos, A.V. Caching strategy based on hierarchical cluster for named data networking. *IEEE Access* **2017**, *5*, 8433–8443. [CrossRef]

25. Lal, K.N.; Kumar, A. A centrality-measures based caching scheme for content-centric networking (CCN). *Multimed. Tools Appl.* **2018**, *77*, 17625–17642. [CrossRef]

26. Amadeo, M.; Campolo, C.; Molinaro, A. NDNe: Enhancing named data networking to support cloudification at the edge. *IEEE Commun. Lett.* **2016**, *20*, 2264–2267. [CrossRef]

27. Hassan, S.; Din, I.U.; Habbal, A.; Zakaria, N.H. A popularity based caching strategy for the future Internet. In Proceedings of the 2016 ITU Kaleidoscope: ICTs for a Sustainable World (ITU WT), Bangkok, Thailand, 14–16 November 2016; pp. 68–74.

28. Dräxler, M.; Karl, H. Efficiency of on-path and off-path caching strategies in information centric networks. In Proceedings of the 2012 IEEE International Conference on Green Computing and Communications, Besancon, France, 20–23 November 2012; pp. 581–587.

29. Zhang, G.; Li, Y.; Lin, T. Caching in information centric networking: A survey. *Comput. Netw.* **2013**, *57*, 3128–3141.

30. Bernardini, C. Stratégies de Cache basées sur la popularité pour Content Centric Networking. Ph.D. Thesis, Lorraine University, University of Lorraine, Nancy, France, 2015.

31. Bernardini, C.; Silverston, T.; Festor, O. SONETOR: A social network traffic generator. In Proceedings of the 2014 IEEE International Conference on Communications (ICC), Sydney, Australia, 10–14 June 2014; pp. 3734–3739.

32. Bernardini, C.; Silverston, T.; Festor, O. MPC: Popularity-based caching strategy for content centric networks. In Proceedings of the 2013 IEEE International Conference on Communications (ICC), Budapest, Hungary, 9–13 June 2013; pp. 3619–3623.

33. Bernardini, C.; Silverston, T.; Festor, O. A comparison of caching strategies for content centric networking. In Proceedings of the 2015 IEEE Global Communications Conference (GLOBECOM), San Diego, CA, USA, 6–10 December 2015; pp. 1–6.

34. Sheather, S.J.; Jones, M.C. A reliable data-based bandwidth selection method for kernel density estimation. *J. R. Stat. Soc. Ser. B (Methodol.)* **1991**, *53*, 683–690. [CrossRef]

35. Badenhop, C.W.; Graham, S.; Ramsey, B.W.; Mullins, B.; Mailloux, L.O. The Z-Wave routing protocol and its security implications. *Comput. Secur.* **2017**, *68*, 112–129. [CrossRef]

36. Dias, J.A.F.F. Performance of Management Solutions and Cooperation Approaches for Vehicular Delay-Tolerant Networks. Available online: http://hdl.handle.net/10400.6/4501 (accessed on 2 September 2019).

37. Beckmann, N.; Chen, H.; Cidon, A. LHD: Improving Cache Hit Rate by Maximizing Hit Density. In Proceedings of the 15th {USENIX} Symposium on Networked Systems Design and Implementation ({NSDI} 18), Renton, WA, USA, 9–11 April 2018; pp. 389–403.

38. Chen, P.; Yue, J.; Liao, X.; Jin, H. Trade-off between Hit Rate and Hit Latency for Optimizing DRAM Cache. *IEEE Trans. Emerg. Top. Comput.* **2018**, *41*, 1. [CrossRef]

39. Tseng, F.-H.; Chien, W.-C.; Wang, S.-J.; Lai, C.F.; Chao, H.C. A novel cache scheme based on content popularity and user locality for future internet. In Proceedings of the 27th Wireless and Optical Communication Conference (WOCC), Hualien, Taiwan, 30 April–1 May 2018; pp. 1–5.

40. Li, S.; Zhang, Y.; Raychaudhuri, D.; Ravindran, R. A comparative study of MobilityFirst and NDN based ICN-IoT architectures. In Proceedings of the 10th International Conference on Heterogeneous Networking for Quality, Reliability, Security and Robustness, Rhodes, Greece, 18–20 August 2014; pp. 189–193.

41. Bai, X.; Liu, S.; Zhang, P.; Kantola, P. ICN: Interest-based clustering network. In Proceedings of the Fourth International Conference on Peer-to-Peer Computing, Zurich, Switzerland, 27 August 2004; pp. 489–497.

42. Modesto, F.; Boukerche, A. A novel service-oriented architecture for information-centric vehicular networks. In Proceedings of the 19th ACM International Conference on Modeling, Analysis and Simulation of Wireless and Mobile Systems, Malta, Malta, 13–17 November 2016; pp. 239–246.

43. Khan, J.A.; Ghamri-Doudane, Y. Saving: Socially aware vehicular information-centric networking. *IEEE Commun. Mag.* **2016**, *54*, 100–107. [CrossRef]

44. Rainer, B.; Petscharnig, S. Challenges and Opportunities of Named Data Networking in Vehicle-To-Everything Communication: A Review. *Information* **2018**, *9*, 264. [CrossRef]

45. Modesto, F.M.; Boukerche, A. Seven: A novel service-based architecture for information-centric vehicular network. *Comput. Commun.* **2018**, *117*, 133–146. [CrossRef]

Article

Reinforcement Learning for Energy Optimization with 5G Communications in Vehicular Social Networks

Hyebin Park and Yujin Lim *

Department of IT engineering, Sookmyung Women's University, Seoul 04310, Korea; hb0390@sookmyung.ac.kr
* Correspondence: yujin91@sookmyung.ac.kr; Tel.: +82-2-2077-7305

Received: 20 February 2020; Accepted: 18 April 2020; Published: 21 April 2020

Abstract: Increased data traffic resulting from the increase in the deployment of connected vehicles has become relevant in vehicular social networks (VSNs). To provide efficient communication between connected vehicles, researchers have studied device-to-device (D2D) communication. D2D communication not only reduces the energy consumption and loads of the system but also increases the system capacity by reusing cellular resources. However, D2D communication is highly affected by interference and therefore requires interference-management techniques, such as mode selection and power control. To make an optimal mode selection and power control, it is necessary to apply reinforcement learning that considers a variety of factors. In this paper, we propose a reinforcement-learning technique for energy optimization with fifth-generation communication in VSNs. To achieve energy optimization, we use centralized Q-learning in the system and distributed Q-learning in the vehicles. The proposed algorithm learns to maximize the energy efficiency of the system by adjusting the minimum signal-to-interference plus noise ratio to guarantee the outage probability. Simulations were performed to compare the performance of the proposed algorithm with that of the existing mode-selection and power-control algorithms. The proposed algorithm performed the best in terms of system energy efficiency and achievable data rate.

Keywords: 5G; D2D communication; vehicle-to-vehicle communication; mode selection; power control; vehicular social network

---

1. Introduction

With the dynamic increase in data traffic in connected vehicles and wireless networks, satisfying cellular data traffic has become relevant. To satisfy these requirements, vehicular social networks (VSNs) have been studied [1,2]. VSNs consist of moving and parked vehicles on roads, and the vehicles communicate with nearby vehicles or infrastructure to report and exchange traffic information. Therefore, one of the challenges is to ensure communication quality and reduce delays in VSNs. To guarantee a high data rate in a VSN, base stations are densely deployed and overlapped. However, densely deployed base stations cause higher system energy consumption. To tackle the energy consumption problem, the traditional base station structure is changed to a centralized structure, such as a heterogeneous cloud radio access network (H-CRAN). In the H-CRAN structure, a traditional base station is divided into a signal-processing part as the baseband unit (BBU) and a signal transceiver part as the radio remote head (RRH). The BBUs are centralized as a BBU pool, and the RRHs are connected to the BBU pool through fronthaul links. The macro base station only serves voice services to users, and the RRHs are densely overlapped in macro cells.

In traditional communication in vehicular networks, vehicles communicate with other vehicles and infrastructure through RRHs. Further, vehicles have high mobility; therefore, they make frequent handovers. This causes heavy loads and energy consumption problems in the system. To solve the energy consumption problem and provide efficient communication between vehicles, device-to-device (D2D) communication has been developed. With D2D communication, nearby wireless devices can communicate directly without an RRH, which can reduce the system energy consumption from the RRH transceivers and fronthaul links. In vehicular networks, D2D communication can facilitate efficient data communication between nearby vehicles. Thus, D2D communication can decrease latency because of the relatively small distance between the transmitter and receiver vehicle. As D2D communication reuses cellular wireless resources, it helps increase the spectral capacity.

However, D2D communication is highly susceptible to interference because of the reuse of cellular resources. Specifically, for D2D communication in H-CRANs, various interference problems are caused by the dense deployment of RRHs. In addition, when traffic congestion occurs in vehicular environments, the high density of devices in D2D communication can cause a critical interference problem [3]. To avoid the interference problem, channel extensions are conducted. However, interference problems remain because of interference from adjacent channels and shared resources [4–6]. To solve these problems, various studies on D2D communication have been conducted. Mode selection is a technique that can solve the interference problem [7–9]. Mode selection allows the devices to select from several communication modes: cellular mode, D2D mode, and detection mode. With mode selection, devices choose modes according to the channel state, which can ensure sufficient quality of service (QoS) for the devices and increase the system capacity. Another solution to solve the interference problem is to use the power-control method [10,11]. Power control is a method that manages interference by controlling the power of the D2D links. Using the power-control method increases the energy efficiency of a device by decreasing the transmission power in low-interference situations and increasing the transmission power in high-interference situations to guarantee the QoS.

In this paper, we propose a joint mode-selection and power-control algorithm using reinforcement learning in a VSN. In our algorithm, the BBU pool uses centralized Q-learning, and the vehicles use distributed Q-learning to achieve improved signal-to-interference noise ratios (SINRs). Centralized Q-learning aims to maximize the system's energy efficiency, while distributed Q-learning aims to maximize the vehicle's achievable data rate. As the energy efficiency of the system and the SINR of a vehicle are in a trade-off relationship, it is important to identify a point that optimizes both objectives. Our algorithm ensures vehicular capacity by considering outage probability as a QoS constraint while maximizing system energy efficiency.

The rest of this paper is organized as follows. In Section 2, we introduce related work. In Section 3, we formulate the system model and the problem definition. The joint mode selection and power-control algorithm using multi-Q-learning is introduced in Section 4. A performance evaluation and discussion of our algorithm are presented in Section 5. We present the conclusions of this study in Section 6.

2. Materials and Methods

To reduce the load and increase the system's energy efficiency, several optimization methods have been developed. In [12], a mode-selection algorithm was developed to maximize the energy efficiency of the device using the transmission rate as a QoS requirement. The mode was determined adaptively, based on various factors of the device, and it could minimize the energy consumption for each successful content delivery. In [13], a mode-selection algorithm that considered the link quality of D2D links was proposed to maximize the system throughput. It estimated the expected system throughput considering the SINR and available resources, so it could maximize the system throughput.

As D2D communication reuses cellular resources, interference management between cellular and D2D links is important. The power-control method that controls the transmission power of D2D links is one method used to manage interference. In [14], a power-control algorithm with variable target SINRs was proposed for application in multicell scenarios. It aimed to maximize the system spectral efficiency using a soft-dropping algorithm to control the transmission power to meet the variable target SINR. A power-control algorithm using stochastic geometry was proposed in [15]. The algorithm can be divided into two types: centralized and decentralized. The centralized type aims to guarantee the coverage probability by solving the optimization power problem. The decentralized type was an interference mitigation method to maximize the sum rates of the devices.

To take advantage of both mode selection and power control, jointly designed algorithms were proposed. In [16], an energy-aware joint power-control and mode-selection algorithm was proposed to minimize the power consumption. It solved the power minimization problem by guaranteeing the QoS constraints and developed a joint strategy under the condition of imperfect channel state information. In [17], a mode-selection and power-control algorithm was proposed to maximize the sum of the achievable data rate. It selected the mode to satisfy the distance and interference constraints from an operator perspective. After mode selection, it first proved that the power-control problem was quasiconvex for the D2D mode and then solved it.

Environments where cellular and D2D modes coexist involve high complexity that reflects the numerous features of network dynamics in the optimization methods. To consider network dynamics, reinforcement learning was used to adapt the optimization of the mode selection and power-control problems. In [18], a mode-selection algorithm based on Q-learning was proposed to maximize QoS and minimize interference. It considered the delay, energy efficiency, and interference to determine the transmission mode. In [19], a mode-selection method based on deep reinforcement learning was proposed to minimize the system power consumption in a fog radio access network. To make optimal decisions, it formulated the energy minimization problem with a Markov decision process by considering the on-off state of processors, communication mode of the device, and precoding vectors of RRH.

Reinforcement learning was implemented in [20] to adapt the power-control algorithm for D2D communication. It consisted of centralized and distributed Q-learning and aimed to maximize the system capacity and guarantee a stable QoS level. In centralized Q-learning, called team Q-learning, the agent in each resource block (RB) managed only one Q-table. In distributed Q-learning, agents in each D2D link learned independently and managed each Q-table. Team Q-learning could reduce the complexity and avoid the overhead from managing the Q-tables in distributed Q-learning. In [21], a joint mode-selection and power-control algorithm with multiagent learning was proposed. The agents in each device managed each Q-table and learned independently. It considered the local SINR information and device modes, such as the cellular mode, D2D mode, and detection mode. It helped D2D transmitters decide on efficient mode selection and power control to maximize the energy efficiency of the D2D links.

In a VSN, vehicles move constantly with high mobility. This high mobility results in frequent handovers in cellular communication and changes the channel states accordingly. However, none of the above studies considered vehicle mobility and changing channel states. In addition, there is a large amount of communication between adjacent vehicles in a VSN. D2D communication has distance limitations that have a significant impact on performance; thus, vehicle mobility should be considered. Therefore, any appropriate mode-selection and power-control method must consider the network dynamics that come with vehicle mobility. This mobility causes signaling overhead and data latency because of frequent changes in the channel information state.

Typical optimization methods incur high complexity if they consider the various relevant features. To make optimal decisions in a dynamic network environment, reinforcement learning can make recommendations according to the various states. In D2D communication applications, deep Q-learning has the advantage of being able to learn directly using network data and process high-dimensional data [15]. However, deep Q-learning is more complex than Q-learning, and the available data for vehicles are limited. So, it brings high complexity and overhead problems.

In this paper, we propose a mode-selection and power-control algorithm using reinforcement learning for the H-CRAN architecture in a VSN. Our algorithm consists of two parts: centralized Q-learning and distributed Q-learning. In the centralized part, the agent in a BBU pool manages one Q-table to maximize the system's energy efficiency and guarantee the QoS constraints. It recommends an appropriate communication mode and transmission power for the vehicles, based on the average

SINR and available resources. To satisfy the outage probability as a QoS constraint, the target SINR that determines the states is adjusted. In the distributed part, the agents in each vehicle manage each table and learn to maximize the received SINR. The agents choose their actions by comparing the actions recommended by the BBU with their own actions.

3. System Model and Problem Definition

In this work, we consider the single-cell scenario in H-CRANs where cellular-mode vehicles and D2D-mode vehicles coexist. The vehicles are expressed as devices in this study. H-CRANs comprise a BBU pool and multiple RRHs, as shown in Figure 1. The total device set can be expressed as U, and it consists of a cellular device set C and D2D device set D. The cellular devices, $C = \{1, \ldots, c\}$, and the number of D2D devices are distributed randomly within the set $D = \{1, \ldots, d\}$. We denote the set of RRHs as $S = \{1, \ldots, s\}$ and RBs as $K = \{1, \ldots, k\}$. We assume that one RB can be allocated to one cellular device and shared with multiple D2D devices. Each device can select the transmission mode between the cellular mode, m_c, and D2D mode, m_d; to select the D2D mode, the distance between the D2D pairs must satisfy the D2D distance threshold.

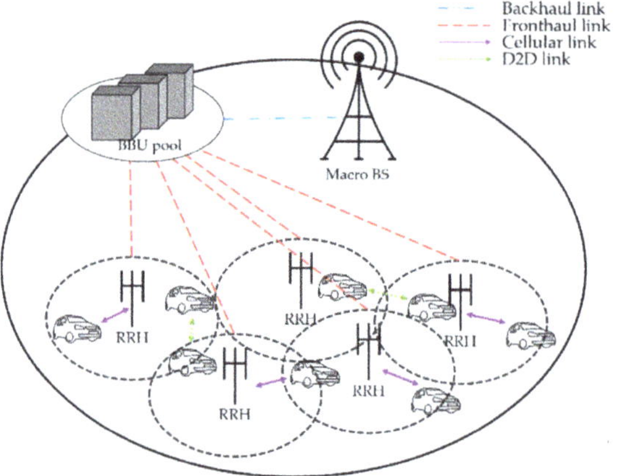

Figure 1. System model.

The SINR of cellular device c at the k-th RB, as reported in [20], is

$$\gamma_k^c = \frac{p_{tr}^0 \cdot g_k^{sc}}{N + \sum_{d \in D} p_k^d \cdot g_k^{cd} + \sum_{s' \in RRH,\ s' \neq s} p_{tr}^0 \cdot g_k^{s'c}}, \tag{1}$$

where p_{tr}^0 is the transmission power of the RRH; g_k^{sc} is the channel gain between the associated RRH s and cellular device c, and N is the noise power spectral density. The channel gain between cellular device c and the associated RRH s can be calculated as $g_k^{sc} = G \varrho^{s,c} \xi^{s,c} L^{s,c-}$, where G is the constant gain from the antenna and amplifier, $\varrho^{s,c}$ is the multipath fading gain with a log-normal distribution, $L^{s,c}$ is the distance between the cellular device c and the associated RRH s, and is the path-loss exponent. The SINR of D2D device d at the k-th RB, as reported in [16], is

$$\gamma_k^d = \frac{p_k^d \cdot g_k^{dd'}}{N + p_k^c \cdot g_k^{cd} + \sum_{j \in D,\ j \neq d} p_k^j \cdot g_k^{jd}}, \tag{2}$$

where p_k^d is the transmission power of D2D device d, $g_k^{dd'}$ is the channel gain between another D2D pair d', p_k^c is the transmission power of cellular device c which shares the k-th RB, and g_k^{cd} is the channel gain between cellular device c and D2D device d. The total system capacity, according to the Shannon capacity, can be expressed as

$$B = W^{total} \sum_{c \in C} \sum_{k \in K} \{ \log_2 (1 + \gamma_k^c) + \sum_{d \in D} \log_2 (1 + \gamma_k^d) \}, \tag{3}$$

where W^{total} is the total bandwidth of the system. The power consumption of the system includes the power consumed by the RRHs and fronthaul devices. The power consumption model of the RRH can be expressed as

$$P_{rrh} = \sum_{s=1}^{S} (\phi_{rrh} + \Delta slope \sum_{c=1}^{C} a_c^s p_{tr}^0), \tag{4}$$

where ϕ_{rrh} is the circuit power of the RRH, $\Delta slope$ is the slope of the load-dependent power consumption of RRH, as reported in [22], and a_c^s is the association indicator of cellular device c, with values of 1 for an association and 0 for a nonassociation. The power consumption model for the fronthaul links was reported in [23] and can be expressed as

$$P_{fronthaul} = \sum_{s=1}^{S} (\phi_{fronthaul} + \ell \cdot t_s), \tag{5}$$

where $\phi_{fronthaul}$ is the circuit power from the fronthaul transceiver and switch, ℓ is the power consumption per bit/s, and t_s is the traffic associated with RRH s. The macro base station provides only voice services; thus, its power consumption is not considered. Therefore, the system power consumption model can be defined as

$$P = P_{rrh} + P_{fronthaul}. \tag{6}$$

The system energy efficiency can be defined as

$$EE = \frac{B}{P}. \tag{7}$$

Our main goal is to maximize the system's energy efficiency, and it can be formulated as

$$\max_{C,D,K} EE, \tag{8}$$

$$s.t. C1 : \gamma \geq \gamma_0 \forall c, d, k,$$
$$C2 : 0 < p_k^d \leq p_{max} \forall d, k,$$
$$C3 : \tau \leq \tau_{max}.$$

where γ is the SINR of users, γ_0 is the SINR constraint, p_{max} is the maximum transmission power, τ is the outage probability, and τ_{max} is the maximum outage probability constraint. The outage probability is the probability that the SINR of the devices is lower than the SINR constraint [24].

4. Proposed Algorithm

In this section, we introduce a mode-selection and power-control algorithm based on Q-learning. When each device has Q-learning agent, the agent cannot get the information to improve system energy efficiency at system level. Even if the system sends the information to the agent, it increases that state space that agent manages and the communication load for data exchange between system and device. So, we proposed two types of Q-learning agent: centralized agent at the system level and distributed agent at each device.

In our previous research, we proposed RRH switching and power-control methods based on the Q-learning mechanism [25]. We considered the available resources and interference levels to maximize the energy efficiency and minimize interference in the cell. However, this cannot account for the QoS of the devices, and cell coverage problems occur because of switching off the RRHs. We need to ensure the QoS of the devices while maximizing the system's energy efficiency, which is in a trade-off relationship. Therefore, in this paper, we determine communication mode and transmission power by using Q-learning mechanism to solve the problems. The centralized agent learns to recommend optimal actions that can maximize system energy efficiency by considering the available resources and interference in the cell. The distributed agent recommends an optimal action that can maximize the SINR of the device by considering the interference of the device. Then, the agent finally selects the optimal action by comparing the expected SINRs of the recommended actions.

Existing algorithms have focused on maximizing the energy efficiency of the devices, which cannot guarantee QoS. Our proposed algorithm learns to maximize the system's energy efficiency and the received SINRs of devices. It also adjusts the target SINR, which is the basis for the proposed Q-learning, according to the interference state. It is important to set the target SINR appropriately because as the target SINR increases, the agent increases the transmission power to increase the SINR. This increases the cellular mode and reduces the system's energy efficiency. Conversely, when the target SINR is reduced, the agent reduces the transmission power to lower the SINR. This increases the system's energy efficiency, but it is likely that the devices cannot guarantee the QoS. The agents of each device are in the transceiver of each device, and we assume that agents get the information of the receiver through delay-free feedback [26].

The operational procedure is described as follows. In Step 1, the device prepares to select the transmission mode. At this step, the transmitting device requests a connection through the BBU pool to communicate with the receiving device, and the BBU pool checks whether the distance between the devices is under the D2D distance threshold. If the device does not satisfy the D2D distance threshold, the device will only be able to select the cellular mode and associate with the RRH that provides the highest SINR. If the device satisfies the D2D distance threshold, it can select the D2D mode and obtain the selectable mode and transmission power from its agent. After that, the BBU pool recommends the selectable mode and transmission power to the device.

In Step 2, the agent calculates the expected SINRs of the recommended actions from its own agent and from the BBU pool. The agent chooses communication mode and transmission power as its action. The agent then informs the BBU pool of the determined action. The BBU pool allocates resources to the device, and the device starts communication based on the determined actions. This procedure is summarized in Figure 2.

To recommend the communication mode and transmission power, Q-learning is used in the BBU pool and in each device. Q-learning is a model-free reinforcement-learning algorithm that learns to find an optimal policy that can maximize the expected reward. Q-learning involves a set of states S, a set of actions A, and a set of rewards R. The agent transits from one state to another state by performing an action a. The agent chooses an action according to the optimal policy π^* in its current state s. The agents manage the Q-table, and the Q-value updating rule is defined as follows:

$$Q^{t+1}\left(s^t, a^t\right) = Q^t\left(s^t, a^t\right) + \alpha\left[r^{t+1} + \beta \max_a Q^t\left(s^{t+1}, a^{t+1}\right) - Q^t\left(s^t, a^t\right)\right], \tag{9}$$

where $Q^{t+1}\left(s^t, a^t\right)$ is the Q-value with state s^t and action a^t at time t, α is the learning rate, r^{t+1} is the expected reward at time $t+1$, and β is the discount factor. The optimal policy π^* with state s can be expressed as

$$\pi^*(s) = \max_a Q(s, a). \tag{10}$$

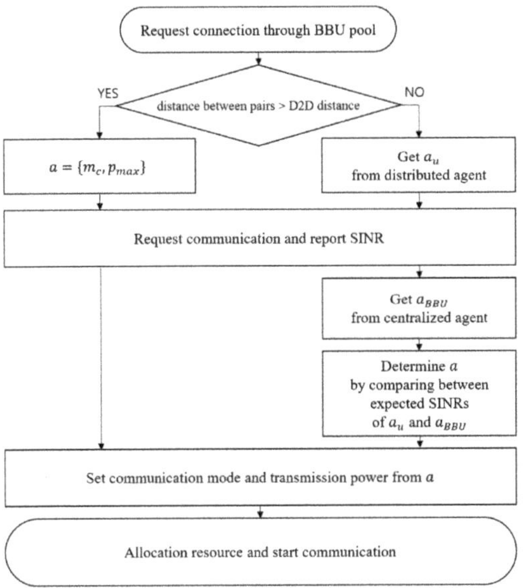

Figure 2. Operation procedure.

4.1. Centralized Q-Learning in the BBU Pool

To recommend the communication mode and transmission power, based on Equation (8), we use a Q-learning agent in the BBU pool. For Q-learning in the BBU pool, the state, action, and reward can be defined as follows. The state of the BBU pool at time t is defined as

$$S^t_{BBU} = \left\{I^t,\ \rho^t\right\}, \tag{11}$$

where I^t is the binary level of the average received SINR of the devices at time t, with a value of 1 when the average SINR is greater than the target SINR, γ_{min}, and 0 when the average SINR is smaller than γ_{min}. ρ^t is the available RB at time t. The action of the BBU pool at time t is defined as

$$A^t_{BBU} = \left\{m^t, p^t\right\}, \tag{12}$$

where m^t is the transmission mode at time t, and p^t is the transmission power at time t. The transmission power p^t is divided into discrete intervals, $(0,\ p_{max}]$. If the transmission mode m^t is the cellular mode m_c, the transmission power p^t will be set to p_{max}. The reward of the BBU pool at time t is defined as

$$R^t_{BBU} = EE. \tag{13}$$

The centralized Q-learning is summarized in Algorithm 1.

Algorithm 1: Pseudocode for centralized Q-learning

Initialization:
   for each $s^t_{BBU} \in S^t_{BBU}$, $a^t_{BBU} \in A^t_{BBU}$ do
      initialize Q-table and policy $\pi^*_{BBU}(s^t_{BBU})$
   end for
Learning:
   loop
      estimate the state s^t_{BBU}
      generate a random real number $x \in [0, 1]$
        if $x < \varepsilon$ // for exploration
          select the action a^t_{BBU} randomly
        else
          select the action a^t_{BBU} according to $\pi^*_{BBU}(s^t_{BBU})$
        recommend action a^t_{BBU} to the devices in the cell
        calculate reward r^t_{BBU}
        update Q-value $Q(s^t_{BBU},\, a^t_{BBU})$ and $\pi^*_{BBU}(s^t_{BBU})$
  end loop

4.2. Distributed Q-Learning in the Devices

According to Equation (8), the BBU pool recommends more D2D modes for the devices. While the D2D mode has the advantage of increasing system capacity, it may not ensure an achievable data rate because of interference. To solve this problem, we use Q-learning agents in each device to maximize the received SINR.

The state of device u at time t is defined as

$$S^t_u = \{I^t_u\}, \tag{14}$$

where I^t_u is the binary level of the received SINR of device u at time t based on current communication mode. If the mode of device is cellular mode, state of device u will be calculated based on received SINR of the cellular link. The action of device u at time t is defined as

$$A^t_u = \{m^t_u,\, p^t_u\}, \tag{15}$$

where m^t_u is the transmission mode of device u at time t, and p^t_u is the transmission power of device u at time t. The reward of device u at time t is defined as

$$R^t_u = \begin{cases} \gamma_u & ,\, I^t_u = 1 \\ -\gamma_u * (p_{max} - p^t_u), & I^t_u = 0,\ \gamma_u \geq 0 \\ \gamma_u * (p_{max} - p^t_u), & I^t_u = 0,\ \gamma_u < 0 \end{cases}, \tag{16}$$

where γ_u is the SINR of device u. In distributed Q-learning, the agent chooses between the action from centralized Q-learning and its own action to find the action that provides a higher expected SINR. Distributed Q-learning is summarized in Algorithm 2.

Algorithm 2: Pseudocode for distributed Q-learning

Initialization:
   for each $s_u^t \in S_u^t$, $a_u^t \in A_u^t$ do
      initialize Q-table and policy $\pi_u^*(s_u^t)$
   end for
Learning:
   loop
      estimate state s_u^t
      generate a random real number $x \in [0,1]$
      if $x < \varepsilon$ // for exploration
         elect action a_u^t randomly
      else
         select action a_u^t according to $\pi_u^*(s_u^t)$
      receive action a_{BBU}^t from algorithm1
      determine action a_u^* by comparing a_u^t and a_{BBU}^t
      execute action a_u^t
      calculate reward r_u^t
      update Q-value $Q(s_u^t, a_u^t)$ and $\pi_u^*(s_u^t)$
   end loop

4.3. Target SINR Updating Algorithm

In our algorithm, the state of Q-learning is determined by the target SINR. To set the target SINR to reflect the interference state, the target SINR at time interval T, denoted as γ_{min}^T, is adjusted as follows:

$$\gamma_{min}^{T+1} = \begin{cases} C * \gamma^+ + (1 - C) * \gamma_{min}^T, & \tau^T \geq \tau_{max} \\ C * \left(\frac{\gamma_{median} + \gamma_0}{2}\right) + (1 - C) * \gamma_{min}^T, & \tau^T \leq \tau_{min} \end{cases}, \tag{17}$$

where C is the weight factor, γ^+ is the largest SINR value of the devices, τ^T is the average outage probability for the time interval T, and τ_{max} is the maximum outage probability. γ_{median} is the median SINR value of the devices; γ_0 is the SINR constraint, and τ_{min} is the minimum outage probability.

The changed target SINR affects the learning, and therefore, the agent can make an optimal decision using the changed target SINR. However, as the optimal actions in the changed target SINR may not be optimal in the original target SINR, those actions may not be selected. To solve this problem, we must adjust the Q-table. At this time, only the Q-table of the BBU is adjusted as follows:

$$Q^{T+1}(s,a) = log_2\big(Q^T(s,a) - 1\big) \, \forall\{s,a\}. \tag{18}$$

5. Results and Discussion

A single-cell H-CRAN environment was considered in this work, and the parameters used in the simulation are summarized in Table 1. We set the parameters and speed requirements for mobility dataset according to the 3GPP (3rd Generation Partnership Project) specifications release 16 [22,27]. Four RRHs and vehicles were distributed randomly in a single macro cell with an intercell distance of 500 m. The mobility dataset used in the simulation was a dataset in an urban area created using a simulation of urban mobility (SUMO) simulator [28]. In the dataset, all vehicles had mobility with a random trip model at the Seoul City Hall in South Korea according to the 3GPP specification release 15 [29]. The datasets consisted of two types of scenarios: light traffic and heavy traffic. The two types were vehicular mobility datasets for urban areas with light traffic and heavy traffic scenarios. In a light traffic scenario, vehicles can move fast and the distance between vehicles increases. This means that fewer vehicles can select the D2D mode because of the D2D distance threshold. In a heavy traffic scenario, vehicles move slowly because of the traffic jam, and the distance between vehicles becomes shorter. This allows the vehicles to select a D2D mode more often. Compared to heavy traffic scenarios, fewer vehicles move and require resources at the same system load situation in light traffic scenarios. The D2D distance threshold affects the system energy efficiency, which can also be affected by the mode selection. To consider and present the effects of the D2D distance threshold, we simulated experiments with various D2D distances, as reported in [27]. We considered Rayleigh fading, log-normal shadowing, and the path-loss model $140.7 + 36.7 \log(distance(km))$, based on [30]. The D2D distance threshold was varied between 250 and 350 m, according to [31,32]. The Q-learning parameters were $\alpha = 0.01$, $\beta = 0.9$, and $\varepsilon = 0.01$. We set the time for the D2D link establishment as 6 ms, as reported in [33], and the size of time unit is 3 s.

Table 1. Parameters used in the simulation.

Parameter	Notation	Value
Noise power spectral density	N	-174 dBm/Hz
Total bandwidth	W^{total}	100 MHz
SINR constraint	γ_0	0.5 dBm
Maximum outage probability constraint	τ_{max}	0.05
Minimum outage probability constraint	τ_{min}	0.01
Circuit power of RRH	ϕ_{rrh}	4.3 W
Slope of RRH	$\Delta slope$	4.0
Circuit power of fronthaul transceiver and switch	$\phi_{fronthaul}$	13 W
Power consumption per bit/s	ℓ	0.83 W
Transmission power of cellular device	p_k^c	23 dBm
Transmission power of RRH	p_{tr}^0	24 dBm

We performed simulations while changing the load of the system and the D2D distance with different traffic scenarios and compared the proposed algorithm with other existing algorithms. First, we denote our proposed algorithm as "proposed algorithm". Second, the condition where no algorithm was applied was used to create a baseline, and we denoted the baseline as "BA". It selects the communication mode by comparing the strength of the expected SINR between cellular and D2D modes. Third, for a comparison with a mode-selection and power-control algorithm with Q-learning, the algorithm in [21] was used. It learned to maximize the device's energy efficiency with a fixed target SINR. The objective that maximizes device energy efficiency can also maximize system energy efficiency by selecting more D2D modes. We denote that as "Compare1". Fourth, we used the algorithm in [20] to compare power control with Q-learning. It also learned to maximize device energy efficiency using a fixed target SINR. We denote that as "Compare2".

The energy efficiencies for different D2D distance scenarios are compared as functions of the system loads with different mobility scenarios, as shown in Figure 3. Figure 3a shows that the system energy efficiency of the proposed algorithm is 83% and 89% higher than those of BA and Compare2, respectively. Compared to Compare1, it presents high energy efficiency, with a difference of approximately 26%. In Figure 3b it can be seen that the proposed algorithm outperformed BA, Compare1, and Compare2 by more than 121%, 26%, and 126%, respectively. Figure 3c shows that the system energy efficiency of the proposed algorithm is 102% and 108% higher than that of BA and Compare2, respectively. Compared to Compare1, it presents higher energy efficiency, with a difference of approximately 21%. Figure 3d shows that the proposed algorithm performed approximately 128%, 27%, and 133% better than BA, Compare1, and Compare2, respectively. As the number of devices capable of D2D communication increases as the D2D distance increases, the difference in system energy efficiency increases. This is because the devices communicate directly rather than through the RRH, which reduces the system's energy consumption. The proposed algorithm is designed to maximize system energy efficiency; thus, it gives the best performance in scenarios with large D2D distances. Furthermore, because D2D-mode vehicles communicate with the same transmission power at different distance thresholds, the achievable data rate and system energy efficiency decrease at longer D2D distance thresholds. This reveals that the overall performance decreases as the D2D distance threshold increases, but the proposed algorithm achieves a better performance compared to other algorithms. In addition, system energy efficiencies in the light traffic scenario get higher-scale results than the heavy traffic scenario. This is because vehicles require more resources in the light traffic scenario compared to the same number of vehicles in a heavy traffic scenario. Required resources affect system capacity, so system energy efficiencies in the light traffic scenario are higher than the heavy traffic scenario in the same system load situation.

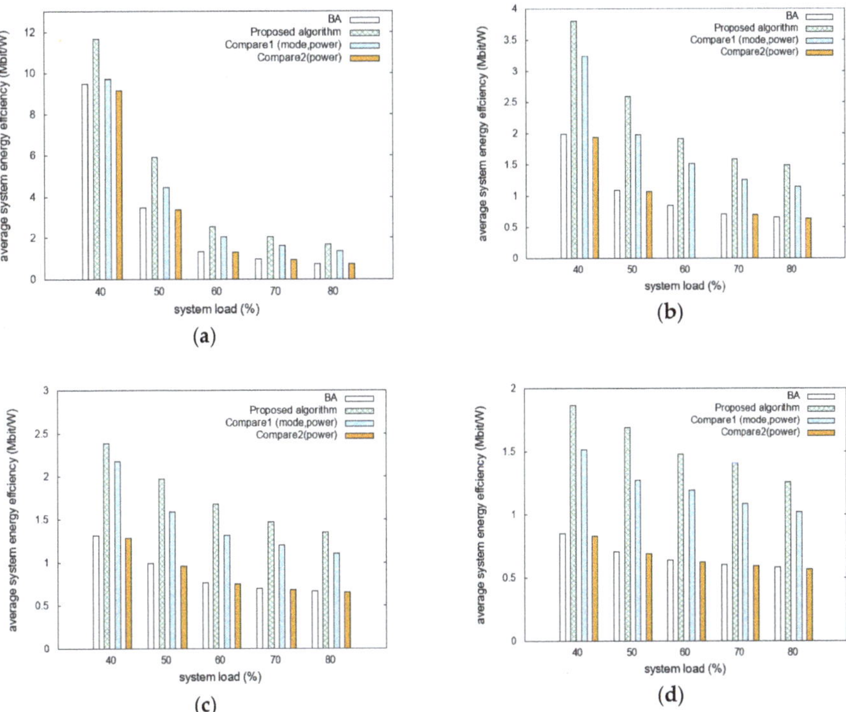

Figure 3. Comparison of system energy efficiencies with various device-to-device (D2D) distances and system loads: (**a**) 250 m with light traffic; (**b**) 350 m with light traffic; (**c**) 250 m with heavy traffic; and (**d**) 350 m with heavy traffic.

The average achievable data rates for the different D2D distance scenarios are compared as functions of system loads with different mobility scenarios, as shown in Figure 4. Figure 4a shows that the proposed algorithm has achievable data rates that are higher than those of BA, Compare1, and Compare2 by approximately 20%, 40%, and 22%, respectively. Figure 4b shows that the achievable data rate of the proposed algorithm is 60% higher than that of Compare1. Compared to BA and Compare2, it presents higher energy efficiency, with differences of approximately 16% and 19%, respectively. In Figure 4c, the proposed algorithm presents approximately 10%, 47%, and 13% higher achievable data rates compared to those of BA, Compare1, and Compare2, respectively. Figure 4d shows that the proposed algorithm outperforms BA, Compare1, and Compare2 by approximately 9%, 65%, and 11%, respectively. As the D2D distance increases, increasing the D2D communication ratio results in reduced energy consumption. However, increasing the D2D distance decreases the achievable data rates because the D2D mode communicates with the same transmission power even in degraded interference situations. In addition, communication without mode-selection algorithms, such as BA and Compare2, can select the transmission mode by simply considering the expected SINR, so it can perform better in terms of achievable data rate than algorithms with mode selection.

The proposed algorithm achieved a higher achievable data rate even in such scenarios by adjusting the target SINR.

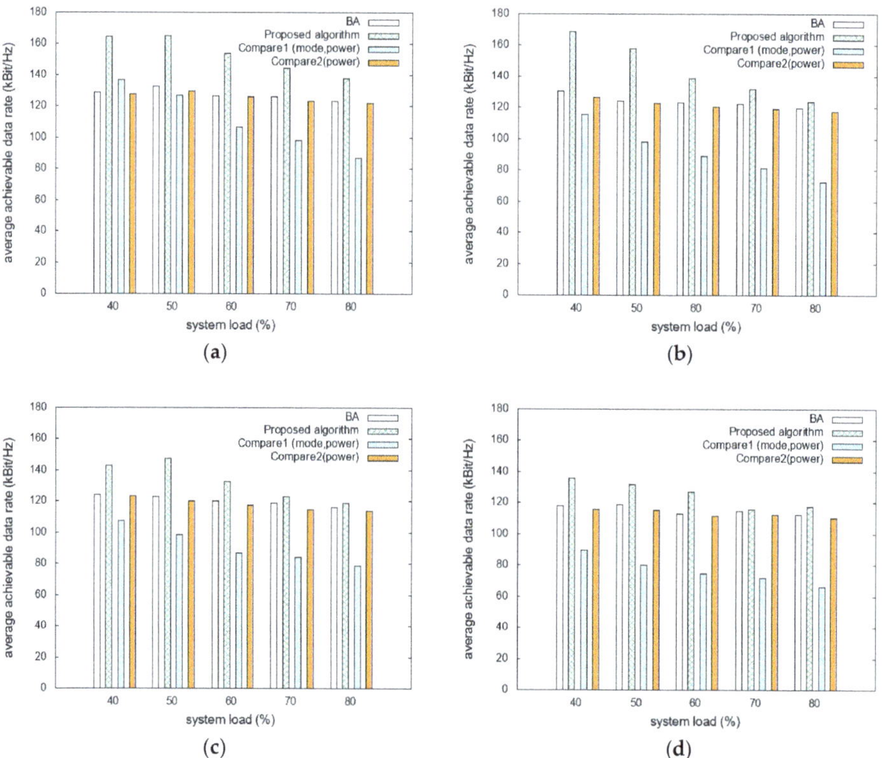

Figure 4. Comparison of average achievable data rate with various D2D distances and system loads: (**a**) 250 m with light traffic; (**b**) 350 m with light traffic; (**c**) 250 m with heavy traffic; and (**d**) 350 m with heavy traffic.

The cumulative distribution functions (CDFs) of SINR received by vehicles for different D2D distance thresholds are compared in Figure 5. Figure 5a–d shows the distribution of the received SINR in each algorithm. In all panels, the proposed algorithm has more vehicles that receive higher SINR compared to BA, Compare1, and Compare2. The proposed algorithm and Compare1 can select the communication mode and transmission power to maximize the system energy efficiency, allowing vehicles to select more D2D modes. It can increase spectral efficiency and reduce system energy consumption, but it can also decrease the received SINR. However, the proposed algorithm is designed to select the mode and transmission power to ensure the received SINR. Therefore, the proposed algorithm has more vehicles, receives higher SINR, and performs better with an achievable data rate compared to Compare1. The BA and Compare2 decide the communication mode based only on the expected SINR of the vehicle, so they can achieve higher performance in a longer D2D distance threshold.

The comparisons of system energy efficiency as a reward function with an increasing number of learning iterations are shown in Figure 6. In Figure 6a it can be seen that as the number of learning iterations increases, the value of system energy efficiency stabilizes after 100 iterations. In heavy traffic scenario, as shown in Figure 6b, the value of the system energy efficiency stabilizes after 100 iterations. The system energy efficiency is overserved with a low value at the beginning of the learning iterations and increases until it reaches a stable point. This means that the proposed algorithm updates its policy optimally and converges faster and in a more stable manner than the other algorithms. The convergence time of the proposed algorithm is longer than those of Compare1 and Compare2. This is because the proposed algorithm has longer decision process and considers more states in decision process. It finally decides an action by comparing two recommended actions. In addition, it considers available resources and interference level as state to make a decision. However, Compare1 and Compare2 consider only interference level as state. Although the convergence time of the proposed algorithm becomes a little bit longer, the proposed algorithm shows better performance in terms of system energy efficiency.

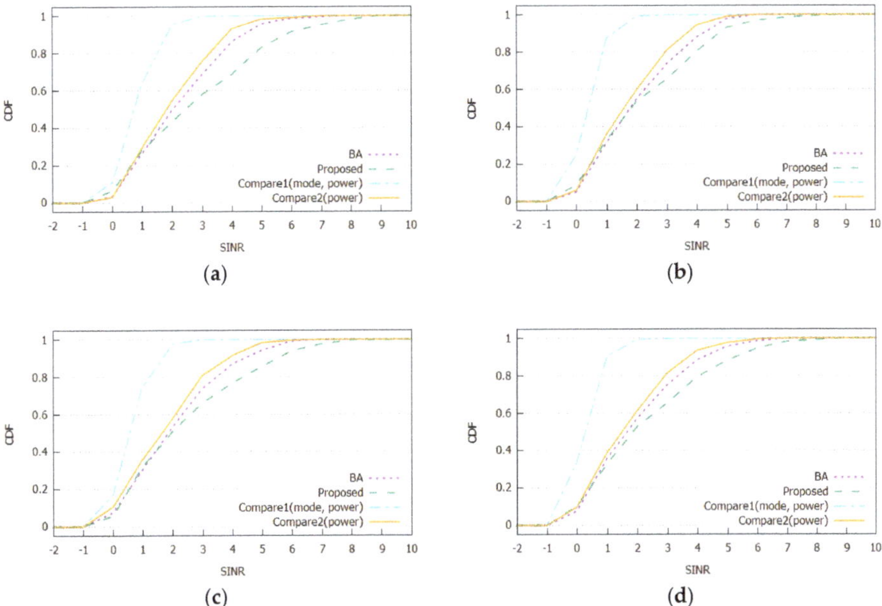

Figure 5. Comparison of signal-to-interference noise ratio (SINR) with various D2D distances and system loads: (**a**) 250 m with light traffic; (**b**) 350 m with light traffic; (**c**) 250 m with heavy traffic; and (**d**) 350 m with heavy traffic.

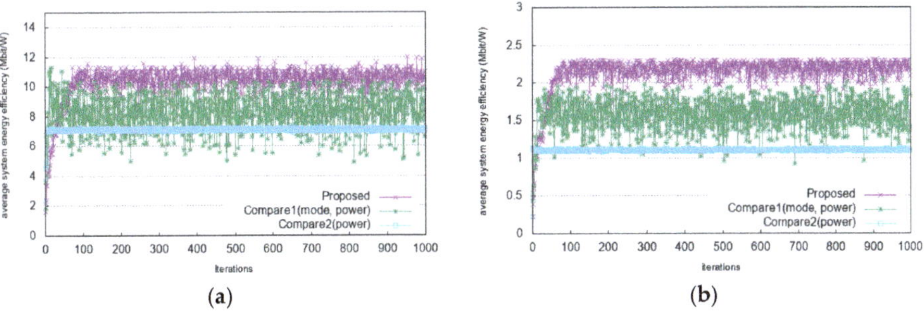

Figure 6. Convergence of system energy efficiency as a reward function with D2D distance = 350 m: (**a**) light traffic scenario; (**b**) heavy traffic scenario.

To show the comparison with the optimal solution, the performance values are shown in Tables 2 and 3. In Table 2, the average system energy efficiencies with system load variance are described.

Based on the optimal solution, the proposed algorithm, BA, Compare1, and Compare2 obtain 44%, 20%, 35%, and 20% of the system energy efficiency, respectively. In Table 3, the average achievable data rate with system load variance is described. Based on the optimal solution, the proposed algorithm, BA, Compare1, and Compare2 obtain 79%, 68%, 50%, and 66% of the achievable data rate, respectively. This shows that the proposed algorithm performs closest to the optimal solution among other compared algorithms because it adopts the system energy efficiency and received SINR as rewards. However, as the system load increases, the differences in performance increase. This is because the interference increases according to increasing system load environment. The proposed algorithm simply models received SINR as whether the average SINR is greater than the target SINR to take account of the interference.

Table 2. Average system energy efficiencies with system load variance, D2D distance = 350 m.

System Load (%)	40	50	60	70	80
Optimal	4.8414	5.2831	4.9594	5.4548	5.5733
BA	1.9857	1.0919	0.8479	0.7086	0.6553
Proposed	3.8026	2.5863	1.9223	1.5863	1.4909
Compare1	3.2371	1.9767	1.5212	1.2576	1.1491
Compare2	1.9429	1.0723	0.8184	0.6992	0.6376

Table 3. Average achievable data rate with system load variance, D2D distance = 350 m.

System Load (%)	40	50	60	70	80
Optimal	183.1308	184.869	181.0795	180.4453	179.5396
BA	130.6982	124.3168	123.1323	122.5992	119.5717
Proposed	168.763	158.1163	138.6988	131.8081	123.8312
Compare1	115.4847	98.18192	88.934	81.2835	72.47508
Compare2	126.4065	122.9883	120.3971	119.4175	117.3408

To evaluate and discuss the performance of the proposed algorithm, we compared it with the BA and the two compared algorithms. The proposed algorithm exhibited the best performance in terms of system energy efficiency, achievable data rate, and SINR in cases of increasing D2D distance thresholds. The proposed algorithm learned to maximize the system energy efficiency while ensuring achievable data rates. The system energy efficiency and achievable data rate have a trade-off relationship, so the proposed algorithm used Q-learning in two ways. To maximize the system energy efficiency, the proposed algorithm sets the system energy efficiency as a reward function of centralized Q-learning. In addition, to ensure an achievable data rate, the proposed algorithm sets the received SINR as a reward function of distributed Q-learning. This implies that the proposed algorithm can achieve the highest energy efficiency when compared to other algorithms. It can also guarantee QoS while increasing the efficiency of the resource and system energy.

6. Conclusions

We proposed a joint mode-selection and power-control algorithm with reinforcement learning to achieve energy optimization in vehicle networks. We defined and formulated the maximization problem for system energy efficiency, subject to SINR and outage probability constraints. We considered cellular and D2D communications coexisting in a single-cell environment and designed a Q-learning algorithm that made optimal transmission-mode-selection and power-control decisions by adjusting the target SINR. The Q-learning algorithm used centralized Q-learning in the BBU pool and decentralized Q-learning in the devices. To show that the proposed algorithm outperformed the other algorithms, we used a SUMO simulator to run various scenarios and compared system energy efficiencies, achievable data rates, and SINRs.

In real network environments, the system may not be fully aware of the channel state information. To solve this problem, clustering and sharing information among neighboring cells is necessary.

In future work, we will consider a cell clustering and sharing method that provides optimal decisions to devices that handover nearby cells in environments with uncertain channel state information.

Author Contributions: H.P.: conceptualization, software, validation, writing-original draft preparation, visualization; Y.L.: conceptualization, methodology, writing-review and editing, supervision, project administration, funding acquisition; All authors have read and agreed to the published version of the manuscript.

Funding: This work was supported by the National Research Foundation of Korea (NRF) grant funded by the Korea government (MSIT) (No. NRF-2018R1A2B6002505)

Conflicts of Interest: The authors declare no conflict of interest.

References

1. Su, Z.; Hui, Y.; Guo, S. D2D-based content delivery with parked vehicles in vehicular social networks. *IEEE Wirel. Commun.* **2016**, *23*, 90–95. [CrossRef]
2. Akinlade, O.M. Adaptive Transmission Power with Vehicle Density for Congestion Control. Master's Thesis, Windsor University, Windsor, ON, Canada, July 2018.
3. Wu, X.; Sun, S.; Li, Y.; Tan, Z.; Huang, W.; Yao, X. A Power control algorithm based on outage probability awareness in vehicular ad hoc networks. *Adv. Multimed.* **2018**, *2018*, 1–8. [CrossRef]
4. Hong, H.; Kim, Y.; Kim, R.; Ahn, W. An Effective Wide-Bandwidth Channel Access in Next-Generation WLAN-Based V2X Communications. *Appl. Sci.* **2019**, *10*, 222. [CrossRef]
5. Liang, L.; Ye, H.; Li, G.Y. Spectrum sharing in vehicular networks based on multi-agent reinforcement learning. *IEEE J. Sel. Areas Commun.* **2019**, *37*, 2282–2292. [CrossRef]
6. Zhang, X.; Peng, M.; Sun, Y. Deep reinforcement learning based mode selection and resource allocation for cellular V2X communications. *IEEE Internet Things* **2019**, 1–12. [CrossRef]
7. Huang, J.; Jinyun, Z.; Cong-Cong, X. Energy-efficient mode selection for D2D communications in cellular networks. *IEEE Trans. Cogn. Commun. Netw.* **2018**, *4*, 869–882. [CrossRef]
8. Akkarajitsakul, K.; Phunchongharn, P.; Hossain, E.; Bhargava, V.K. Mode selection for energy-efficient D2D communications in LTE-advanced networks: A coalitional game approach. In Proceedings of the IEEE international conference on communication systems, Singapore, 21–23 November 2012; pp. 488–492.
9. Zuo, J.; Chao, Z.; Nan, B. Mode selection for energy efficient D2D communications underlaying C-RAN. In Proceedings of the International Conference on Information Technology, Singapore, 27–29 December 2017; pp. 287–291.
10. Huang, J.; Liao, Y.; Xing, C. Efficient power control for D2D with SWIPT. In Proceedings of the Proceedings of the Conference on Research in Adaptive and Convergent Systems, Honolulu, HI, USA, 9–12 October 2018; pp. 106–111.
11. Peng, S.; Kang, G.; Hailin, Z.; Liang, H. Transmit power control for D2D-underlaid cellular networks based on statistical features. *IEEE Trans. Veh. Technol.* **2017**, *66*, 4110–4119.
12. Xu, Y.; Wang, S. Mode selection for energy efficient content delivery in cellular networks. *IEEE Commun. Lett.* **2016**, *20*, 728–731. [CrossRef]
13. Klaus, D.; Chia-Hao, Y.; Cassio, B.R.; Peckka, J. Mode selection for device-to-device communication underlaying an LTE-advanced network. In Proceedings of the IEEE Wireless Communication and Networking Conference, Sydney, Australia, 18–21 April 2010; pp. 1–6.
14. Yuri, V.L.M.; Rodrigo, L.B.; Carlos, F.M.S.; Tarcisio, F.M.; Jose, M.B.S.; Francisco, R.P.C. Uplink power control with variable target SINR for D2D communications underlaying cellular networks. In Proceedings of the 20th European Wireless Conference, Barcelona, Spain, 14–16 May 2014; pp. 1–5.
15. Lee, N.; Lin, X.; Andrews, J.G.; Heath, R.W. Power control for D2D underlaid cellular networks: Modelling, algorithms, and analysis. *IEEE J. Sel. Areas Commun.* **2015**, *33*, 1–13. [CrossRef]
16. Bei, M.; Hailin, Z.; Zhaowei, Z. Joint power allocation and mode selection for D2D communications with imperfect CSI. *China Commun.* **2015**, *12*, 73–81.
17. Yifei, H.; Ali, A.N.; Salman, D.; Xiangyun, Z. Mode selection, resource allocation, and power control for D2D-enabled two-tier cellular network. *IEEE Trans. Commun.* **2016**, *64*, 3534–3547.
18. Arifur, R.; Youngdoo, L.; Insoo, K. An efficient transmission mode selection based on reinforcement learning for cooperative cognitive radio networks. *Hum. Cent. Comput. Inf. Sci.* **2016**, *6*, 1–14.

19. Yaohua, S.; Mugen, P.; Shiwen, M. Deep reinforcement learning-based mode selection and resource management for green fog radio access networks. *IEEE Internet Things J.* **2019**, *6*, 1960–1971.

20. Shiwen, N.; Zhiqiang, F.; Ming, Z.; Xinyu, G.; Lin, Z. Q-learning based power control algorithm for D2D communication. In Proceedings of the IEEE 27th Annual International Symposium on Personal, Indoor, and Mobile Radio Communications, Valencia, Spain, 4–8 September 2016; pp. 1–6.

21. Yimingm, Q.; Zelin, J.; Yonghao, Z.; Guanghao, M.; Gang, X. Joint mode selection and power adaption for D2D communication with reinforcement learning. In Proceedings of the 15th International Symposium on Wireless Communication Systems, Lisbon, Portugal, 28–31 August 2018; pp. 1–6.

22. Auer, G.; Giannini, V.; Desset, C.; Godor, I.; Skillermark, P.; Olsson, M.; Imran, M.A.; Sabella, D.; Gonzalez, M.J.; Blume, O.; et al. How much energy is needed to run a wireless network? *IEEE Wirel. Commun.* **2011**, *18*, 40–49. [CrossRef]

23. Xiaojian, L.; Shaowei, W. Efficient remote radio head switching scheme in cloud radio access network: A load balancing perspective. In Proceedings of the IEEE Conference on Computer Communications, Atlanta, GA, USA, 1–4 May 2017; pp. 1–9.

24. Arnob, G.; Laura, C.; Eitan, A. Nash equilibrium for femto-cell power allocation in HetNets with channel uncertainty. In Proceedings of the IEEE Global Communications Conference, San Diego, CA, USA, 6–10 December 2015; pp. 1–7.

25. Park, H.; Lim, Y. Adaptive Power Control using Reinforcement Learning in 5G Mobile Networks. In Proceedings of the International Conference on Information Networking, Barcelona, Spain, 7–10 January 2020; pp. 409–414.

26. Nasir, Y.S.; Guo, D. Multi-agent deep Reinforcement Learning for Distributed Dynamic Power Allocation in Wireless Networks. *arXiv* **2018**, arXiv:1808.00490.

27. 3GPP. *TR 37.885, Technical Specification Group Radio Access Networks, Study on Evaluation Methodology of New Vehicle-to-Everything (V2X) use cases for LET and NR, Rel. 16*; Sophia-Antipolis: Valbonne, France, 2018.

28. Daniel, K.; Jakob, E.; Michael, B.; Laura, B. Recent development and applications of SUMO-simulation of urban mobility. *Int. J. Adv. Syst. Meas.* **2012**, *5*, 128–138.

29. 3GPP. *TR 22.886, Technical Specification Group Radio Access Networks, Study on Enhancement of 3GPP Support for 5G V2X Services, Rel. 15*; Sophia-Antipolis: Valbonne, France, 2017.

30. YooSeung, S.; Hyungkyun, C. Analysis of V2V Broadcast Performance Limit for WAVE Communication Systems Using Two-Ray Path Loss Model. *ETRI J.* **2017**, *39*, 213–221.

31. 3GPP. *TR 36.872, Technical Specification Group Radio Access Network, Small Cell Enhancements for E-UTRA and E-UTRAN Physical Layer Aspects, Rel. 11*; Sophia-Antipolis: Valbonne, France, 2013.

32. Khaled, S.H.; Engy, M.M. Device-to-Device communication distance analysis in interference limited cellular networks. In Proceedings of the ISWCS 2013; The Tenth International Symposium on Wireless Communication Systems, Ilmenau, Germany, 27–30 August 2013; pp. 1–5.

33. Seok, B.; Sicato, J.C.S.; Erzhena, T.; Xuan, C.; Pan, Y.; Park, J.H. Secure D2D Communication for 5G IoT Network Based on Lightweight Cryptography. *Appl. Sci.* **2020**, *10*, 217. [CrossRef]

Article

Secure and Blockchain-Based Emergency Driven Message Protocol for 5G Enabled Vehicular Edge Computing

Lewis Nkenyereye [1], **Bayu Adhi Tama** [2], **Muhammad K. Shahzad** [3] and **Yoon-Ho Choi** [4,*]

[1] Department of Computer and Information Security, Sejong University, Seoul 05006, Korea; nkenyele@sejong.ac.kr
[2] Department of Mechanical Engineering, Pohang University of Science and Technology, Pohang 37673, Korea; btama@acm.org
[3] Department of Computing, National University of Sciences and Technology (NUST), Islamabad 44000, Pakistan; mkhuram.shahzad@seecs.edu.pk
[4] School of Computer Science and Engineering, Pusan National University, Busan 46241, Korea
* Correspondence: yhchoi@pusan.ac.kr

Received: 5 November 2019; Accepted: 23 December 2019; Published: 25 December 2019

Abstract: Basic safety message (BSM) are messages that contain core elements of a vehicle such as vehicle's size, position, speed, acceleration and others. BSM are lightweight messages that can be regularly broadcast by the vehicles to enable a variety of applications. On the other hand, event-driven message (EDM) are messages generated at the time of occurrence such as accidents or roads sliding and can contain much more heavy elements including pictures, audio or videos. Security, architecture and communication solutions for BSM use cases have been largely documented on in the literature contrary to EDM due to several concerns such as the variant size of EDM, the appropriate architecture along with latency, privacy and security. In this paper, we propose a secure and blockchain based EDM protocol for 5G enabled vehicular edge computing. To offer scalability and latency for the proposed scenario, we adopt a 5G cellular architecture due to its projected features compared to 4G tong-term evaluation (LTE) for vehicular communications. We consider edge computing to provide local processing of EDM that can improve the response time of public agencies (ambulances or rescue teams) that may intervene to the scene. We make use of lightweight multi-receiver signcryption scheme without pairing that offers low time consuming operations, security, privacy and access control. EDM records need to be kept into a distributed system which can guarantee reliability and auditability of EDM. To achieve this, we construct a private blockchain based on the edge nodes to store EDM records. The performance analysis of the proposed protocol confirms its efficiency.

Keywords: vehicle edge computing; 5G cellular networks; blockchain; multi-receiver signcryption; security; privacy

---

1. Introduction

Abrupt situations on roads, such as car accidents, slippery roads, or land sliding can be reported by the vehicles using the inbuilt sensors and devices. Reporting emergency situations can be an effective approach in situations where the evident proofs are required such as car accidents or reckless driving. Those reports can also provide additional materials like pictures, videos, audios to the rescue departments for an efficient and timely intervention [1]. Emergency warning can be divided in three categories; (1) a vehicle-to-vehicle (V2V) warning dissemination such as hazardous or slippery road where the vehicles in the vicinity need to slow down or take further precautions,

(2) vehicle-to-infrastructure (V2I) warning such as car accidents or road sliding where the rescue teams (police or medical teams) can use the multimedia proof for an effective and timely response, (3) the last warning dissemination is the combination of the two scenarios. We do focus on the second scenario in this work and those files are called event-driven messages (EDM). Moreover, the EDM files would be sent to remote servers which will require a heavy bandwidth with excessive response delay. In big cities with millions of vehicles running on the road every day, the amount of data to be processed would be massive [2–4].

Edge computing was introduced as a new paradigm that takes the computing tasks to the network edge. The edge nodes collect data from the vehicles and process them rather than sending them to a central cloud server. This paradigm offers a number of benefits such as geo-distribution and low latency and can be applied in vehicular networks to offer real time services such as emergency warning, road surface monitoring and navigations [5–7]. However, despite the merits of edge computing in vehicular communications, the proposed solutions and applications have not been deployed worldwide due to lack of scalability and adequate communication supports. Recently, the researchers have raised the limitations of IEEE 802.11p due to its lacks of mobility support, also the long-term evolution (LTE) of fourth generation networks (4G) cannot offer effective latency that suits the vehicular networks based applications [8–10]. To overcome these limitations, 5G cellular networks were adopted as the ultimate architecture which could help a real deployment of vehicular based technologies. For instance, Uber made a successful test of driverless vehicle using 5G cellular networks recently. The cellular networks offer higher mobility support, reduced latency and massive connectivity that are core requirements for vehicular applications [11–13]. Meanwhile, security and privacy issues are also a considerable concern for vehicular communications. For example, vehicles reporting the emergency warnings might not need to disclose their locations expect to authorized third parties. The identities of the vehicles participating in the sending the warnings, their destinations and itinerary are highly sensitive information that need to be carefully handled. Currently, the misuse of vehicles' users data have been reported massively in the press where untrusted third parties and malicious users leaked the vehicles' sensitive data [14].

In the literature, a considerable number of articles has been published on secure message dissemination for vehicle networks and can be categorized in three groups: (1) Security for beacons messages also knows as basic safety message (BSM) for the US. These are periodic messages containing vehicle's position, speed and direction, etc. [15–17]; (2) event-driven message (EDM) that are generated at the time of occurrence such as accident alerts or emergency reports [18]. There are many solutions that addressed BSM based scenarios for privacy and security as shown in this recent survey [19]. However, these schemes cannot be directly applied to EDM scenarios for the following reasons. First, the schemes are not built based on 5G-enabled architecture which offers low latency and mobility support for vehicular communications. Second, the current literature such as [20] mainly suggests anonymous authentication schemes and message encryption for secure communications and the central cloud or edge nodes are mainly supposed to be secure. However, if the central cloud is compromised, the rescue services can not retrieve the important files needed for their services, thus data auditability and reliability is very crucial and one of the solutions to achieve data auditability would be through a private blockchain maintained by the edge devices [21,22]. Third, most of the schemes is built using expensive bilinear pairing techniques which are expensive and time consuming operations that degrade the overall protocol performance. To the best of our knowledge, there is one relevant article for emergency message dissemination for vehicular communications [23]. The authors basically presented a fog assisted architecture, highlighted limitations of relevant schemes in the literature and concluded with open research discussions.

Thus, a privacy preserving, secure yet auditable protocol for emergency message in vehicle edge computing is appealing. The contributions of this paper are three folds:

- We describe a novel architecture for emergency warning dissemination using edge computing and private blockchain. The proposed architectures uses 5G network technologies for communication.

In our model, we design a secure and privacy preserving model that protect the sensitive data (identity, location, shared data, etc) of the vehicles participating in emergency warning dissemination. We assume that the edge nodes and cloud are semi trusted, therefore our architecture proposes a private blockchain using edge nodes to record the EDM in an immutable and verifiable ledger to guarantee EDMs auditability.

- We design a secure and blockchain based EDM protocol for 5G enabled vehicle edge computing using the private blockchain technique to provide EDM auditability. We make use of lightweight multi-receiver signcryption scheme without pairing that offer low time consuming operations, security, privacy and access control.
- We provide an analysis of security and privacy features of the proposed protocol and evaluation in respect of private blockchain construction, computational and communication costs.

The remainder of this paper is as follows. We review the related work on 5G enabled vehicular edge computing and secure emergency message dissemination in Section 2. We design the system model and the preliminaries of the core cryptographic schemes used in our protocol in Section 3. Section 4 describes the proposed scheme and we provide security, privacy and performance analysis in Section 5. Finally, the concluding remarks are given in Section 6.

2. Related Work

This section first describes the basic concepts of vehicular edge computing, then outlines the basic notions of private blockchain technology and concludes with a review of the current schemes on EDM schemes in vehicular networks.

2.1. 5G Enabled Vehicular Edge Computing

Vehicular edge computing (VEC) or vehicular edge computing networks (VECONs) are extended from the conventional VANETs. The main difference is that VEC are made by an additional edge layer [24,25]. VEC is basically made by three layers, first the vehicle layer where the embedded sensors in the vehicle collect the data and send them to the edge layer using the onboard unit (OBU). The edge layer is a cluster made by several roadside units (RSUs) within a given distance. The RSU keeps or processes the data provided by the vehicles in the edge cluster. The following layer is called the cloud layer that manages the edge layers. The cloud layer can store massive data and make complex delay tolerant operations on the data provided by the edge layers. The cloud layer can be a data center or intelligent transportation system or regional trusted authority. Although 4G technology can be used, there are inherent drawbacks that has been raised by the research community for an effective vehicular communications networks using 4G technologies. A number of attacks performed on the international subscriber identity for the 4G LTE networks revealed its weakness to provide integrity, non-repudiation, accountability on user data. In addition, IEEE 802.11p can not be relied on for VEC due to its lack of mobility support [8–10]. In the forthcoming fifth-generation (5G) cellular-based vehicle networks, the use of denser and smaller cells are anticipated to offer a high transmission rate for the vehicles users. This will enable a range of application starting from the safety related applications to entertainment use cases. The nature of vehicular networks presents a different requirement from the conventional mobile networks. This is basically due to the volatile mobility of the vehicles in the network, the speed of the vehicles along with the topology of the dynamic wireless networks. Therefore, the use of 5G cooperative small cells is discussed in the literature as a promising recommendation [26–28]. We adopt in this paper a 5G enabled edge vehicular computing model as underlying network architecture for the proposed protocol.

2.2. Blockchain

Blockchain technology is a distributed ledger technique that was first introduced for the financial domain starting with Bitcoin crypto currency. It offers data auditability using authenticated blocks added on the system. It is also a decentralized network since it does not require a centralized entity because it is a peer to peer network [21,22]. In order to add a new block, a consensus algorithm need to be agreed upon by the entities in the chain. Private blockchain was introduced for restricted environment such as businesses or companies where public readability can not be applied. A private blockchain is a network where the participants require a permission to join. In this work, we consider a proof of stake consensus algorithm that randomly choose one of the entities proportionally to each node's stake to run the process [29].

2.3. Secure Schemes for Emergency Warning in VEC

A considerable number of researchers have documented security and privacy related solutions for BSM based scenario as shown in this survey [18]. Nevertheless, these schemes are not directly applicable to EDM scenarios for the following reasons. The protocols do not consider a 5G-enabled architecture which offers low latency and mobility support for vehicular communications. Then, the schemes in the literature such as [20] mainly suggest anonymous authentication techniques and EDM/BSM encryption for secure communications while the central cloud or edge nodes are most of the time supposed to be secure. As mentioned earlier, this raises a huge issue with billions of connected devices, the regional or center cloud or edge devices can be compromised, thus data auditability and reliability need to be taken into consideration. One alternative way of achieving data auditability would be through a private blockchain maintained by the edge devices [21,22]. Also, most of the scheme such as [20] are built using expensive bilinear pairing techniques which are very heavy for mobile and ad hoc networks. In [23], the authors proposed an emergency message dissemination for vehicular communications. Though the paper specifically target EDM, the authors mainly presented fog assisted architecture, highlighted limitations of relevant schemes in the literature and concluded with open research discussions. Their protocol do not address security issues, latency sensitive architecture, distributed environment and EDM reliability and auditability. Thus, we are appealed in this paper to investigate on secure communication for EDM scenario taking into consideration that EDM could be very heavy (heavy videos, audio), while the privacy of the vehicles' users is not neglected, yet EDM reliability and auditability are guaranteed.

3. System Model

In this section we first present the system architecture of the proposed protocol and outline the basic concepts of cryptographic techniques used to construct our protocol.

3.1. Main Entities

Our proposed system model is made by a main regional overviewer called RTA, the road side units (RSCs) that make a edge cluster and the vehicles that provide EDM files collected through their sensors as shown in Figure 1. We outline the role of each entity in the following:

- Regional Transportation Authority (RTA): RTA is considered as a trusted agency that offers the registration of all the entities within the proposed system (vehicles and edge nodes) and generate cryptographic materials to the entities during the system setup.
- RSU edge nodes: Similar to a sever with limited capabilities, edge nodes are devices placed on the roads with efficient computing, communication and also storage aptitude. Their principal role is the collection of EDM provided by the vehicles, verifies the validity of EDM through designcryption and share the EDM to RTA or any entity that might need the EDM. In real life applications, the EDM could be needed by rescue services such as police or medical centers. We did not explicitly add these entities but we assumed that they have servers in the cloud which are connected to RTA servers as shown in Figure 1. We assume that edge nodes are connected to a source that generate electricity power.

- Vehicles: The vehicles are assumed to be equipped with several sensors and devices such as camera. The onboard units (OBU) in the vehicle gather all the those data in form of EDM files, sends them to edge nodes using different communication means such as D2D or mmWave communications. All vehicles need to register with the RTA at the time of periodic inspection. Besides the well known identifiers of vehicles such as the Electronic License Plate (ELP) or the electronic chassis number (ECN), every vehicle is given a 5G unique identifier (5GID), which is similar to subscriber identification module (SIM) as it is for 3G and 4G cellular networks.

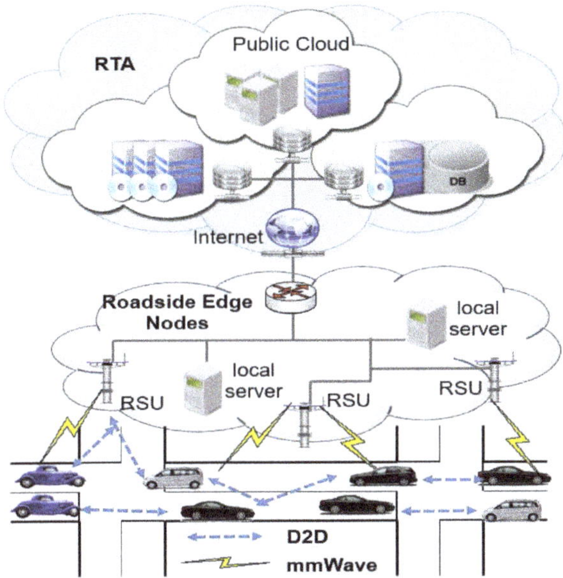

Figure 1. System architecture.

3.2. Communication Model

Motivated by the 5G cellular networks architecture, the proposed 5G enabled vehicle edge computing is made by the following components:

- Heterogeneous networks: This network aims at achieving high data rate and network capacity for the 5G-enabled vehicle edge computing. Therefore, two alternative techniques may help to get the mentioned capacities through smaller cells which increase the spectral efficiency. In addition, using the mmWave spectrum would offer high data rates since it operates within the range of 30–300 GHz and 1–10 mm for the spectrum and wavelength respectively [11].
- D2D Communications: D2D communication would enable the vehicles to communicate with each edge device within the licensed cellular bandwidth without considering the base stations. In the 5G edge based vehicular networks, the communication between the vehicles and edge devices can be done through D2D communication or mmWave technology.

3.3. Adversary Model

In this section, we describe the main attacks which an malicious user might conduct in the absence of this protocol EDM reporting scheme.

- A malicious vehicle can try to send EDM files when he is not enrolled for participation.
- A malicious user or vehicle can try to know the identity of the vehicles that reported the EDM file.

- A malicious vehicle can try to get the raw content of EDM which were sent through the network.
- A malicious vehicle can try to attack one or several edge nodes and try to process the EMD by impersonating a given edge node.
- A number of attackers (within or without the participating group) can try to jeopardize the whole network through a denial of service attack.

3.4. Security Objectives

We outline the security goals which the proposed scheme needs to achieve:

- Identity privacy preservation: the identities of the vehicles that report the EDMs should be preserved.
- Authentication: each vehicle that is involved in sending the EDMs should be authenticated before it is allowed to join the system.
- Confidentiality and integrity: the EDMs files generated and sent through the network should not be intercepted and modified during the communication.
- Key escrow resilience: the keys of the entities (vehicles) participating in EDM reporting should not be generated by a single entity. Thus, even if the RTA is comprised, the attackers can not disclose the signing keys of the vehicles.
- Access control: only the entities with matching policies should be able to retrieve the contents of EDMs.
- Non-repudiation and traceability: a vehicle should not deny any participation in the EDM reporting. In addition, RTA should be able to disclose the true identity of any entity if needed.
- Auditability: the EDMs records that are saved in the system should be securely kept and easily verifiable. Even if one node in the chain is compromised, the malicious user should not be able to modify and upload any EDM content.

3.5. Preliminaries

In this section, we describe the two main cryptographic techniques used to build our protocol. We first outline a lightweight signcrpytion technique which is not built on pairing operations, we also describe the underlying concepts for constructing a private blockchain.

3.5.1. Signcryption Scheme without Bilinear Pairings

This scheme is made by six sub protocol namely Setup, SecretValue, Partialkeypair, keypair, Signcryption and DeSigncryption [30].

- $Setup(1^\lambda)$: using a parameter λ, RTA runs the system to generate a master secret key mk and the parameters $params$.
- $SecretValue(ID, params)$: a user runs the algorithm to return a secret value V_{ID} using his/her identity ID.
- $Partialkeypair(params, V_{ID}, ID)$: RTA runs the algorithm and returns the partial private key y_{ID} and partial public key D_{ID} using the user identity ID and the secret value V_{ID}
- $Keypair(D_{ID}, y_{ID}, params)$: the user generates the key pairs (PK_i, SK_i) using the partial key pairs (D_{ID}, y_{ID}).
- $Signcrypt(L, m, SK_i, params)$: the user target a group of authorized receivers' public keys $L = \{PK_1, PK_2, PK_3, .., PK_n\}$ where n is a positive integer. Output a ciphertext δ on the message m.
- $Designcrypt(params, \delta, SK_i)$: using the system parameters $params$, the receiver's private key SK_i and the ciphertext δ, an authorized receiver recovers the message m.

3.5.2. Private Blockchain

The private blockchain concept used in the paper is made by the following sub-phases, namely setup, initial stage, leader selection and block generation [21,22]:

- Setup: in this phase, different slot $\{ts_1, ts_2, ts_3, \cdots\}$ are generated and a private ledger is attached with a one block for every time slot ts_i. In addition, a leader selection algorithm $F(.)$ is assigned to each edge node.
- Initial stage: this is a first stake distribution phase when the first block also called genesis block is generated. The genesis block includes the edge nodes identities, public keys and stakes. The first block is assumed to have an empty blockheader and signcryption is generated on it.
- Leader selection: taking each time slot ts_i, the edge nodes identities, their public key, the probability of an edge node corresponding to its stake, this function output the node leader.
- Blockgeneration: the chosen leader generates a new block which is made by a block header, its stake, the number of EDM recorded. Note that the blockheader is made by a blockheader number, hash of previous blockheader, a merkle hash root along with a time stamp. For interested readers, the overall details can be found in [21,22].

4. Protocol Description

Our proposed protocol is made by five main sub-protocols: setup, participation agreement, EDM reporting, EDM collection and private blockchain generation

4.1. Protocol Setup

Our protocol assume that a regional traffic authority (RTA) manages the reporting of EDM messages, therefore both the vehicles and the edge nodes in the region are registered to the RTA. RTA first runs $Setup(1^\lambda)$ to generates the parameters parameters $(\mathbb{G}_1, q, P)$ with the $\mathbb{G}_1$ being a cyclic additive group of order q and a generator P over an elleptic curve that is defined on finite field F_w where w is an integer chosen by RTA. RTA then selects a random $s \in \mathbb{Z}_q^*$ as a master secret and generates the public key of RTA as $P_{RTA} = sP$. RTA selects five hash functions $H_1 : \{0,1\}^* \to \mathbb{G}_1$, $H_2 : \{0,1\}^* \to \mathbb{G}_1, H_3 : \{0,1\}^* \to \mathbb{G}_1, H_4 : \{0,1\}^* \to \mathbb{G}_1, H_5 : \{0,1\}^* \to \mathbb{G}_1$ and publishes the public parameters $(\mathbb{G}_1, H_1, H_2, H_3, H_4, H_5, P_{RTA}, P, q)$ to all the entities.

4.2. Participation Agreement

In order to participate in EDM reporting, the vehicles and the edge nodes are registered by the RTA. The registration of these entities is done as follows

- Step 1. Assume there is a vehicle inspection within a given period (12 or 18 months), the vehicle owner or user can express its desire to be part of EDM reporters. In this case, the vehicles does the following:

  1. A vehicle v_i with its identity $5GID_{v_i}$ selects a secret $t_i \in \mathbb{Z}_q^*$ and computes $V_{v_i} = t_1 P$ and send $(5GID_{v_i}, V_{v_i})$ to RTA.
  2. Upon receiving $(5GID_{v_i}, V_{v_i})$, RTA choose a pseudonym for $5GID_{v_i}$ as $P5GID_{v_i}$, and keep the mapping table securely. RTA selects $d_i \in \mathbb{Z}_q^*$ and computes $y_i = H_1(P5GID_{v_i}, V_{v_i}, d_i) + s(mod)w$ and $D_i = H_1(P5GID_{v_i}, V_{v_i}, d_i)P$. Then RTA returns (D_i, y_i) to v_i
  3. v_i receives (D_i, y_i) and checks if the equation $y_i P = D_i + P_{RTA}$ is correct. If yes, v_i generates its public key $PK_{v_i} = D_i + H_2(P5GID_{v_i}, V_{v_i})V_{v_i}$.
  4. v_i generates its private key $SK_{v_i} = H_2(P5GID_{v_i}, PK_{v_i})(y_i + H_2(5GID_{v_i}, V_{v_i})t_i)(mod)w$. The key pair of the vehicle v_i is (PK_{v_i}, SK_{v_i}).

- Step 2. In the same way, RTA registers the edge nodes as follows:

1. A edge node Ed_i with its identity ID_{Ed_i} selects a secret $m_i \in \mathbb{Z}_q^*$ and computes $V_{Ed_i} = m_1 P$ and send (ID_{Ed_i}, V_{Ed_i}) to RTA.
2. After receiving $(5GID_{Ed_i}, V_{Ed_i})$, RTA selects $f_i \in \mathbb{Z}_q^*$ and computes $a_i = H_1(ID_{Ed_i}, V_{Ed_i}, f_i) + s(mod)w$ and $F_i = H_1(ID_{Ed_i}, V_{Ed_i}, f_i)P$. Then RTA returns (F_i, a_i) to v_i
3. Ed_i receives (F_i, a_i) and check if the equation $a_i P = F_i + P_{RTA}$ is correct. If yes, Ed_i generates its public key $PK_{Ed_i} = F_i + H_2(ID_{Ed_i}, V_{Ed_i})V_{Ed_i}$.
4. Ed_i generates its private key $SK_{Ed_i} = H_2(ID_{Ed_i}, PK_{Ed_i})(a_i + H_2(ID_{Ed_i}, V_{Ed_i})m_i)(mod)w$. The key pair of the edge node Ed_i is (PK_{Ed_i}, SK_{Ed_i}).

4.3. Emergency Driven Message Reporting

Whenever an emergency event such as land sliding occurs, v_i performs the following:

- Composes EMD file as $M = \{Dt, Ts, loc, file\}$ representing the date, the time, the location and main file which has been captured. $file$ could be a multimedia item such as pictures or audio files.
- v_i generates a list of edge nodes that can recover the message, and in this case we adopt proximity protocol based on the location as described in [31]. v_i generates $L = \{ID_{Ed_1}, ID_{Ed_2}, ID_{Ed_3}, \cdots, ID_{Ed_n}\}$ and make the signcryption on the event message as follows

1. Computes $Q_i = PK_{Ed_i} + P_{RTA}$ with $i = 1, 2, 3, \cdots, n$
2. Selects a integer $x \in \mathbb{Z}_q^*$ and computes $X = xP$ and $C_i = xH_2(PID_{Ed_i}, PK_{Ed_i})Q_i$ and $\alpha_i = H_3(C_i, X)$ where $i = 1, 2, ..n$
3. Selects an integer $\xi \in \mathbb{Z}_q^*$ and computes the polynomial $f(v) = \prod_{i=1}^{n}(v - \alpha_i) + \xi(mod)w$, which equals to $a_0 + a_i v + .. + a_{n-1}v^{n-1}$ for $a_1 \in \mathbb{Z}_q^*$
4. Computes $k = H_4(\xi)$, $J = Enc_k(m||P5GID_{v_i})$ and $h = H_5(m||P5GID_{v_i}, \xi, a_0, a_1, .., a_{n-1}, X)$
5. Generates h^{-1} that satisfy $hh^{-1} \equiv 1(mod)w$ and computes $z = h^{-1}(SK_{v_i} + x)(mod)w$
6. Generates the cipher text $CT = <J, X, z, h, a_0, a_1, \cdots, a_{n-1}>$ and send it to edge nodes.

4.4. Emergency-Driven Message Collection

Upon receiving the cipher text CT, an edge node Ed_i does the following to recover the emergency warning

- Compute $C_i = SK_{Ed_i}X$ and $\alpha = H_3(C_i, X)$
- Then computes $f(v) = a_0 + a_i v + .. + a_{n-1}v^{n-1} + v^n$ and $\xi = f(\alpha_i)$
- Computes $k = H_4(\xi)$ and retrieve the message trough the decryption $Dec_k(J) = m||P5GID_{v_i}$
- Also compute $h' = H_5(P5GID_{v_i}, \xi, a_0, \cdots, a_{n-1}, X)$ and verifies if the equation $h = h'$ is correct. Otherwise, the emergency message is rejected
- Upon receiving the vehicle public PK_{v_i}, Ed_i checks if the equation $hzP = H_2(P5GID_{v_i}, PK_{v_i})(PK_{v_i} + P_{RTA})$. The correctness is as follows:

$$= hh^{-1}(SK_{v_i} + x)P$$
$$= SK_{v_i}P + W$$
$$= H_2(P5GID_{v_i}, PK_{v_i})(y_i + H_2(P5GID_{v_i}, V_{v_i})t_i)P + X$$
$$= H_2(P5GID_{v_i}, PK_{v_i})(D_i + H_2(P5GID_{v_i}, V_{v_i})V_{v_i} + P_{RTA}) + X$$
$$= H_2(P5GID_{v_i}, PK_{v_i})(PK_{v_i} + P_{RTA}).$$

If yes, Ed_i keeps the EDM message.

4.5. Private Blockchain Generation

We do assume that every edge node Ed_i is a stakeholder having its proper stake that could be the number of valid EDM that Ed_i has received. A private blockchain *Chain* is constructed as follows:

1. Assume that the time is divided into time slot $\{ts_1, ts_2, \cdots, ts_1\}$ in which a block is attached to the ledger for each time sequence.

2. The initial block also called genesis block is generated as the first state distribution and it contains the edge nodes identities, their public keys, their stakes as $B_{gen} = <\{ID_{Edi=1}{}^R\}, \{PK_{Ed}\}_{i=1}^R$ and $\{ST_{Ed}\}_{i=1}^R >$. We assume that first blockheader B_{gen} to be empty.

3. Therefore, in a given area, each edge node Ed_i set $C = B_0$ where B_0 is the genesis block

4. An edge node Ed_i collects n EDM and verifies each EDM as shown in Section 4.4 by running $Designcrypt(params, \delta, SK_i)$. To choose a leader edge node L_{Ed_i}, the probability p_i for being chosen should be relative to its stake that are in previous block.

5. Ed_i runs a leader selection protocol $F(.)$ [32] that input $< \{ID_{Edi=1}{}^R\}, \{PK_{Ed}\}_{i=1}^R, p_{Ed_i}, st_i >$ representing respectively the edge nodes identities, their public key, the probability of the leader and the corresponding time slot with $p_{Ed_i} = st_i / \sum_{j=1}^{Ed_i} st_{Ed_i}$.

6. $F(.)$ outputs a leader edge node $L_{Ed_i} \in \{Ed_i, Ed_2, Ed_3, \cdots, Ed_n\}$

7. To generate a block, the selected edge node $L_E d_i$ output a block B_{ts_i} that corresponds to the time slot ts_i with $B_{ts_i} = \{Num_{ts_i}, H_{ts_i}, MHR_{ts_i}, t_{ts_i}\}$ representing respectively the number of the block, a hash corresponding to previous blockheader, merkle hash root corresponding to a merkle tree built using n EDM.

8. Ed_i performs the update of its stake st_{ts_i} and generates a signcryption on the entire message.

9. Finally add the block to the chain and send a notification to the entire network

5. Performance

This section is made by the security analysis of the proposed protocol, the experiment on private blockchain, the computational and communication cost along with the simulation.

5.1. Security Analysis

We provide in this section the analysis in regards to the security goals for the proposed scheme.

5.1.1. Privacy Preservation

The communication within the proposed protocol is entirely based on anonymous interactions. While the vehicles engaged in reporting EDM are sending their messages, they make use of pseudonyms. RTA is the one and only entity that can map the real identity of a vehicle participating in the EDM reporting to its pseudonyms. As described in Section 4, when v_i requests partial key pair by running the function $Partialkeypair(params, V_{ID}, ID)$, it sends its real identity to RTA which generates a pseudonymous $P5GID_{v_i}$. Therefore it is infeasible for any entity inside or outside the network to know the real identity of the EDM participant except the RTA. This would require the adversary to access the database that maps the vehicles real identities and their pseudonyms.

5.1.2. Authentication

In the proposed protocol, bad actors or malicious vehicles or any entity can not successfully engage in forging an EDM report because the authentication between a vehicle v_i and an edge node Ed_i is achieved through the signcryption function $Signcrypt(L, m, SK_i, params)$ that is made on each message. Once an EDM is generated, v_i makes signcryption of the message $M = \{Dt, Ts, loc, file\}$ by running $Signcrypt(L, m, SK_i, params)$. Any entity needs to possess a valid private key SK_i to be able to verify the correctness of the equation $hzP = H_2(P5GID_{v_i}, PK_{v_i})(PK_{v_i} + P_{RTA})$. It is hard for an adversary to have the full private key of an edge node because it is partial generated both by the RTA and the edge node through the functions $Partialkeypair(params, V_{ID}, ID)$ and $Keypair(D_{ID}, y_{ID}, params)$. The signcryption technique that was used to build the proposed protocol achieves unforgeability through the strong existential unforgeability against chosen plain text, selvi2008efficient. Therefore, we confirm that the proposed protocol achieves authentication property.

5.1.3. Confidentiality and Integrity

The proposed protocol achieves the confidentiality and integrity of the EDM messages that are sent by the vehicles to edge nodes through the two in one technique that both provide encryption and digital signature in a single step. As shown in Section 4, the cipher $CT =< J, X, z, h, a_0, a_1, \cdots, a_{n-1} >$ accomplishes the duties of message encryption and signature. A malicious user can not tamper with the integrity of the EDM because the signcryption phase transform the data into hash values $C_i = xH_2(PID_{Ed_i}, PK_{Ed_i})Q_i$ and $\alpha_i = H_3(C_i, X)$ as described in Section 4. Thus, we guarantee that the proposed scheme achieves data integrity and confidentiality because the underlying technique is fully proved to satisfy security under adaptively chosen ciphertext [33].

5.1.4. Key Escrow Resilience

The 4th industrial revolution projects a massive connectivity of devices to offer diversified services such as EDM reporting using the inbuilt vehicle sensors. In additional, 5G cellular networks were adopted in this work to provide effective latency. Security wise, key escrow resilience property needs to be achieved for applications within a massive connectivity environment. In the proposed scheme, the entities first generate a secret value by running $SecretValue(ID, params)$ and RTA will then provide partialkey pair to the entities by computing $Partialkeypair(params, V_{ID}, ID)$ function. The entities in our system compute their key pairs through the function $Keypair(D_{ID}, y_{ID}, params)$. Therefore, the proposed protocol achieves key escrow resilience.

5.1.5. Access Control

In the current era with millions of devices connection, a single point failure should be avoided as much as possible. While the EDM are categorized to be safety related messages that contain sensitive data, multi receiver property (or access control) is a key point that need to be considered. In the proposed protocol, a vehicle v_i selects a number of valid edge nodes, in this case, even in a scenario where a number of edge nodes have been compromised, the probability that the EDMs at least get to one receiver is higher. Therefore v_i generates $L = \{ID_{Ed_1}, ID_{Ed_2}, ID_{Ed_3}, \cdots, ID_{Ed_n}\}$ and run $Signcrypt(L, m, SK_i, params)$ to signcrypt the messages. Only valid receiver within the L can recover the EDM. Therefore, the proposed scheme achieves fine-grained access control by using attribute based encryption.

5.1.6. Traceability and Non Repudiation

In the proposed system, when a valid user sends a fake EDM (probably for criminal profit), the edge node will discard the message because the signcryption correctness $hzP = H_2(P5GID_{v_i}, PK_{v_i})(PK_{v_i} + P_{RTA})$ will not hold. However, Ed_i will keep a log of pseudo identity of the vehicles. Thus, the vehicle can not deny its own pseudo identity. Additionally, in case of legal disputes, RTA consults its database that maps the identities and their pseudo identities to reveal the real identity of the vehicle. Therefore, we do confirm that the proposed protocol can achieve traceability and non repudiation of misbehaving entities.

5.1.7. Auditability

The proposed scheme does achieve data auditability by building a private blockchain between the edges nodes. The achievement of this property can be summarized in three steps:

- The blockchain that is built in this scheme is private, any participant requires a permission or an invitation to join the private chain. In this case, it is infeasible for an malicious user to add bogus block to the chain.
- Each participant in the private chain keeps a replica of any appended ledger of emergency warning messages. This is crucial in case a crash occurs in any of the remote servers where the EDM are kept.

- Transactions immutability: It is hard for a malicious entity to tamper the EDM that is exchanged between the vehicles and the edges. In case of a legal dispute that require the thorough auditability of the EDM, the transaction immutability of blockchain can strengthen such services.

5.1.8. Secure against Known Attacks

We describe in this section few well known attacks within the vehicular networks and how our proposed scheme can overcome them.

- Impersonation attack: as mentioned earlier, the malicious vehicles cannot succeed to impersonate a legitimate vehicle because the authentication between a vehicle v_i and an edge node Ed_i is achieved through the signcryption function $Signcrypt(L, m, SK_i, params)$ that is made on each message. Once an EDM is generated, v_i makes signcryption of the message $M = \{Dt, Ts, loc, file\}$ by running $Signcrypt(L, m, SK_i, params)$. Every participating vehicle needs to possess a valid private key SK_i to be able to verify the correctness of the equation $hzP = H_2(P5GID_{v_i}, PK_{v_i})(PK_{v_i} + P_{RTA})$. Based on the hardness of the DL problem, the signature provided on the message cannot match the verification and the message will be discarded. Thus, it is almost impossible to perform an impersonation attack in our proposed scheme
- Masquerade attack: suppose a malicious user eavesdrops an EDM message and tries to know the EMD contents. That malicious user can not tamper with the integrity of the EDM because the signcryption phase transforms the data into hash values $C_i = xH_2(PID_{Ed_i}, PK_{Ed_i})Q_i$ and $\alpha_i = H_3(C_i, X)$ as described in Section 4. Therefore, the malicious user cannot learn any useful information from the eavesdropped message nor reveal the identity of the message owner.
- DDoS attack: our scheme is able to resist against DDoS attacks either launched by legitimate or illegitimate vehicles. Assume an illegitimate vehicle tries to send multiple EDM to a given edge node, as demonstrated in the impersonation attack, those EDM will be discarded by the edge node because the message verification will not hold. In addition, assume a legitimate vehicle is generating excessive EDM to cause a DDoS attack, in that scenario, the edge nodes will use the time stamp on any EDM given message to predict the frequency of message compared to other users because every EDM message contains a time stamp as shown in message content as $M = \{Dt, Ts, loc, file\}$ representing respectively the date, the time, the location and $file$ which could be a multimedia item such as pictures or audio files. Therefore, the messages from the suspicious user can be discarded.

5.2. Computational Cost

In this section we provide the analysis of the proposed protocol in terms of generating the EDM by the vehicle and the recovery of the message by the dedicated node. We performed the benchmark using a desktop of Core i7 3.5-GHz,16GB RAM with a crypto ++ library [34] with 6 as the embedded degree and $\mathbb{G}$ and q equivalent respectively to 161 bits and 160 bits. We mainly focused on the following main operations; point scalar multiplication, modular exponentiation and bilinear pairing. These operations dominate the process of sending and receiving the emergency messages. Table 1 shows the cost of the main cryptographic operations. A vehicle v_i after generating the EDM, it performs $(T+1)T_m + nT_p$ to signcrypt a emergency message while the designcrypt operation requires $2T_m + T_p$ as shown in Table 2. As mentioned, there are several articles that addressed security solutions for BSM messages but few have addressed the EDM. BSM content being a predefined with limited content, the size of the BSM is supposed to be small and constant. As shown in this recent survey on secure protocol for vehicular communications [19], we compared the proposed protocol with the protocol in [20] as shown in Table 3.

We further considered two main elements than can effect the complexity of the whole scheme. First we investigated the number of attributes that can be associated with a given policy. Assume a user v_i wants to share his emergency files with five governmental agencies. For instance, the emergency files contains few pictures of a land sliding scene. Those pictures can both be used by the ambulance team, the evacuation, the police and any other. Therefore the access policy might contain a number of attributes. In our simulation scenario we considered a range of attributes varying from 0 or 50, $range = [0 - 50]$. We then investigate the time needed for signcryption and designcryption based on the number of attributes in a given access policy. It is obvious that the obtained results are increasing gradually based on the number of attributes. For an average number of 30 attributes, the designcryption time was 17 s as shown in Figure 2a. Though the results are not very competitive, they are still feasible especially for edge nodes that have considerable computing capabilities. Also, we investigated the time needed for signcryption and designcryption when the number of files is fixed. The cost of encryption for one of two files was constant since we assume that the files (assume three separate images taken from different angles) are encrypted using a similar access policy. On the other hand, the decryption phase took much more time due to reconstruction of the secret value using Lagrange algorithm. As shown in Figure 2b, for a maximum folder of 10 files, we have a decryption cost of around 32 s. Since the decrypting devices could be servers or computing gadgets with sufficient communication power, the obtained decryption is acceptable for non real-time scenarios such as EDM reporting.

Table 1. Measurement of cryptographic operations.

$Notation$	Operations	Time (ms)
T_b	Bilinear pairing	4.5
T_m	Point scalar multiplication	0.6
T_p	Point adddition on ECC	0.047
T_e	Exponentiation	3.9
T_{as-dec}	Asymmetric decryption	0.61
T_{s-enc}	Symmetric encryption	0.51
T_{s-dec}	Symmetric decryption	0.55
T_h	Execution time of a general hash function	0.0001

Table 2. Computational cost of signcrypt and designcript (n is number of receiver).

Phase	Operation
Signcryp an EDM	$(T + 1)T_m + nT_p$
Designcryp an EDM	$2T_m + T_p$

Table 3. Comparison Performance of Proposed Work and literature.

Scheme	lightweight	Traceability	Tamper Proof	Privacy	Decentralization	IoT Friendly
Liu et al., [20]	Low	YES	Low	Yes	NO	NO
Proposed Framework	High	YES	HIGH	YES	HIGH	HIGH

There are a considerable number of schemes within [19], however, these protocols are built based on expensive operations that compromise their efficient even if these protocols address BSM scenarios. For EDM, the size can be very important with the multimedia contents that can be added, thus a lightweight protocol could be more efficient. In additional, the protocols in the survey are not decentralized to offer immutability of transactions, however we achieve this property in our scheme by using private blockchain.

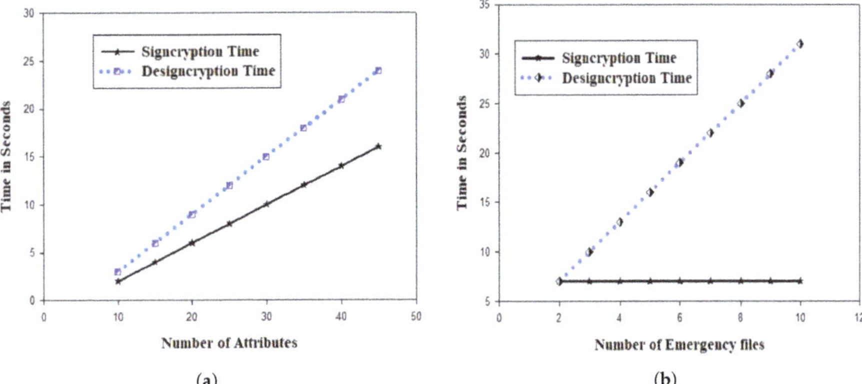

Figure 2. Signcryption/Designcryption time versus Number of Attributes (**a**) and Signcryption/Designcryption time versus Number Emergency Files (**b**).

5.3. Communication Cost

In this section, we provide the communication cost of the proposed protocol. We first computed the overhead caused by the additional cryptographic primitives that were added on the raw message. We did not consider the element in the EDM since this can vary in real life application based on the multimedia content within an EDM. As mentioned in [35], in pairing operations, the size of elements equals to $64 \times 2 = 128$ bytes while for the ECC based operations, the size of the elements are equal to $20 \times 2 = 40$ bytes. As seen in the construction of the proposed, our scheme is not built based on pairing operations, and as described in Table 2, the size caused by security primitives are 80 bytes for the proposed protocol.

5.4. Private Blockchain Evaluation

We perform the experiments for the private blockchain that was built based on the edges nodes. Our experiment considers seven settings. As shown in Table 4, we considered a scenario that can generate 5 to 35 blocks. In the three first settings, we assumed 10 edge nodes while we considered 15 edges nodes in the four last settings. We computed the average time within our seven steps including the time required from system setup to the generation of the block. As seen in Table 4, the additional cost caused by the generation of block is not very heavy for a 5G cellular network, in the same time this technique offers immutable transactions with EDM auditability even if one or several edge nodes crash. We can see from the table that an edge node can create a new block to the added on the private blockchain with a cost of 0.056 s during the seventh step.

Table 4. Analysis of Edge node made private blockchain (/second).

Phase	No Trans/Block	No of ED	Initialization	Request	Response	Matching	Updating
1	5	10	0.022	0.44	0.15	1.89	0.0022
2	10	10	0.022	0.61	0.26	2.56	0.0089
3	15	10	0.022	0.98	0.63	2.29	0.014
4	20	15	0.045	1.44	0.89	3.51	0.031
5	25	15	0.045	1.79	1.25	4.01	0.056
6	30	15	0.045	2.14	1.67	4.98	0.17
7	35	15	0.045	4.12	3.90	6.67	0.56

5.5. Simulation

In this section we provide the simulation that focus on the network performance of the proposed protocol. To achieve this, we made use of VANETSIM 2.02 that offers simulation for vehicle mobility and NS-3 was considered as a tool for network simulation. In our simulation, we considered a 5G functional network that can achieve a connection speed of 1.2Gb/s as reported and confirmed in several reports [36]. We focused on analyzing the performance of well known block ciphers techniques that vehicles can choose as symmetric encryption k as shown in Section 4. These algorithms were TWOFISH/CTR with 256 bit key and speed of 147 MB/s, then SERPENT/CTR that has 256 bit key with a 65 MB/s speed and lastly the famous AES/CBC of 256 bit key for a 455 MB/s as speed.

The rest of the parameters that were considered in our simulation are described in Table 5. Our simulation mainly focus on the size of the EDM because till now we cannot tell what would be the real size of EDM, therefore using a 5G benchmarked connection, we investigated the performance of signcryption of EDM based on different sizes. The size of an EDM message varies between 1 to 6 Gigabytes. Figure 3b shows that the time needed by a vehicle to signcrypt an EDM, ranges between 20 to 40 s for an EDM that has a size of 2 GB. In addition, we investigated the overall time to signcrypt and designcrypt an EMD as shown in Figure 3a, we still found that as long as an EDM does not go beyond 2 GB of size, the overall time is not that much when we consider the 5G projected features. In this case, the highest record which corresponds to Serpent/CTR algorithm is around 100 s.

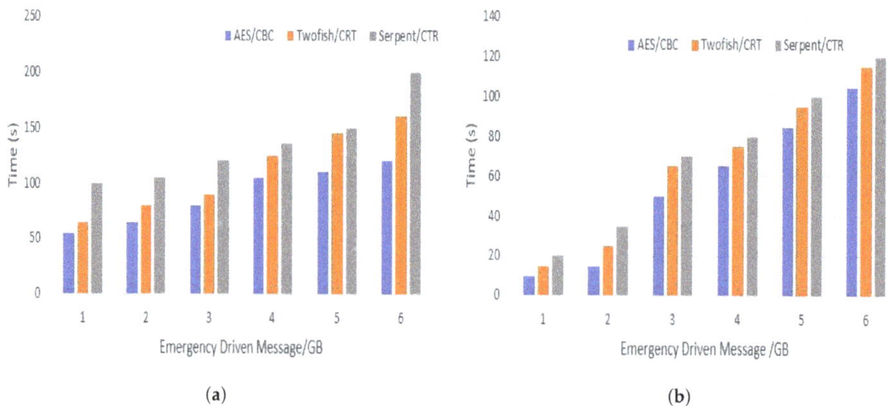

Figure 3. Overall time overhead (**a**) and signcryption time overhead (**b**).

Table 5. Setting of simulation parameters.

Tools/Parameter	Value/Specification
Mobility generation tool	VANETSIM 2.02
Network Simulation tool	ns-3
Data Rate	1.2 GBps
Number-of-vehicle	200
Number-of-edge nodes	40
Distance between two edge nodes	150 m
Simulation time	100 min
Wireless protocol	802.11a
Departure interval	180 s
RSU/Edge radius	800 m
mobility model	shortest path
Range of EDM size	(1–6 GB)

6. Conclusions

In this paper, we presented a secure and blockchain based EDM protocol for 5G-enabled vehicular edge computing. To provide scalability and latency for the proposed scheme, we adopted a 5G cellular architecture due to its projected features compared to 4G long-term evaluation (LTE) for vehicular communications. We considered an edge computing architecture to provide local processing of EDM in order to improve the response time. We made use of lightweight multi-receiver signcryption scheme without pairing that offers lightweight consuming operations, security, privacy and access control. To keep EDM records into a distributed system for reliability and auditability, we constructed a private blockchain using the edge nodes. The performance analysis of the proposed protocol in terms of security analysis, communication, computational and simulation confirms the efficiency of the protocol.

Author Contributions: Conceptualization, L.N.; methodology, L.N.; validation, B.A.T.; writing—original draft preparation, L.N.; writing—review and editing, M.K.S.; supervision, Y.-H.C.; funding acquisition, Y.-H.C. All authors have read and agreed to the published version of the manuscript.

Funding: This work was supported by BK21PLUS, Creative Human Resource Development Program for IT Convergence, by basic science research program through national research foundation korea (NRF) funded by the ministry of science, ICT and future planning (NRF-2018R1D1A3B07043392).

Acknowledgments: This work was supported by BK21PLUS, Creative Human Resource Development Program for IT Convergence, by basic science research program through national research foundation korea (NRF) funded by the ministry of science, ICT and future planning (NRF-2018R1D1A3B07043392).

Conflicts of Interest: Authors declare that there is no conflict of interest.

References

1. Sanguesa, J.A.; Fogue, M.; Garrido, P.; Martinez, F.J.; Cano, J.C.; Calafate, C.T. A survey and comparative study of broadcast warning message dissemination schemes for VANETs. *Mob. Inf. Syst.* **2016**, *2016*, 8714142. [CrossRef]
2. Karagiannis, G.; Altintas, O.; Ekici, E.; Heijenk, G.; Jarupan, B.; Lin, K.; Weil, T. Vehicular networking: A survey and tutorial on requirements, architectures, challenges, standards and solutions. *IEEE Commun. Surv. Tutorials* **2011**, *13*, 584–616. [CrossRef]
3. Kamouch, A.; Chaoub, A.; Guennoun, Z. Mobile big data in vehicular networks: The road to internet of vehicles. In *Mobile Big Data*; Springer: Cham, Switzerland, 2018; pp. 129–143.
4. Nkenyereye, L.; Park, Y.; Rhee, K.H. Secure vehicle traffic data dissemination and analysis protocol in vehicular cloud computing. *J. Supercomput.* **2018**, *74*, 1024–1044. [CrossRef]
5. Lavanya, R. Fog Computing and Its Role in the Internet of Things. In *Advancing Consumer-Centric Fog Computing Architectures*; IGI Global: Hershey, PA, USA, 2019; pp. 63–71.
6. Nkenyereye, L.; Liu, C.H.; Song, J. Towards secure and privacy preserving collision avoidance system in 5G fog based Internet of Vehicles. *Future Gener. Comput. Syst.* **2019**, *95*, 488–499. [CrossRef]
7. Yi, S.; Li, C.; Li, Q. A survey of fog computing: Concepts, applications and issues. In Proceedings of the 2015 Workshop on Mobile Big Data, Hangzhou, China, 21 June 2015; pp. 37–42.
8. Mir, Z.H.; Filali, F. LTE and IEEE 802.11 p for vehicular networking: A performance evaluation. *EURASIP J. Wirel. Commun. Netw.* **2014**, *2014*, 89.
9. Vinel, A. 3GPP LTE versus IEEE 802.11 p/WAVE: Which technology is able to support cooperative vehicular safety applications? *IEEE Wirel. Commun. Lett.* **2012**, *1*, 125–128. [CrossRef]
10. Bellalta, B.; Belyaev, E.; Jonsson, M.; Vinel, A. Performance evaluation of IEEE 802.11 p-enabled vehicular video surveillance system. *IEEE Commun. Lett.* **2014**, *18*, 708–711. [CrossRef]
11. Shen, X. Device-to-device communication in 5G cellular networks. *IEEE Netw.* **2015**, *29*, 2–3. [CrossRef]
12. Tehrani, M.N.; Uysal, M.; Yanikomeroglu, H. Device-to-device communication in 5G cellular networks: Challenges, solutions, and future directions. *IEEE Commun. Mag.* **2014**, *52*, 86–92. [CrossRef]
13. Schneider, P.; Horn, G. Towards 5G security. In Proceedings of the 2015 IEEE Trustcom/BigDataSE/ISPA, Helsinki, Finland, 20–22 August 2015; Volume 1, pp. 1165–1170.
14. Park, J.; Kim, J.; Lee, B. Are uber really to blame for sexual assault?: Evidence from New York city. In Proceedings of the 18th Annual International Conference On Electronic Commerce: e-Commerce in Smart Connected World, Suwon, Korea, 17–19 August 2016; p. 12.

15. Darus, M.Y.; Bakar, K.A. Review of Congestion Control Algorithm for Event-Driven Safety Messages in Vehicular Networks. *Int. J. Comput. Sci. Issues* **2011**, *8*, 49.
16. Djahel, S.; Ghamri-Doudane, Y. A robust congestion control scheme for fast and reliable dissemination of safety messages in VANETs. In Proceedings of the 2012 IEEE Wireless Communications and Networking Conference (WCNC), Shanghai, China, 1–4 April 2012; pp. 2264–2269.
17. Zhang, W.; Festag, A.; Baldessari, R.; Le, L. Congestion control for safety messages in VANETs: Concepts and framework. In Proceedings of the 2008 8th International Conference on ITS Telecommunications, Phuket, Thailand, 24 October 2008; pp. 199–203.
18. Ma, X.; Kanelopoulos, G.; Trivedi, K.S. Application-level scheme to enhance VANET event-driven multi-hop safety-related services. In Proceedings of the 2017 international conference on computing, networking and communications (ICNC), Santa Clara, CA, USA, 26–29 January 2017; pp. 860–864.
19. Ali, I.; Hassan, A.; Li, F. Authentication and privacy schemes for vehicular ad hoc networks (VANETs): A survey. *Veh. Commun.* **2019**, *16*, 45–61. [CrossRef]
20. Liu, Y.; Wang, L.; Chen, H.H. Message authentication using proxy vehicles in vehicular ad hoc networks. *IEEE Trans. Veh. Technol.* **2015**, *64*, 3697–3710. [CrossRef]
21. Kosba, A.; Miller, A.; Shi, E.; Wen, Z.; Papamanthou, C. Hawk: The blockchain model of cryptography and privacy-preserving smart contracts. In Proceedings of the 2016 IEEE Symposium on Security and Privacy (SP), San Jose, CA, USA, 22–26 May 2016; pp. 839–858.
22. Gao, F.; Zhu, L.; Shen, M.; Sharif, K.; Wan, Z.; Ren, K. A blockchain-based privacy-preserving payment mechanism for vehicle-to-grid networks. *IEEE Netw.* **2018**, *32*, 184–192. [CrossRef]
23. Ullah, A.; Yaqoob, S.; Imran, M.; Ning, H. Emergency message dissemination schemes based on congestion avoidance in VANET and vehicular FoG computing. *IEEE Access* **2019**, *7*, 1570–1585. [CrossRef]
24. Zhang, K.; Mao, Y.; Leng, S.; He, Y.; Zhang, Y. Mobile-edge computing for vehicular networks: A promising network paradigm with predictive off-loading. *IEEE Veh. Technol. Mag.* **2017**, *12*, 36–44. [CrossRef]
25. Feng, J.; Liu, Z.; Wu, C.; Ji, Y. AVE: Autonomous vehicular edge computing framework with ACO-based scheduling. *IEEE Trans. Veh. Technol.* **2017**, *66*, 10660–10675. [CrossRef]
26. Wang, C.X.; Haider, F.; Gao, X.; You, X.H.; Yang, Y.; Yuan, D.; Aggoune, H.M.; Haas, H.; Fletcher, S.; Hepsaydir, E. Cellular architecture and key technologies for 5G wireless communication networks. *IEEE Commun. Mag.* **2014**, *52*, 122–130. [CrossRef]
27. Ge, X.; Cheng, H.; Mao, G.; Yang, Y.; Tu, S. Vehicular communications for 5G cooperative small-cell networks. *IEEE Trans. Veh. Technol.* **2016**, *65*, 7882–7894. [CrossRef]
28. Ge, X.; Li, Z.; Li, S. 5G software defined vehicular networks. *IEEE Commun. Mag.* **2017**, *55*, 87–93. [CrossRef]
29. Kiayias, A.; Konstantinou, I.; Russell, A.; David, B.; Oliynykov, R. A Provably Secure Proof-of-Stake Blockchain Protocol. *IACR Cryptol. EPrint Arch.* **2016**, *2016*, 889.
30. Pang, L.; Kou, M.; Wei, M.; Li, H. Efficient Anonymous Certificateless Multi-Receiver Signcryption Scheme Without Bilinear Pairings. *IEEE Access* **2018**, *6*, 78123–78135. [CrossRef]
31. Zheng, Y.; Li, M.; Lou, W.; Hou, Y.T. Location based handshake and private proximity test with location tags. *IEEE Trans. Dependable Secur. Comput.* **2017**, *14*, 406–419. [CrossRef]
32. Kiayias, A.; Russell, A.; David, B.; Oliynykov, R. Ouroboros: A provably secure proof-of-stake blockchain protocol. In *Annual International Cryptology Conference*; Springer: Cham, Switzerland, 2017; pp. 357–388.
33. Selvi, S.S.D.; Vivek, S.S.; Shukla, D.; Chandrasekaran, P.R. Efficient and provably secure certificateless multi-receiver signcryption. In *International Conference on Provable Security*; Springer: Berlin/Heidelberg, Germany, 2008; pp. 52–67.
34. Wei, D. Crypto++ Library 5.6.5, a Free C++ Class Library of Cryptographic Schemes. 2019. Available online: http://www.cryptopp.com (accessed on 29 August 2019).
35. He, D.; Zeadally, S.; Xu, B.; Huang, X. An efficient identity-based conditional privacy-preserving authentication scheme for vehicular ad hoc networks. *IEEE Trans. Inf. Forensics Secur.* **2015**, *10*, 2681–2691. [CrossRef]
36. BBCNews. 5G Researchers Manage Record Connection Speed, 2015. 2019. Available online: http://www.bbc.co.uk/news/technology-31622297 (accessed on 29 August 2019).

Article

Fast, Resource-Saving, and Anti-Collaborative Attack Trust Computing Scheme Based on Cross-Validation for Clustered Wireless Sensor Networks

Chuanyi Liu [1,2] and Xiaoyong Li [3,*]

1   Harbin Institute of Technology (Shenzhen), School of Computer, Shenzhen 518055, China;
     liuchuanyi@hit.edu.cn
2   Cyberspace Security Research Center, Peng Cheng Laboratory, Shenzhen 518000, China
3   The Key Laboratory of Trustworthy Distributed Computing and Service, Ministry of Education,
     Beijing University of Posts and Telecommunications, Beijing 100876, China
*   Correspondence: lixiaoyong@bupt.edu.cn

Received: 9 February 2020; Accepted: 9 March 2020; Published: 12 March 2020

Abstract: The trust computing mechanism has an increasing role in the cooperative work of wireless sensor networks. However, the computing speed, resource overhead, and anti-collaborative attack ability of a trust mechanism itself are three key challenging issues for any open and resource-constrained wireless sensor networks. In this study, we propose a fast, resource-saving, and anti-collaborative attack trust computing scheme (FRAT) based on across-validation mechanism for clustered wireless sensor networks. First, according to the inherent relationship among three network entities (which are made up of three types of network nodes, namely base stations, cluster heads, and cluster members), we propose the cross-validation mechanism, which is effective and reliable against collaborative attacks caused by malicious nodes. Then, we adopt a fast and resource-saving trust computing scheme for cooperation between between cluster heads or cluster members. This scheme is suitable for wireless sensor networks because it facilitates resource-saving. Through theoretical analysis and experiments, the feasibility and effectiveness of the trust computing scheme proposed in this study are verified.

Keywords: cross-validation; anti-collaborative attack; resource-saving; trust computing; wireless sensor networks

---

1. Introduction

Wireless sensor networks (WSNs [1–5]) are widely used in several fields such as intelligent perception, military, disaster warning, medical care, etc. The main application of WSN is to sense the surrounding environment and send the obtained information to the base station (BS) for subsequent processing. For clustered WSNs such as EEHC [6], EC [7], HEED [8], TRAST [9], and LDTS [10], clustering algorithms can significantly improve the performance and efficiency of wireless sensor networks [11]. The clustering algorithm is used to divide nodes into multiple clusters. In each cluster, a node with powerful computing capability is selected as the cluster head (CH). Multiple CHs together form a higher level information transmission network. This layered network structure helps increase the speed of data collection and can limit network operations that consume large amounts of bandwidth [10,12]. Many applications in WSNs require coordination through wireless communications between participating nodes for interactive operations such as task collaborations and data transmissions [10,13–15].

However, the inherent security issues of WSNs also arise in the cooperation between participating nodes. WSNs are usually highly accessible in existing applications, which makes them very vulnerable to malicious attacks. Therefore, it becomes very important to provide a secure and trusted collaboration mechanism for WSNs. The trust mechanism and network entity behavior are an important factors that WSNs must consider [10,16–20]. The trust mechanism can be used to detect the reliability and security of the cooperative nodes (or to identify the faulty nodes) or to assist in the decision-making process, such as whether a node needs to choose a partner to complete the data transmission task [21–26].

1.1. Challenges and Motivations of This Work

The trust mechanism is an important element in any network computing environment [27–33]. There are many advantages to introducing a trust mechanism in clustered WSNs [16–20] and selecting a CH to detect failed or malicious nodes in the cluster [34]. In a multi-hop cluster environment [8], the trust mechanism supports the selection of a trusted routing node (usually a CH), and a cluster member (CM) can send the collected data to the CH. In communication between clusters, the trust mechanism also supports the selection of trustworthy routing gateway nodes or other trustworthy CHs through which the sender forwards data to the BS [10]. The BS is a powerful device that can process the information collected from the CM and interact with the user.

However, due to its high resource consumption (such as memory, time, and communication overhead), this makes traditional trust computing solutions developed for wired and wireless ad hoc networks unsuitable for sensor networks. The computing speed and resource-saving problem of a trust system are the most key requirements for resource-constrained WSNs. At the same time, the anti-collaborative attack ability of a trust computing mechanism itself is another challenging issue for any WSN (including clustered WSNs). Currently, there is a lack of a universal trust computing solution designed for clustered WSNs that can simultaneously achieve computing speed, resource efficiency, and resistance to collaborative attacks.

- *Most studies do not consider both computational speed and resource overhead issues of the trust computing scheme itself.* The trust mechanism should be fast and save resources to serve a large number of resource-constrained nodes in terms of accuracy, calculation speed, storage overhead, and communication overhead [17,18,35]. Currently, many representative works have been proposed for clustering WSNs, such as the group-based trust computing mechanism (GTMS) [18], the belief-based trust evaluation mechanism (BTEM) [34], the trust and reputation scheme (ATRM) [36], the trust-based cluster head election mechanism (TCHEM) [37]. However, most of these studies do not simultaneously consider the computational speed and resource overhead issues of the trust computing scheme itself. Most of these studies use complex trust calculation algorithms at each CM or CH, which will greatly affect the applicability of the trust model.
- *Most studies do not consider the anti-collaborative attack ability of the trust computing scheme itself.* The malicious nodes may cooperate to provide false feedback information to attack the trust computing system. The anti-collaborative attack ability of the trust computing scheme should have the ability to identify cooperative attacks. WSNs are usually deployed in insecure and highly complex network environments, and nodes may be attacked and cooperatively spoofed. Furthermore, an attacker can disrupt communication or spread misleading sensor values through sensor nodes that have been compromised. In the traditional WSN trust mechanism, the trust system collects remote feedback and then aggregates such feedback to generate a global trust for the node, which can be used to evaluate the overall trust of the node (such as TCHEM [37] and HTMP [34]). There are a large number of malicious nodes in an open or hostile WSN environment. The ratings of these malicious nodes may produce erroneous results. Organized cooperative attacks are the most important threat to trust mechanisms in WSNs [10,12,18]. However, most previous studies lacked consideration of malicious cooperative attacks, which severely affected the security, availability, and reliability of the system.

By identifying the inherent relationship among CMs, CHs, and BSs, we propose a fast, resource-saving, and anti-collaborative attack trust computing scheme based on the cross-validation mechanism. This scheme can effectively eliminate collaborative attacks initiated by large collaborative groups and large-scale malicious nodes. Different from the previous trust computing methods, in the proposed trust mechanism, feedback comes not only from CMs, but also from CHs and BSs. This cross-validation mechanism can effectively reduce malicious feedback and improve system security.

1.2. Main Idea and Contributions

To the best of our knowledge, this study is the first to construct a fast, resource-saving, and anti-collaborative attack trust computing scheme based on an innovative cross-validation mechanism. Compared with existing methods, the main contributions of this paper are as follows:

- A cross-validation trust aggregating mechanism, which has anti-collaborative attack ability against garnished and collaborative attacks caused by malicious nodes. In the proposed cross-validation trust aggregating mechanism, feedback not only comes from CMs, but also from CHs and BSs. The feedback information from multiple sources confirms each and constitutes a cross-validation mechanism. Such CM-level trust computing, with three trust factors, including CM-to-CM direct trust, CM-to-CM feedback, and CH-to-CM feedback, constitutes a cross-validation relationship. For CH-level trust computing, three trust factors, CH-to-CH direct trust, CH-to-CH feedback, and BS-to-CH feedback, also constitute a cross-validation relationship. This cross-validation mechanism can effectively reduce the risk of the system, while improving system reliability and security. We investigated representative trust schemes in clustered WSNs, such as LDTS [10], GTMS [18], DST [19], BTEM [34], ATRM [36], and TCHEM [37]. We found that many of these studies lacked considerations of the anti-collaborative attack ability of the trust scheme itself. We extended the traditional trust schemes in clustered WSNs and proposed a cross-validation trust mechanism based on multiple trust factors, which has a stronger anti-collaborative attack ability against collaborative attacks compared with existing trust mechanisms.
- A fast and resource-saving trust computing scheme for cooperation between CMs or between CHs, which is suitable for resource-constrained WSNs. The computational speed and resource-saving of a trust system are the most fundamental requirements for resource-constrained WSNs. However, most of these studies (such as LDTS [10], GTMS [18], DST [19], BTEM [34], ATRM [36], and TCHEM [37]) failed to consider the resource efficiency issue of the trust computing scheme itself. In this study, the number of successful transmissions was considered as the key credential to determine the trustworthiness of a node. We adopted fast algorithms and a resource-saving mechanism to compute the trust value between nodes, which was suitable for resource-constrained WSNs with large-scale nodes.

Together, these innovative designs made the fast, resource-saving, and anti-collaborative attack trust computing scheme (FRAT) solution a fast, resource-efficient, and cooperative attack-resistant solution that could be used in a clustered WSN environment. This study provided the theoretical basis and experimental results for verifying the design of FRAT. Theoretical analysis and experimental results showed that compared with the existing methods, FRAT had superior performance.

The main contents of the rest of this study are as follows: Section 2 provides an overview of related work. The cross-validation mechanism for trust computing for clustered WSNs is described in Section 3. Section 4 gives the details of the trust scheme in the FRAT scheme. Sections 5 and 6 respectively provide the theoretical and experimental analyses of FRAT. Section 7 is the conclusion of this paper.

2. Related Work

Desai et al. proposed a trust evaluation method that used node's internal resources to evaluate node-level trust [12]. Using the suggested self-test algorithm, this method helped nodes trust themselves after booting by ensuring reliable system memory. This algorithm was a completely intermediary technology and had nothing to do with the network topology and auxiliary information.

In [18], a group-based trust computing mechanism (GTMS) was proposed for clustered WSNs. Compared with traditional trust schemes that always focus on the trust value of a single node, GTMS evaluates the trust of a group of nodes. This approach provides a benefit to WSNs, which requires less memory to store trust records on each node. GTMS helps to reduce the costs associated with trust evaluation of remote nodes significantly.

In [34], a belief-based trust evaluation mechanism (BTEM) was developed for wireless sensor networks. The proposed mechanism could resist against various network attacks such as on-off attacks, bad-mouth attacks, and DoS (denial of service) attacks. Simulation-based experimental results showed that the trust mechanism could not only successfully identify and isolate malicious nodes to a certain extent, but also improve the detection rate of malicious behaviors.

In [37], a trust-based cluster head (CH) election mechanism (TCHEM) was proposed. Its basic framework was proposed based on the clustered network model. In this network collaboration model, all nodes had unique local IDs. This method could reduce the possibility of malicious or damaged nodes becoming CHs. This mechanism discouraged sharing trust information between sensor nodes. Therefore, this method reduced the impact of cooperative attacks.

In [36], a trust and reputation scheme (ATRM) based on a distributed agent mechanism was proposed for WSNs. With the help of a mobile agent running on each network node, ATRM collects trusted information and calculates the node's trust. The benefits of local management schemes for trust and reputation are that there is no need for a centralized repository, and the node itself can provide its own reputation information when needed. As a result, there are no network-wide floods or acquisition delays when performing reputation calculations and propagation.

In [35], the authors proposed a robust trust-aware routing framework (TARF) for dynamic WSNs. Because there was no consideration of tight time synchronization, TARF provided a reliable and energy-saving trust scheme. Facts have proven that TARF could effectively prevent harmful attacks due to identity spoofing; through simulation and empirical experiments on large WSNs in various scenarios including mobile and RF shielded network conditions, verifying the flexibility of TARF through extensive evaluation. In addition, the authors implemented a low-overhead TARF module. This implementation could be incorporated into existing routing protocols with minimal changes.

In [10], the authors proposed LDTS, a lightweight and highly reliable trust system for collecting data through wireless sensor networks. First, a lightweight trust decision scheme based on node identity (role) in a clustered wireless sensor network was proposed. This scheme is suitable for such wireless sensor networks because it is conducive to energy saving. Because feedback between cluster members (CM) or the cluster head (CH) is eliminated, this method can greatly improve system efficiency while reducing attacks on the system by malicious nodes. Considering that the CH undertook a large number of data forwarding and communication tasks, this study defined the cooperation of reliability enhancement trust assessment methods between CHs. This method could effectively reduce the network consumption caused by malicious or selfish CHs.

The research on the trust mechanism for WSNs has received extensive attention from scholars. In WSNs, how to identify malicious nodes accurately is a challenging problem that has aroused widespread concern in academia and industry. Table 1 concludes about the features of the trust computing mechanism for WSNs mentioned. From Table 1, we can find that, in view of the security and trustworthiness of WSNs, some feasible and rich solutions were proposed [10,12,18,34–37], but an efficient trust computing mechanism designed for clustered WSNs from the simultaneous achievement of overhead-saving and anti-collaborative attack ability is still necessary.

Table 1. The comparison of the trust computing mechanisms for WSNs. FRAT, fast, resource-saving, and anti-collaborative attack trust computing scheme.

Model	Clustered WSNs	Cross-Validation	Resource-Saving	Anti-Collaborative Attack	Trust Evaluation Approach
Desai model [12]	No	No	No	No	Subjective
Shaikh model [18]	Yes	No	Yes	No	Subjective
Raja model [34]	No	No	No	Yes	Subjective
Crosby model [37]	Yes	No	Yes	No	Subjective
Boukerche model [36]	No	No	Yes	No	Subjective
Zhan model [35]	No	No	No	No	Subjective
Li model [10]	Yes	No	Yes	No	Adaptive
FRAT in this paper	Yes	Yes	Yes	Yes	Adaptive

3. Cross-Validation Mechanism for Trust Computing

In this section, we first present the conceptual model and formal definitions based on the cross-validation mechanism, which is employed by FRAT. We then establish a trustworthy clustered WSN environment based on the trust relationship among network entities (CMs, CHs, and BSs). We will also analyze possible attack patterns that threaten to build trust relationships.

3.1. Three-Tier Network Architecture Model

The clustering algorithm provides one of the most feasible solutions for communication in WSNs due to its inherent resource-saving characteristics and its suitability for highly scalable networks. The FRAT solution is based on a clustered WSN with a backbone, and its core function is to build a reliable and efficient data aggregation network.

As shown in Figure 1, according to its characteristics, the nodes in a clustered WSN environment can be identified as the CH or CM [10]. The CM in the cluster can communicate directly with its CH. The communication between the CM and the BS can only be performed through the CH. In each cluster, only the CH can forward data directly to the BS. The CH collects, aggregates, and forwards data from the CM to the BS. The BS, CH and CM form a three-layer network architecture model (Figure 1).

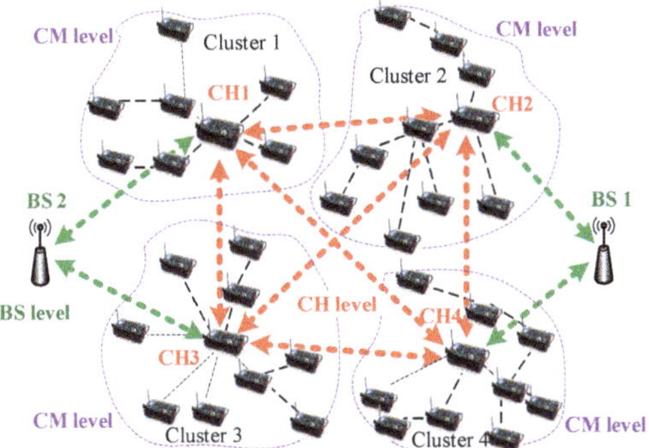

Figure 1. Three-tier network architecture model, in which a member can be identified as a BS, a CH, or a CM according to their features.

One of the key tasks of this study is to construct a trust-based network topology model that can reduce the possibility of malicious members being selected as cooperative partners in data forwarding. Through the cooperation of other CHs, one CH can forward the collected data to the central BS node. It was assumed that the members were divided into multiple clusters based on existing clustering algorithms, such as [8,10]. We also assumed that each node had a unique ID, which could be used to distinguish it from other nodes, similar to the assumptions in [18,34,37]. Once the cluster was formed, they would maintain the same CMs unless a CM was blacklisted or dead or a new node joined the sensor network.

3.2. Formal Definitions of Trust Based on Cross-Validation

Based on the inherent relationship among CMs, CHs, and BSs, this paper first systematically studies and constructs a cross-validation trust computing scheme for clustered WSN environment. In Figure 2, following the functions of the network nodes in the cluster WSN, a total of three network entities exists, namely CMs, CHs, and BSs. Thus, a total of three collaborative groups can be formed: a CM group ($\{CM_1, CM_2, \cdots, CM_i, \cdots, CM_I\}$, where i is the unique identity of a CM, I is the total number of CM nodes in the system); a CH group ($\{CH_1, CH_2, \cdots, CH_j, \cdots, CH_J\}$, where j is the unique identity of a CH, J is the total number of CHs); and a BS group ($\{BS_1, BS_2, \cdots, BS_k, \cdots, BS_K\}$, where k is the unique identity of a BS, K is the total number of BSs). There are two basic trust relationships between these network entities. One is the trust relationship between two CMs. This is the most basic trust relationship in a clustered WSN environment [12]. The other is the trust relationship between two CHs, and this is a special trust relationship in the clustered wireless sensor network environment. This is a crucial factor in encouraging cooperation between CHs and is highly important in successfully deploying a trustworthy clustered WSN. Referring to the methods in [10,33,38], we then provide the cross-validation definitions of the trust relationship used in the clustered WSN environment.

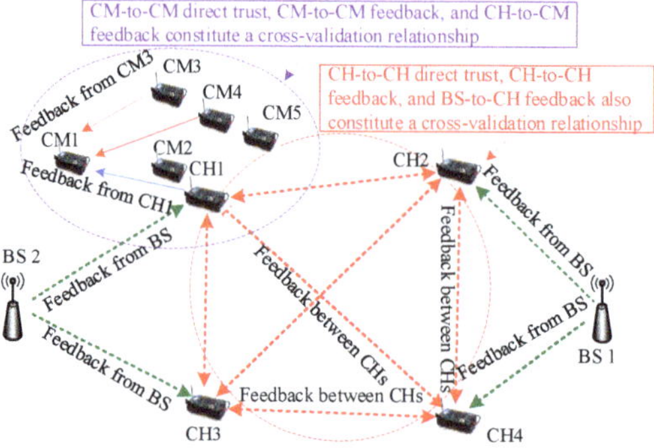

Figure 2. Cross-validation mechanism for trust computing based on the three-tier network architecture model.

The main innovation of the cross-validation approach is embodied in the following two aspects:

- First, three trust factors constitute a cross-validation relationship in the CM-level (or CH-level) trust computing. For CM-level trust computing, three trust factors, including CM-to-CM direct trust, CM-to-CM feedback, and CH-to-CM feedback, constitute a cross-validation relationship. For CH-level trust computing, three trust factors, CH-to-CH direct trust, CH-to-CH feedback, and BS-to-CH feedback, also constitute a cross-validation relationship. Different from the previous trust computing methods, in the proposed trust mechanism, feedback not only comes from CMs, but also from CHs and BSs. This cross-validation mechanism can effectively reduce the risk of the system, while improving system reliability and security.
- Second, relying on the theory of standard deviation analysis [39,40], we used an aggregating method for the overall trust degree in which three trust factors were further cross-validated with one another. In statistics, deviation analysis refers to the absolute difference between any number

in a set and the mean of the set [40]. Different from traditional methods, our mechanism based on the theory of deviation analysis is a cross-validated trust calculation mechanism based on multiple trust factors (in CM-level trust computing, including CM-to-CM direct trust, CM-to-CM feedback, and CH-to-CM feedback; in CH-level trust computing, including CH-to-CH direct trust, CH-to-CH feedback, and BS-to-CH feedback). The trust factor with a larger deviation compared with the other two values is eliminated from the overall trust aggregation process. At the same time, this removal solves the adaptive aggregation problem caused by malicious nodes (malicious CMs or CHs).

Definition 1. *Trust relationship between two CMs based on the cross-validation mechanism (called CM-to-CM overall trust). The CM-to-CM overall trust is a quantifiable value of the competence of another CM (the CM to be evaluated) to complete the task of the CM, based on the CM's direct evaluation and the feedback of CHs and other CMs. As the CH feedback information is integrated into CM-to-CM overall trust computing, this CM-to-CM overall trust computing approach is a cross-validation mechanism.*

Definition 1 and Figure 2 shows that the overall trust degree from CM to CM is the result of fusion calculation through three trust factors, namely CM-to-CM direct evaluation, CH-to-CM feedback, and CM-to-CM feedback. Due to feedback from two sources, this trust computing approach is called the cross-validation mechanism. In traditional feedback-based trust calculation mechanisms, such as in [12], feedback information mainly comes from the CMs, which could cause many problems, such as malicious attacks and coordinated deception. In a clustered WSN, a CH is usually selected by CMs according to its reliability, such as power, data forwarding success rate, and trust. Thus, feedback from the CHs should have higher reliability. From this point of view, the cross-validation mechanism can minimize system risks and improve the security of clustered WSNs.

Definition 2. *Trust relationship between two CHs based on the cross-validation mechanism (called CH-to-CH overall trust). The CH-to-CH overall trust is a a quantifiable value in the judgment of another CH (the CH to be evaluated) to complete the task of the CH, based on the CH's direct evaluation, and the feedback of BSs and other CHs (as the BS feedback information is integrated into CH-to-CH overall trust calculation, this CH-to-CH overall trust computing approach is a cross-validation mechanism).*

Similar to CM-to-CM overall trust, the overall trust of CH-to-CH is the result of aggregation calculation through three trust factors, namely CH-to-CH direct evaluation, CH-to-CH feedback, and BS-to-CH feedback. To integrate more reliable feedback from BSs, existing BS equipment is usually managed by a reputable ISP. The CH-to-CH overall trust is significantly enhanced. In addition, the basic function of BSs allows for dynamic monitoring of the forwarding behavior of CHs. Thus, each BS could provide feedback based on real monitoring data, which could then partly solve the problem of malicious feedback from CHs.

Definition 3. *Feedback between two CMs or between two CHs (called CM-to-CM feedback or CH-to-CH feedback). Feedback between two CMs (or between two CHs) is a rating based on the CM or CH history behavior. After the data forwarding task is completed, the CM or CH will calculate the real-time trust. When another CM or CH requests it, the CH provides the value to the requester.*

Definition 4. *Feedback of a CH to a CM (called CH-to-CM feedback). CH's feedback on the CM is an objective rating based on the historical behavior of the CM. A CH dynamically monitors the CM behavior during the data forwarding. After the data forwarding task is completed, the CH calculates the real-time trust of the CM. When another CM requests it, the CH provides the value to the requester.*

Definition 5. *Feedback of a BS to a CH (called BS-to-CH feedback). BS's feedback on the CH is an objective evaluation based on the historical behavior of the CH. A BS dynamically monitors the CH behavior* *during the data forwarding. After the data forwarding task is completed, the BS will calculate the real-time trust of the CH. When another CH requests it, the BS provides the value to the requester.*

According to Definitions 1 and 2, the FRAT scheme needs to maintain two levels of trust relationship: CM-to-CM overall trust and *CH-to-CH* overall trust. In this paper, CM-to-CM overall trust is represented by $T_{CM_x,CM_y}(\Delta t)$, and CH-to-CH overall trust is represented by $T_{CH_i,CH_j}(\Delta t)$. Since trust is a dynamic value that changes over time, we added a timestamp *Deltat* to the expression. Likewise, the FRAT scheme needs to maintain four levels of feedback relationship: (1) feedback between a CM and a CM, (2) feedback of a CH to a CM, (3) feedback of a CH to a CH, and (4) feedback of a BS to a CH. We use $F_{CM_x,CM_y}(\Delta t)$ to represent CM-to-CM feedback (use $F_{CH_i,CH_j}(\Delta t)$ to represent *CH-to-CH* feedback), $F_{CH_i,CM_x}(\Delta t)$ to represent *CH-to-CH* feedback, $F_{CH_i,CH_j}(\Delta t)$ to represent *CH-to-CH* feedback, and $F_{BS_z,CH_j}(\Delta t)$ to represent *BS-to-CH* feedback. $D_{CM_x,CM_y}(\Delta t)$ and $D_{CH_i,CH_j}(\Delta t)$ are the direct trust between two CMs or CHs. These are different from traditional trust computing methods (such as LDTS [10], GTMS [18], DST [19], BTEM [34], ATRM [36], and TCHEM [37]), in which the feedback comes from a single source. In summary, in the proposed FRAT scheme, the feedback information comes from multiple mutual cross-validation sources. Three types of feedback relationship form a cross-validation mechanism, and this mechanism has a protective ability against collaborative attacks caused by malicious nodes through the theory of deviation analysis.

To clarify the cross-validation mechanism, we provide the following example. Consider the case in Figure 3, where CM-to-CM overall trust is computed based on the cross-validation mechanism. In this case, if CM1 wants to compute the overall trust of CM2, CM1 first asks for the feedback of CM2 in two ways (CMs and its CH). When the CM transmits data, all other CMs in the cluster are listening. Each CM can hear the transmission of all other CMs within its broadcast range, and these CMs are generally neighbor nodes. The neighbor nodes of CM1 (including CM3, CM4, and CM5) will send their feedback to CM1 (including $F_{CM_3,CM_2}(\Delta t)$, $F_{CM_4,CM_2}(\Delta t)$, and $F_{CM_5,CM_2}(\Delta t)$). The CH1 will send its feedback $F_{CH_1,CM_1}(\Delta t)$ to CM1. Then, integrating its direct trust $D_{CM_1,CM_2}(\Delta t)$, CM1 can obtain an overall trust $T_{CM_1,CM_2}(\Delta t)$ for CM2 based on a fusion calculation method. CM-to-CM direct trust, CM-to-CM feedback, and CH-to-CM feedback constitute a cross-validation relationship.

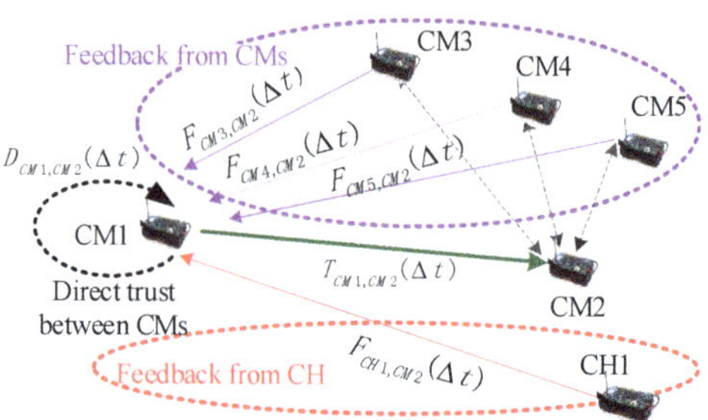

Figure 3. CM-to-CM overall trust computing based on the cross-validation mechanism. CM-to-CM direct trust, CM-to-CM feedback, and CH-to-CM feedback constitute a cross-validation relationship.

A similar example of CH-to-CH overall trust computing based on the cross-validation mechanism is depicted in Figure 4. We can easily understand how to compute the overall trust in the CH-to-CH case from the CM-to-CM overall trust example in Figure 3. In this case, if CH1 needs to compute the overall trust of CH4 (CH1-to-CH4 overall trust $T_{CH_1,CH_4}(\Delta t)$), CH1 will ask for the feedback of CM4 in two ways (CHs and its BS). In the case of Figure 4, CH2 and CH3 provide their CH-to-CH feedback $F_{CH_2,CH_4}(\Delta t)$ and $F_{CH_3,CH_4}(\Delta t)$ to CH1, and BS1 provides its BS-to-CH feedback $F_{BS_1,CH_4}(\Delta t)$ to CH1. At the same time, CH1 needs to compute the direct trust of CH4 (CH1-to-CH4 direct trust

$D_{CH_1,CH_4}(\Delta t)$). After the collection of trust information, CH1 uses the theory of standard deviation analysis to perform the fusion calculation of overall trust $T_{CH_1,CH_4}(\Delta t)$.

In the proposed FRAT scheme, evaluation methods are different for these trust (or feedback) relationships. Both $T_{CM_x,CM_y}(\Delta t)$ and $T_{CH_i,CH_j}(\Delta t)$ are trust decision credentials (or trust authorization credentials), and they can directly act as authorization credentials for node selection in data aggregation, fusion, and higher level transmission. However, $D_{CM_x,CM_y}(\Delta t)$, $D_{CH_i,CH_j}(\Delta t)$, $F_{CM_x,CM_y}(\Delta t)$, $F_{CH_i,CH_j}(\Delta t)$, $F_{CH_i,CM_x}(\Delta t)$, and $F_{BS_z,CH_j}(\Delta t)$ are trust evaluation factors. Each of these factors is one-sided and cannot fully reflect the interactive relationship of nodes in the entire system. Therefore, these factors cannot act as the authorization credential directly. We need to perform fusion calculations on these trust factors in order to obtain a more adequate and accurate overall trust. As mentioned in Section 1, in terms of accuracy, calculation speed, storage overhead, and communication overhead, the trust mechanism should be fast and resource-saving in order to provide services for a large number of resource-constrained nodes. In this work, we propose a series of fast and resource-saving trust computing methods for cooperation between CMs or between CHs. The calculation methods for these nodes' trust (or feedback) relationships are introduced in Section 4.

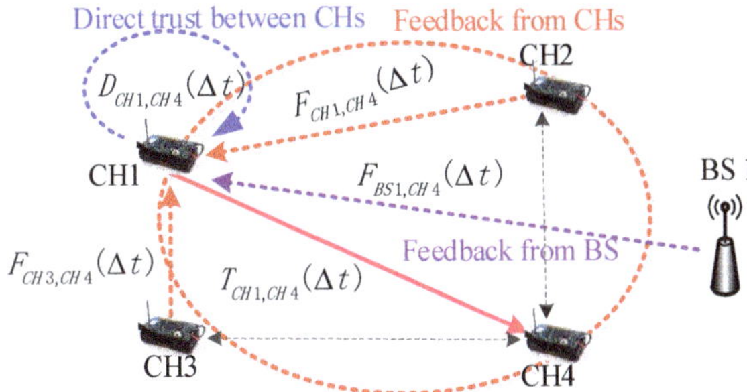

Figure 4. CH-to-CH overall trust computing based on the cross-validation mechanism. CH-to-CH direct trust, CH-to-CH feedback, and BS-to-CH feedback also constitute a cross-validation relationship.

The content of the feedback mainly includes three types of trust, that is the CM-to-CM overall trust degree, or CH-to-CM feedback trust, or CH-to-CH overall trust degree. In Sections 4.1 and 4.2, we introduce the calculation methods of these three types of trust. According to the calculation methods in Sections 4.1 and 4.2, the information transmitted during feedback should be a positive integer between one to 10.

3.3. Attack Pattern Analysis in the FRAT Scheme

In a clustered WSN, the ultimate goal of a trust system is to obtain accurate and reliable functionality against selfish or collaborative network attacks [10]. An effective trust computing system should have a good defense against malicious attacks, that is it should be able to resist selfish or cooperative attacks from the CH and CM. In a clustered WSN environment, network attacks may originate from both malicious CHs and CMs [41].

Definition 6. *Collaborative attacks from CMs or CHs. As long as feedback is considered, a malicious CM or CH will provide dishonest feedback to structure a good CM or CH and/or increase the trust of its stakeholders. This type of attack is called a collaborative attack and is the most direct type of attack in a clustered WSN environment.*

The feedback from the cooperative nodes may produce incorrect trust evaluation results and how to adopt a defense mechanism to prevent cooperative attacks by malicious nodes is the key task of this work. After determining the attack methods of malicious nodes, we can create an effective trust calculation method to prevent malicious entities from achieving their goals by evaluating the behavior of malicious entities, thereby resisting such attacks. However, directly identifying collaborative attacks is a daunting task. In this study, we adopt an adaptive fusion computing method to eliminate false feedback based on the theory of bias analysis, in which the three trust factors are further cross-validated with each other. Compared with the traditional method, our mechanism based on the theory of bias analysis is a cross-validation trust computing mechanism. Compared with the other two values, the biased feedback is eliminated from the entire trust aggregation process.

4. Trust and Feedback Calculation in FRAT

As shown in Figures 2–4, there are two types of direct trust relationship and four types of indirect feedback relationship in the clustered WSN environment. These trust factors have different computing systems because their attributes are completely different. In this section, we introduce related computing mechanisms for these trust factors.

4.1. CM-to-CM Overall Trust Calculation

CM-to-CM direct trust calculation. As mentioned earlier, the problem of saving overhead is the most basic WSN that requires resource constraints. In probability theory and statistics, the beta distribution is a series of continuous probability distributions defined on the interval [0, 1]. It is parameterized by two positive shape parameters, which are indexed by random variables. The form appears and controls the shape of the distribution. In [42], a beta trust system based on statistical theory was proposed. The system had the characteristics of flexibility and high resource efficiency. Inspired by the innovative work in [10,42], we used an improved betaprobability density function to calculate the CM's direct trust in the CM. The direct trust calculation on the CM is defined by the following formula:

$$
\begin{aligned}
&D_{CM_x,CM_y}(\Delta t) \\
&= \left\lceil 10 \times E(\varphi(p|S_{CM_x,CM_y}(\Delta t), U_{CM_x,CM_y}(\Delta t))) \right\rceil
\end{aligned}
\tag{1}
$$

where Δt is a time window. The length Δt can be shorter or longer depending on the network analysis scheme. Therefore, as time goes by, the window forgets the old experience, but adds new experiences. The operation $\lceil \cdot \rceil$ is the closest integer function, such that $\lceil 0.82148 \rceil = 8$. The symbol p reflects the posterior probability of the binary event $(S_{CM_x,CM_y}(\Delta t), U_{CM_x,CM_y}(\Delta t))$, and $S_{CM_x,CM_y}(\Delta t)$ is the total number of successful data communications between nodes CM_x with CM_y during time Δt. $U_{CM_x,CM_y}(\Delta t)$ is the total number of unsuccessful data communications between nodes CM_x with CM_y during time Δt. $E(\varphi(p|S_{CM_x,CM_y}(\Delta t), U_{CM_x,CM_y}(\Delta t)))$ is the expected probability of the beta distribution $\varphi(p|S_{CM_x,CM_y}(\Delta t), U_{CM_x,CM_y}(\Delta t))$:

$$
\begin{aligned}
&\left\lceil E(\varphi(p|S_{CM_x,CM_y}(\Delta t), U_{CM_x,CM_y}(\Delta t)) \right\rceil \\
&= \left\lceil \frac{10 \times (S_{CM_x,CM_y}(\Delta t) + 1)}{S_{CM_x,CM_y}(\Delta t) + \alpha * U_{CM_x,CM_y}(\Delta t) + 2} \right\rceil
\end{aligned}
\tag{2}
$$

where positive integer $\alpha \in [1 - N]$ is a punitive factor that reflects the punitive nature of failed interactions. In special cases, if $S_{CM_x,CM_y}(\Delta t) + U_{CM_x,CM_y}(\Delta t) = 0$, which denotes no interactions between node CM_x with CM_y during time Δt. According to Equation (1), the value of $D_{CM_x,CM_y}(\Delta t) = 5$. If $S_{CM_x,CM_y}(\Delta t) \neq 0$ and $U_{CM_x,CM_y}(\Delta t) = 0$, then the value of $D_{CM_x,CM_y}(\Delta t)$ is a positive increasing value with the increase in the number of successful interactions. Figure 5 depicts the evolution trend of CM-to-CM direct trust. We can observe that the value of CM-to-CM direct trust quickly reduces with the increase in the number of failed interactions, which reflects the strictly punitive nature of the proposed trust mechanism for the failure of interactions.

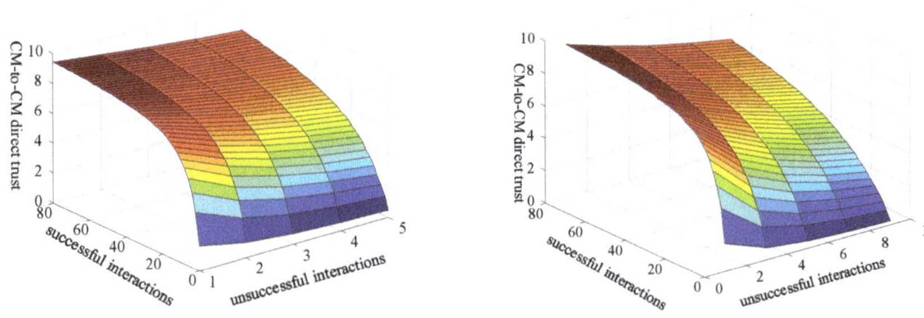

(**a**) The number of unsuccessful interactions is 0 to 5, and that of successful interactions is 0 to 80.

(**b**) The number of unsuccessful interactions is 0 to 20, and that of successful interactions is 0 to 80.

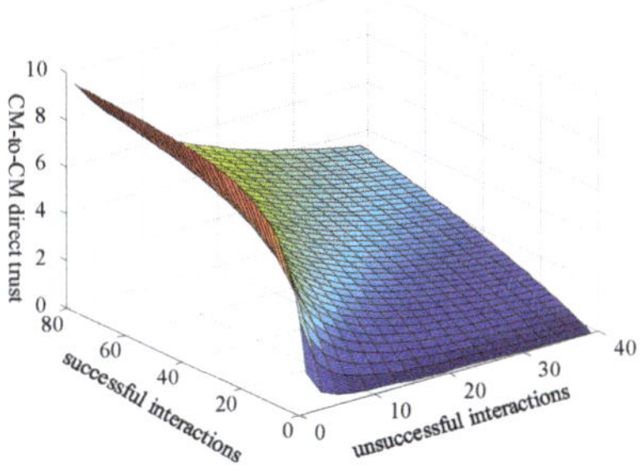

(**c**) The number of unsuccessful interactions is 0 to 40, and that of successful interactions is 0 to 80.

Figure 5. The value of CM-to-CM direct trust with penalty factor $\alpha = 4$.

Compared with the original method proposed by [42], the main difference of the improved beta probability density function is the penalty factor α to be introduced. If $\alpha = 1$, then our approach falls back to [42]. In $\alpha > 1$, then our approach reflects the punitive nature of the failure of interaction. We use Figure 6 for quantitative analysis of CM-to-CM direct trust under different values of penalty factor α. In Figure 6, S is the number of successful interactions, and U is the number of unsuccessful interactions. From Figure 6, the value of CM-to-CM direct trust shows a downward trend with increasing α, which reaches our design goal for punishment of failed interactions. In WSN systems with high security requirements, we should choose the value of α to approach 10.

CM-to-CM feedback calculation. As mentioned earlier, feedback is an important task for both CMs and CHs. It also provides information and key performance indicators for trust assessment. There are many collaborative CMs in a clustered WSN environment, and the feedback from these CMs is considered as a social rating and should have a high reference value for node trust evaluation. We used the improved beta probability density function with a strict punitive nature to compute $F_{CM_x,CM_y}(\Delta t)$. As a result, the calculation efficiency was improved.

$$F_{CM_x,CM_y}(\Delta t) = \left\lceil \frac{10 * (\xi(CM_y) + 1)}{(\xi(CM_y) + \alpha * \gamma(CM_y) + 2)} \right\rceil \tag{3}$$

where the positive integer $\alpha \in [1 - N]$ is a penalty factor, which reflects the penalty nature for malicious feedback. $\xi(CM_y)$ is the number of positive feedbacks (>0.5) toward CM_y from other CMs in the cluster, whereas $\gamma(CM_y)$ is the number of negative feedbacks (<0.5) from other CMs.

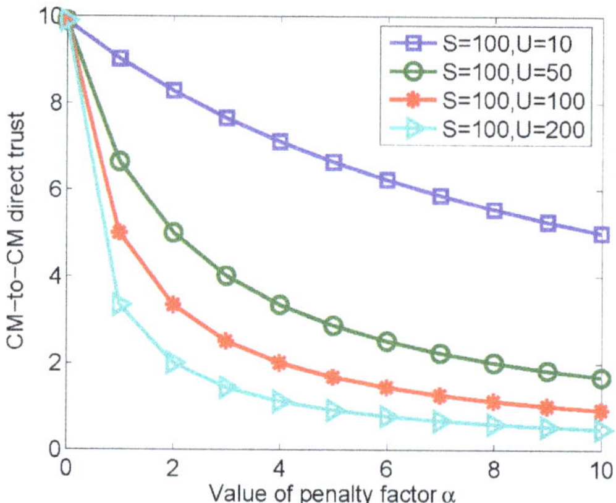

Figure 6. Analysis of CM-to-CM direct trust under different values of α.

CH-to-CM feedback calculation. As shown in Figure 3, different from CM-to-CM feedback, the CH-to-CM feedback is a value based on the CH rating. We assumed that I CMs existed in a cluster. The CH would broadcast request packets in the cluster periodically. In response, all CMs in the cluster would forward their direct trust values to other CMs to the CH. CHs would then maintain these trust values in the matrix f_{CH_i}, as follows:

$$f_{CH_i} = \begin{pmatrix} D_{CM_1,CM_1} & D_{CM_1,CM_2} & \cdots & D_{CM_1,CM_I} \\ D_{CM_2,CM_1} & D_{CM_2,CM_2} & \cdots & D_{CM_2,CM_I} \\ & & \ddots & \\ D_{CM_I,CM_1} & D_{CM_I,CM_2} & \cdots & D_{CM_I,CM_I} \end{pmatrix} \tag{4}$$

where $D_{CM_i,CM_y}(i \in [1, I], y \in [1, I])$ is the direct trust of a network member CM_i for CM_y. In addition, if $i = y$, this means that the value is the node's feedback for itself. In this study, an improved beta probability density function is used to calculate $F_{CH_i,CM_y}(\Delta t)$.

$$F_{CH_i,CM_x}(\Delta t) = \left\lceil \frac{10 * (g(CM_y) + 1)}{(g(CM_y) + \alpha * b(CM_y) + 2)} \right\rceil \tag{5}$$

where positive integer $\alpha \in [1 - N]$ is the penalty factor, which reflects the penalty function of malicious feedback. $g(CM_y)$ is the number of positive feedbacks (>0.5) toward CM_y from other CMs in the cluster, whereas $b(CM_y)$ is the number of negative feedbacks (<0.5) from other CMs. Analyzing Equations (4) and (5), we find that both feedback aggregation mechanisms are resource-saving methods with simple formulas and are suitable for resource-constrained wireless sensor networks with large sensor nodes.

The feedback value is a positive integer between one and 10. Thus, we can define how a CH/CM detects that a received feedback is positive or negative. If the value is less than or equal to five, we consider this feedback to be negative. If the value is more than five, we consider this feedback to be positive.

CM-to-CM overall trust aggregating calculation based on standard deviation analysis. As indicated in Definition 1, the *CM-to-CM* overall trust is evaluated based on three factors: $D_{CM_x,CM_y}(\Delta t)$, $F_{CM_x,CM_y}(\Delta t)$, and $F_{CH_i,CM_y}(\Delta t)$. Therefore, aggregating these trust factors into a single value in an unbiased manner is a challenging problem. In statistics, standard deviation analysis means the absolute difference between any number in a set and the mean of the set [39,40]. The basic idea of standard deviation analysis is (1) to eliminate the number with a larger deviation than the other numbers and (2) to calculate the average of the remaining numbers.

We suppose that $\mu(\Delta t)$ is the summation value of the three trust factors at time stamp Δt. $f_{max}(\Delta t)$ is the maximum value of the three trust factors at time stamp Δt. $f_{min}(\Delta t)$ is their minimum value at the same time stamp Δt. $\gamma(\Delta t)$ is the average value of the three trust factors. Then, the standard deviation of the three trust factors is defined as follows:

$$\delta(\Delta t) = \sqrt{\Omega(\Delta t)/3} \qquad ,$$
$$\Omega(\Delta t) = (D_{CM_x,CM_y}(\Delta t) - \gamma(\Delta t))^2 +$$
$$(F_{CM_x,CM_y}(\Delta t) - \gamma(\Delta t))^2 +$$
$$(F_{CH_i,CM_y}(\Delta t) - \gamma(\Delta t))^2 \tag{6}$$

From a statistical perspective, the standard deviation of a dataset is a measure of the amount of deviation between the observations contained in the dataset. Relying on the theory of bias analysis, we adopted an aggregation method for the overall trust, which could overcome the limitations of the traditional trust computing system [39,40]. The traditional trust mechanism weighs the attributes of the trust manually or subjectively.

$$T_{CM_x,CM_y}(\Delta t) = \begin{cases} \frac{\mu(\Delta t) - f_{max}(\Delta t)}{2}, & f_{max}(\Delta t) > (\gamma(\Delta t) + \delta(\Delta t)) \\ \frac{\mu(\Delta t) - f_{min}(\Delta t)}{2}, & f_{min}(\Delta t) < (\gamma(\Delta t) - \delta(\Delta t)) \\ \gamma(\Delta t) & , otherwise \end{cases} \tag{7}$$

Compared with the traditional methods, our mechanism in Equation (7) performs adaptive trust calculation. The trust factor with a larger deviation compared with the other two values is eliminated from the overall trust aggregation process using Equations (5), (6), and (7). This removal solves the adaptive aggregation problem caused by collaborative attack CMs.

4.2. CH-to-CH Overall Trust Calculation

CH-to-CH direct trust calculation. We used a similar mechanism to calculate the direct trust from CH-to-CH, that is the direct trust from CM-to-CM. The direct trust assessment method on CHs is defined by the following formula:

$$D_{CH_i,CH_j}(\Delta t) = \left\lceil \frac{10 \times (\psi_{CH_i,CH_j}(\Delta t) + 1)}{\psi_{CH_i,CH_j}(\Delta t) + \alpha * \beta_{CH_i,CH_j}(\Delta t) + 2} \right\rceil \tag{8}$$

where $\alpha \in [1 - N]$ is a penalty factor. Δt is a window of time. $\psi_{CH_i,CH_j}(\Delta t)$ is the total number of successful data forwards. $\beta_{CH_i,CH_j}(\Delta t)$ is the total number of unsuccessful data forwards of node CH_i with CH_j during time Δt.

CH-to-CH feedback calculation. In this study, we use an improved beta probability density function to calculate $F_{CH_i,CM_j}(\Delta t)$.

$$F_{CH_i,CM_j}(\Delta t) = \left\lceil \frac{10 * (\omega(CH_j) + 1)}{(\omega(CH_j) + \alpha * \theta(CH_j) + 2)} \right\rceil \tag{9}$$

where $\alpha \in [1 - N]$ is a penalty factor. $\omega(CH_j)$ is the number of positive ratings (>0.5) toward CH_j from other CHs, whereas $\theta(CH_j)$ is the number of negative ratings (<0.5) from other CHs.

BS-to-CH feedback calculation. As shown in Figure 4, the BS-to-CH feedback is a value based on the BS rating. We assumed the existence of J CHs that interacted with a BS. The BS periodically broadcast the request packet for feedback. In response, all CHs forwarded their direct trust values to other CHs to the BS. The BS then maintained these trust values in the matrix c_{CH_i}:

$$h_{CH_i} = \begin{pmatrix} D_{CH_1,CM_1} & D_{CH_1,CM_2} & \cdots & D_{CH_1,CM_J} \\ D_{CH_2,CM_2} & D_{CH_2,CM_2} & \cdots & D_{CH_2,CM_J} \\ & & \ddots & \\ D_{CH_J,CM_1} & D_{CH_J,CM_2} & \cdots & D_{CH_J,CM_J} \end{pmatrix} \tag{10}$$

where $D_{CH_i,CH_j}(i \in [1,J], j \in [1,J])$ is the direct trust of node CH_i for CH_j. In this study, we use an improved beta probability density function to calculate $F_{BS_z,CH_j}(\Delta t)$.

$$F_{BS_z,CH_j}(\Delta t) = \left\lceil \frac{10 * (o(CH_j) + 1)}{(o(CH_j) + \alpha * p(CH_j) + 2)} \right\rceil \tag{11}$$

where $\alpha \in [1 - N]$ is the penalty factor. $o(CH_j)$ is the number of positive ratings (>0.5) toward CM_y from other CHs, whereas $p(CH_j)$ is the number of negative ratings (<0.5) from other CHs.

CH-to-CH overall trust aggregating calculation based on standard deviation analysis. We used an aggregation method for overall trust based on deviation analysis theory, which could overcome the limitations of traditional trust computing mechanisms where trusted attributes were weighted manually or subjectively [39]. We supposed that $v(\Delta t)$ was the summation value of the three trust factors ($D_{CH_i,CH_j}(\Delta t)$, $F_{CH_i,CH_j}(\Delta t)$, and $F_{BS_z,CH_j}(\Delta t)$) at time stamp Δt. $s_{max}(\Delta t)$ is the maximum value; $s_{min}(\Delta t)$ is its minimum value; and $\rho(\Delta t)$ is the average value. The standard deviation of the three trust factors is defined as follows:

$$\omega(\Delta t) = \sqrt{\Psi(\Delta t)/3} \quad ,$$
$$\Psi(\Delta t) = (D_{CH_i,CH_j}(\Delta t) - \rho(\Delta t))^2 +$$
$$(F_{CH_i,CH_j}(\Delta t) - \rho(\Delta t))^2 + \tag{12}$$
$$(F_{BS_z,CH_j}(\Delta t) - \rho(\Delta t))^2$$

Then, the overall trust degree based on deviation analysis is defined as follows:

$$T_{CH_i,CH_j}(\Delta t) = \begin{cases} \frac{v(\Delta t) - s_{max}(\Delta t)}{2}, & s_{max}(\Delta t) > (\rho(\Delta t) + \omega(\Delta t)) \\ \frac{v(\Delta t) - s_{min}(\Delta t)}{2}, & s_{min}(\Delta t) < (\rho(\Delta t) - \omega(\Delta t)) \\ \rho(\Delta t), & otherwise \end{cases} \tag{13}$$

This trust aggregation in Equation (13) is an adaptive trust calculation mechanism. The trust factor with a larger deviation compared with the other two values is eliminated from the overall trust aggregation process using Equations (12) and (13).

5. Performance Analysis

In this section, we analyze the proposed trust mechanism from three aspects: (1) the attacker's ability to resist collaborative attacks and the trust computing scheme itself, (2) time complexity, and (3) communication overhead (the latter two can reflect the computing speed and resource efficiency of the trust computing solution).

5.1. Time Complexity Analysis

We took some resource-saving steps to calculate the trust value between nodes, which was suitable for WSNs because it helped to save resources. In addition, we used an improved beta probability density function to calculate the overall trust value. It was found that this mechanism was a method to save resources and was suitable for resource-constrained nodes in large-scale sensor networks. Because the calculation of all these trust factors was a statistical operation, the computational overhead of the calculation could be ignored.

Theorem 1. *Using the proposed trust evaluation scheme, the total time complexity of CM-to-CM overall trust computing was no more than* $O(g) + O(m) + O(k) + O(1)$.

Proof. In the period of CM-to-CM direct trust calculation (from Equation (1) to Equation (2)), the time complexity is $O(g)$, and $g = S_{CM_x,CM_y}(\Delta t) + U_{CM_x,CM_y}(\Delta t)$. In the period of CM-to-CM feedback calculation (Equation (3)), the time complexity is $O(m)$ and $m = \xi(CM_y) + \gamma(CM_y)$. In the period of CH-to-CM feedback calculation (from Equation (4) to Equation (5)), the time complexity is $O(k)$ and $k = g(CM_y) + b(CM_y)$. In the period of CM-to-CM overall trust aggregating calculation (from Equation (6) to Equation (7)), the time complexity is $O(1)$. Thus, the time complexity is $O(g) + O(m) + O(k) + O(1)$. $\square$

Theorem 2. *Based on the proposed trust evaluation scheme, the total time complexity of CH-to-CH overall trust computing was no more than* $O(q) + O(w) + O(r) + O(1)$.

Proof. In the period of CH-to-CH direct trust calculation (Equation (8)), the time complexity is $O(q)$ and $q = \psi_{CH_i,CH_j}(\Delta t) + \beta_{CH_i,CH_j}(\Delta t)$. In the period of CH-to-CH feedback calculation (Equation (9)), the time complexity is $O(w)$ and $w = \omega(CH_j) + \theta(CH_j)$. In the period of CS-to-CH feedback calculation (from Equation (10) to Equation (11)), the time complexity is $O(r)$ and $r = o(CH_j) + p(CH_j)$. In the period of CH-to-CH overall trust aggregating calculation (from Equation (12) to Equation (13)), the time complexity is $O(1)$. Thus, the time complexity is $O(q) + O(w) + O(r) + O(1)$. $\square$

In the period of trust factor measurement based on improved beta probability density functions (from Equation (1) to Equation (13)), the computing time complexity was no more than $O(g) + O(m) + O(k) + O(1)$ (or $O(q) + O(w) + O(r) + O(1)$), which showed that the computing complexity of the proposed trust computing scheme was far superior to those of some existing schemes, such as the fuzzy-based trust models [11]), whose time complexity was $O(n^3 log_2 n)$. In traditional trust computing schemes, if $n \to \infty$, trust aggregation calculations would become extremely slow. In this study, we used a time-saving computer system that greatly increased the speed of trust calculation, which made the trust calculation scheme very suitable for large WSNs.

5.2. Communication Overhead Analysis

In order to analyze the communication overhead of the FRAT mechanism under full load conditions, we assumed that in the worst case, each CM wanted to communicate with other CMs in the cluster and each CH wanted to communicate with other CHs in the cluster. In addition, each CH needed to collect feedback from other CMs, and the BS must collect feedback reports from other CHs.

Theorem 3. *Supposed that the network consists of J clusters and that the average size of clusters is I (including the CH of the cluster). Based on the proposed trust computing scheme, the maximum communication overhead is:* $2I^2 + 2J^2 + 2I * J$.

Proof. (1) From Figure 3, in the cross-validation-based CM-to-CM overall trust calculations, feedback came from three sources. First, when node CM_i wanted to collect feedback from node CM_x, the node CM_i sent at most one CM feedback request, and this node CM_i received a response. Second, CM_i sent a feedback request to its CH and obtained feedback from the CH. Finally, CM_i used its self-feedback information, which required no communication overhead. Therefore, if node CM_i wanted to collect feedback from all nodes in the cluster, the maximum communication overhead became $2[(I-1)+1] = 2I$. If all nodes wanted to transfer data to each other, the maximum communication overhead was $2I * I = 2I^2$.

(2) From Figure 3, in the cross-validation-based CH-to-CH overall trust calculation, the feedback came from three sources. First, when CH_j wanted to collect feedback from CH_y, CH_j sent a maximum of one CH feedback request, for which CH_i received one response. Second, CH_j sent one feedback

request to its BS and received one feedback from the BS. Lastly, CH_j used its self-feedback information, which did not require communication overhead. Therefore, if CH_j wanted to collect feedback from all CHs in the network, the maximum communication overhead became $2[(J-1)+1] = 2J$. If all CHs wanted to communicate with one another, then the maximum communication overhead was $2J * J = 2J^2$. In addition, in the trust calculation from CH to CH, when the CH wanted to collect feedback from its I members, it sent a I request and received a I response, thus resulting in a total communication overhead of $2I$. Therefore, the overall trust of the largest communication overhead CH-to-CH was calculated as $2J^2 + 2I * J$. $\square$

5.3. Anti-Collaborative Attack Ability Analysis

In Figures 2–4, according to the inherent relationship among the three network entities, we propose the cross-validation mechanism, which is effective and reliable against collaborative attacks caused by malicious nodes. In this sub-section, we analyze the anti-collaborative attack ability of the FRAT scheme against collaborative attacks on the trust mechanism.

Theorem 4. *Equations (1)–(3) consider not only the number of positive (or negative) ratings ($\xi(CM_y)$ and $\gamma(CM_y)$), but also the punitive nature for failed transactions. The feature of Equations (1)–(3) can effectively prevent collaborative attacks from accomplice CMs.*

Proof. If $\gamma(CM_y) > \xi(CM_y)$, then $F_{CM_x,CM_y}(\Delta t) \geq 5$, which covers a collaborative scenario where individual CMs attempt to lie about a bad CM [10,18]. We must prove that when $\gamma(CM_y) > \xi(CM_y)$, then $F_{CM_x,CM_y}(\Delta t) < 5$. From Equation (3), feedback from CMs can be calculated using the following improved beta probability density functions:

$$F_{CM_x,CM_y}(\Delta t) = \left\lceil \frac{10 * (\xi(CM_y) + 1)}{(\xi(CM_y) + \alpha * \gamma(CM_y) + 2)} \right\rceil \implies$$

$$F_{CM_x,CM_y}(\Delta t) \leq \frac{10 * (\xi(CM_y) + 1)}{(\xi(CM_y) + \alpha * \gamma(CM_y) + 2)}$$

Under the case that $\gamma(CM_y) > \xi(CM_y)$, we must prove that $F_{CM_x,CM_y}(\Delta t) < 5$, that is,

$$\frac{(\xi(CM_y) + 1)}{(\xi(CM_y) + \alpha * \gamma(CM_y) + 2)} < \frac{1}{2}$$

Under the condition $\gamma(CM_y) > \xi(CM_y)$, a negative feedback exceeds a positive feedback. Thus, we only need to prove the following:

$$2(\xi(CM_y) + 1) < \xi(CM_y) + \alpha * \gamma(CM_y) + 2 \implies$$

$$2\xi(CM_y) + 2 < \xi(CM_y) + \alpha * \gamma(CM_y) + 2 \implies$$

$$\xi(CM_y) < \alpha * \gamma(CM_y)$$

Due to $\gamma(CM_y) > \xi(CM_y)$ and $\alpha > 1$, $\xi(CM_y) < \alpha * \gamma(CM_y)$ must exist, which proves Theorem 4. $\square$

Through Theorem 4, we proved that our trust system at the CM level had a protective ability against collaborative attacks from malicious nodes because this system could prevent such nodes from fulfilling their objectives.

Theorem 5. *Equation (9) considers not only the number of positive (or negative) feedbacks from CHs, but also the punitive nature for failed transactions. The feature of Equation (7) can effectively prevent collaborative attacks caused by accomplice CHs.*

Proof. We assumed that $\varpi(CH_j)$ was the number of positive feedbacks and $\theta(CH_j)$ was the number of negative feedbacks in CH-to-CH trust computing. If $\varpi(CH_j) < \theta(CH_j)$, then $F_{CH_i,CH_j}(\Delta t) \geq 5$, which covers a collaborative attack scenario where individual CHs attempt to lie about a bad CH. We must prove that when $\varpi(CH_j) < \theta(CH_j)$, then $F_{CH_i,CH_j}(\Delta t) < 5$. From Equation (9), feedback from CHs can be calculated using the following improved beta probability density functions:

$$F_{CH_i,CM_j}(\Delta t) = \left[\frac{10 * (\varpi(CH_j) + 1)}{(\varpi(CH_j) + \alpha * \theta(CH_j) + 2)} \right] \Longrightarrow$$

$$F_{CH_i,CM_j}(\Delta t) \leq \frac{10 * (\varpi(CH_j) + 1)}{(\varpi(CH_j) + \alpha * \theta(CH_j) + 2)}$$

We must prove that $F_{CH_i,CM_j}(\Delta t) < 5$, that is,

$$\frac{(\varpi(CH_j) + 1)}{(\varpi(CH_j) + \alpha * \theta(CH_j) + 2)} < \frac{1}{2}$$

Under the condition $\varpi(CH_j) < \theta(CH_j)$, the number of negative feedbacks exceeds the number of positive feedbacks. Thus, we only need to prove the following:

$$2(\varpi(CH_j) + 1) < \varpi(CH_j) + \alpha * \theta(CH_j) + 2 \Longrightarrow$$

$$2\varpi(CH_j) + 2 < \varpi(CH_j) + \alpha * \theta(CH_j) + 2 \Longrightarrow$$

$$\varpi(CH_j) < \alpha * \theta(CH_j)$$

Based on known conditions, existing $\theta(CH_j) > \varpi(CH_j)$, and $\alpha > 1$, thus $\varpi(CH_j) < \alpha * \theta(CH_j)$ must be established, which proves Theorem 5. $\square$

Through Theorem 5, we proved that our trust system at the CH level had a protective ability against collaborative attacks from malicious nodes because this system could prevent such nodes from fulfilling their objectives.

6. Experiment-Based Analysis and Evaluation

In this section, we first describe how to set up the experimental method in a simulated WSN environment, including how to deploy the recommended trust scheme on the simulated environment and how to set up the experimental configuration. The experimental results are then reported.

6.1. Experimental Methods and Parameters

Extensive experiments were conducted by using the NetLogo event simulator [10,43–45] to validate the effectiveness of FRAT. For comparison, we also added GTMS [18] and ATRM [36] into the simulator because both of them are independent of any specific routing mechanism.

In order to make the experiment closer to the real WSN environment, three types of nodes were deployed in the simulator according to their identity, namely the CM, CH, and BS [10]. The CM could be one of two types: good CM (GCM) and bad CM (BCM). The GCM always provided successful cooperation, while the BCM provided unsuccessful cooperation. The behavior of a CM as a feedback provider could be one of two types: honest CM (HCM) and malicious CM (MCM). The HCM always provided correct feedback to any CM, while the MCM always provided feedback to other CMs contrary to actual data. Similar to the CM, the GCH always provided successful cooperation, while the BCH provided unsuccessful cooperation. The HCH always provided the correct feedback, while the MCH always provided the opposite feedback of the actual data of the other CHs.

In the proposed trust computing scheme based on the cross-validation mechanism, the main threat was caused by malicious feedback. We designed several performance mechanisms for a comprehensive trust assessment scheme. Due to the limitation of the paper length, we mainly evaluated the performance of FRAT based on the following two aspects: the successful packet transmission rate under different MCMs and the successful packet transmission rate under different MCHs.

Table 2 lists the simulation parameters and default values used in the experiment. A total of 1000–10,000 nodes were deployed in the simulator, and an average of 100 CMs were deployed in each cluster. The penalty factor α was set at two to reflect a double punitive factor for selfish nodes or failed collaborators. The total time step of the simulation run was 1000, and the time window for trust calculation was 20. The percentage of the HCM was 30–100%, and the percentage of the HCH was 50–100%.

Table 2. Parameters and their possible values. HCM, honest CM; HCH, honest CH.

Symbol	Description	Possible Values
$N = I \times J$	total number of CMs	1000–10,000
I	number of CMs in a cluster	100
J	total number of CHs	100–1000
K	total number of BSs	10
t	time-step of simulation runs	1000
Δt	time-window for trust computing	20
H_{CM}	percentage of HCMs	30–100%
H_{CH}	percentage of HCHs	50–100%
α	penalty factor	2

6.2. Evaluation under Different MCMs

We computed the packet successful delivery ratio (PSDR) [10] to reflect the reliability of the trust computing systems. A higher PSDR indicated higher reliability. In this set of experiments, we assumed that most CHs in the WSN environment were trusted, of which MCHs only accounted for 10%. This WSN environment was very similar to the actual situation, and most CHs were honest and trustworthy.

Figure 7 illustrates the PSDR comparison at different percentages of the MCM. In this set of experiments, we assumed that the WSN environment was a trusted network community, of which 90% of the CHs were honest. The remaining 10% of CHs were malicious feedback providers. We set the percentage of MCMs to 10%, 20%, 30%, 50%, 60%, and 70%, which indicated that the cluster environment was fully honest (10%), honest (20%), relatively honest (30%), partly dishonest (50%), dishonest (60%), and fully dishonest (70%), respectively. Figure 7a shows a fully honest WSN environment, where the percentage of MCMs was only 10%. All three kinds of trust mechanisms had high PSDR values beyond 92%. These results reflected that the three kinds of trust mechanisms exhibited high reliability under an honest WSN community.

A robust trust mechanism should have a strong ability to counteract malicious behavior from MCMs. To evaluate the performance of the trust system under a more complex network environment, we gradually increased the proportion of malicious nodes. In Figure 7b–f, the proportion of MCMs were set to 20%, 30%, 50%, 60%, and 70%, and the results indicated larger differences compared with MCMs set to 10%. With the increase in the percentage of MCMs, the performance of GTMS and ATRM exhibited a marked decline; the PSDR of GTMS dropped to 93%, and the PSDR of ATRM dropped to 90%. The performance degradation may be mainly due to the usage of a one-way feedback mechanism. Relatively, FRAT exhibited robust performance in a complex network environment with a larger number of MCMs. These results were consistent with the actual situation, that is in a dishonest network environment, the MCM may conduct cooperative attacks, which may seriously affect the performance of the WSN environment. In order to improve the reliability of the proposed trust management mechanism, we adopted the idea that the overall trust of CM-to-CM was an adaptive combined value of bidirectional feedback (CM-to-CM feedback and CH-to-CH feedback). This new feedback mechanism could significantly improve the reliability of the proposed trust mechanism.

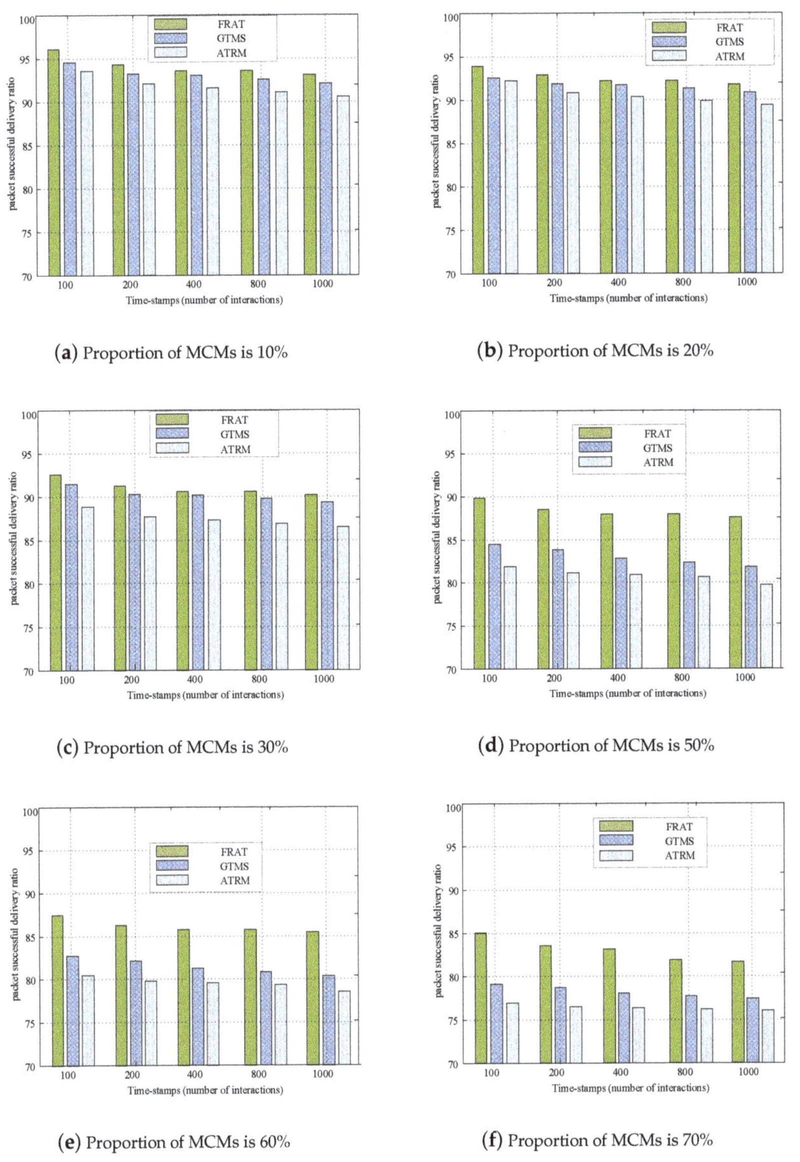

(**a**) Proportion of MCMs is 10%

(**b**) Proportion of MCMs is 20%

(**c**) Proportion of MCMs is 30%

(**d**) Proportion of MCMs is 50%

(**e**) Proportion of MCMs is 60%

(**f**) Proportion of MCMs is 70%

Figure 7. PSDR comparison with different percentages of malicious CMs (MCMs), where the proportion of malicious CHs is 10% GTMS, group-based trust computing mechanism; ATRM, a trust and reputation scheme.

6.3. Evaluation under Different MCHs

To evaluate the performance of the proposed trust mechanism at different MCH percentages, in this set of experiments, we assumed that each cluster environment was honest, and the MCM ratio was 20%. We set the proportion of MCHs to 10%, 20%, 30% 50%, 60%, and 70%, respectively. When the proportion of MCHs was set to 10%, the WSN environment was trustworthy. Most CHs in this network could keep their commitment and provide consistent stable feedbacks. When the proportion of MCHs was set to 20% or 30%, the WSN environment was relatively untrustworthy. More than half of the CHs in this WSN environment could keep their commitment and provide a consistently stable feedback. When the proportion of MCHs was set to 50%, the WSN environment was highly untrustworthy. Over half of the CHs in this WSN environment provided contrary feedback of the actual data for other CHs. Figure 8 shows a comparison of PSDR with different MCH percentages. A reliable trust computing system should have a strong ability to resist malicious behavior from MCHs.

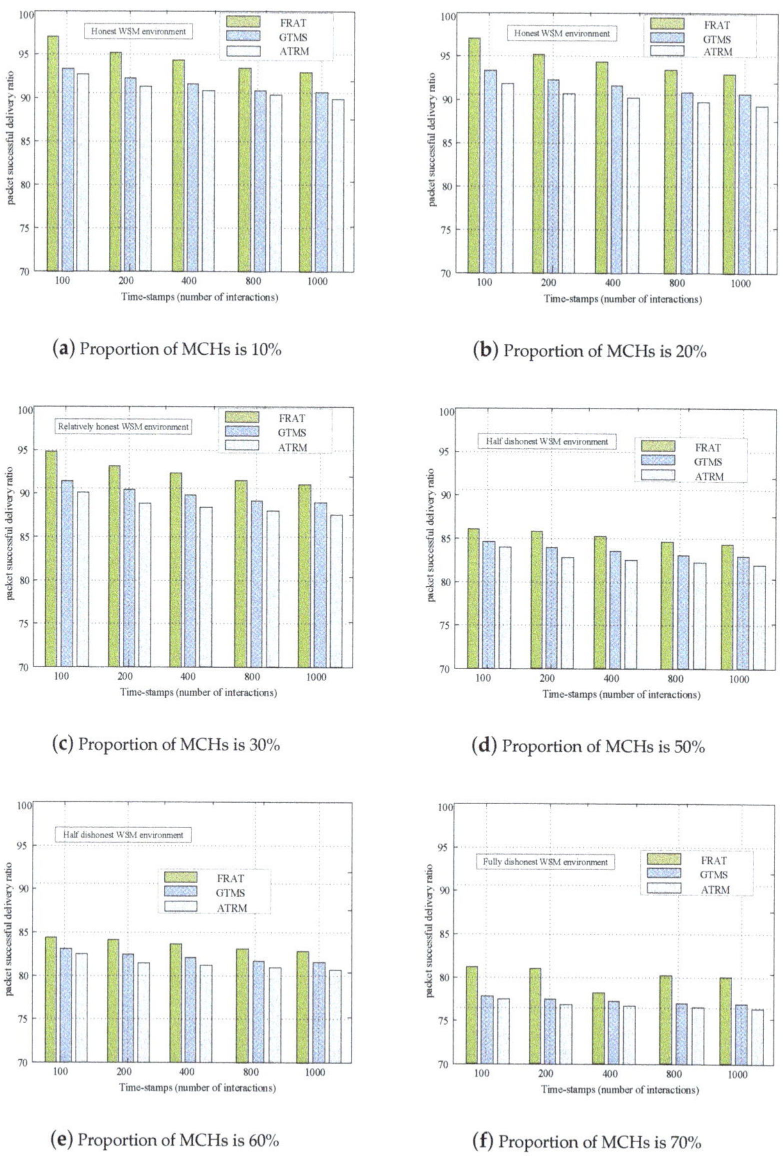

(**a**) Proportion of MCHs is 10%

(**b**) Proportion of MCHs is 20%

(**c**) Proportion of MCHs is 30%

(**d**) Proportion of MCHs is 50%

(**e**) Proportion of MCHs is 60%

(**f**) Proportion of MCHs is 70%

Figure 8. PSDR comparison with different percentages of MCHs, where the proportion of malicious CMs is 20%.

In order to evaluate the performance of trust mechanisms in more complex network environments, we gradually increased the proportion of malicious CHs in the system, and the proportion of MCHs was set to 10%, 20%, 30%, 50%, 60%, and 70% in Figure 8a–8f. Figure 8a shows an honest WSN environment, where the percentage of MCHs was only 10%. All three kinds of trust mechanisms had a high PSDR under this WSN environment, in which all values fluctuated around 90%. These results reflected that the three kinds of trust mechanisms exhibited high reliability under an honest WSN community.

With the increase in the percentage of MCHs, the WSN environment rapidly evolved from honest to fully dishonest. Figure 8d–f show that the performance of GTMS and ATRM exhibited a marked decline; the PSDR of GTMS dropped from 92% to 83%, and the PSDR of ATRM dropped from 90% to 82%. The performance degradation may be mainly due to the usage of a one-way feedback mechanism in GTMS and ARTM. Relatively, FRAT exhibited a more reliable performance in a complex network environment with a larger number of MCHs. These results were consistent with the actual situation, that is, in a dishonest network environment, MCHs may conduct cooperative attacks, which may seriously affect the performance of the WSN environment. To improve the reliability of the proposed trust management mechanism, we adopted the idea that the CH-to-CH overall trust was an adaptively merged value by the cross-validation feedback mechanism: CH-to-CH feedback and BS-to-CH feedback. This cross-validation feedback mechanism could significantly improve the anti-collaborative attack ability of the proposed trust mechanism. Thus, FRAT had a more robust reliability than GTMS and ATRM under five kinds of WSN environment, i.e., honest, relatively honest, partly dishonest, half dishonest, and fully dishonest, and it was suitable for trust computing under an open WSN.

6.4. Overhead Evaluation

To evaluate the performance in a large-scale network environment, we adopted different cluster numbers and different cluster sizes. Figure 9 shows the compared results of communication overhead under different network scales. Six types of network environments were evaluated: (a) the network consisted of 10,000 clusters, and each cluster included 20 nodes; (b) the network consisted of 10,000 clusters, and each cluster included 50 nodes; (c) the network consisted of 10,000 clusters, and each cluster included 100 nodes; (d) the network consisted of 10,000 clusters, and each cluster included 200 nodes; (e) the network consisted of 10,000 clusters, and each cluster included 300 nodes; and (f) the network consisted of 10,000 clusters, and each cluster included 500 nodes. We compared our mechanism with GTMS [18], ATRM [36], and UWSN [20].

As the value of each feedback was a positive integer between one and 10, one byte was required for each feedback information. Table 3 lists the communication overhead (bytes) under full-load conditions. When the number of nodes in each cluster was relatively small (Figure 9a–c), we could observe that the communication overhead of FRAT was far below that of the other two trust mechanisms, GTMS and ATRM, but slightly larger than UWSN. The reason was that UWSN adopted a flat wireless sensor networks and did not require the overhead of the CH node. When the number of nodes in each cluster was relatively larger (Figure 9d–f), we could see that the communication overhead of FRAT was far below those of GTMS and ATRM. The communication overhead of FRAT gradually approached that of UWSN. According to Theorem 3. and Figure 9, the proposed trust computing scheme based on the cross-validation mechanism needed less communication overhead, and it was suitable for large-scale resource-constrained WSNs.

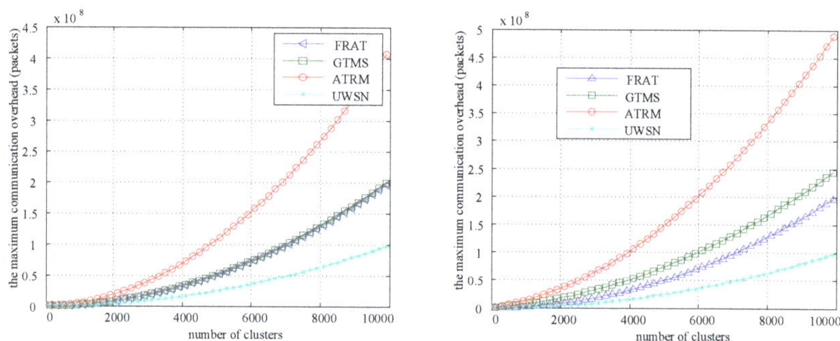

(**a**) network consists of 10,000 clusters, each cluster includes 20 nodes

(**b**) network consists of 10,000 clusters, each cluster includes 50 nodes

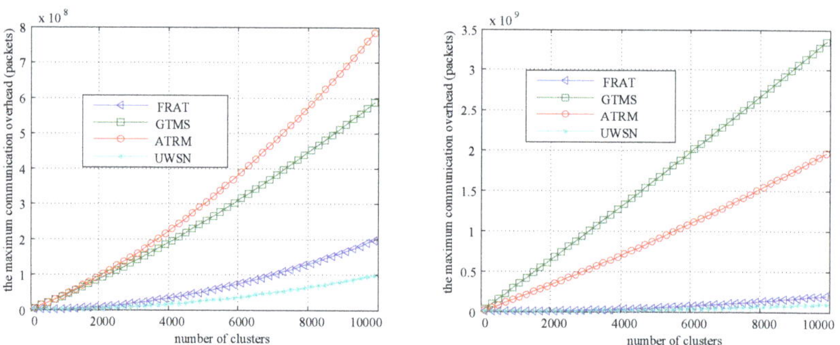

(**c**) network consists of 10,000 clusters, each cluster includes 100 nodes

(**d**) network consists of 10,000 clusters, each cluster includes 200 nodes

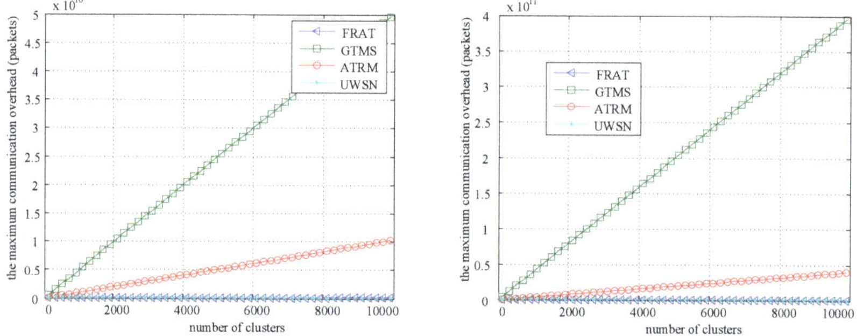

(**e**) network consists of 10,000 clusters, each cluster includes 300 nodes

(**f**) network consists of 10,000 clusters, each cluster includes 500 nodes

Figure 9. Comparing results of communication overhead under different network scales.

Table 3. Communication overhead under full-load conditions.

	Communication Overhead
FRAT	$2I^2 + 2J^2 + 2I * J$
GTMS	$2J(I(I-1)(I-2) + (J-1)$
ATRM	$4J(I(I-1) + (J-1))$
UWSN	$2J^2$

7. Conclusions

In this study, we proposed a trust computing scheme based on a cross-validation mechanism for clustered WSNs. Based on the theory of standard deviation analysis, this mechanism could remove the biased factor from multiple feedback sources. The theoretical analysis and experimental results provided useful insights. In a highly complex WSN environment with large percentages of malicious and selfish nodes, the proposed trust computing scheme based on the cross-validation mechanism may be insignificant, and thus, it should be given considerable attention in practical WSN applications. However, future work can pursue the following research directions:

- In a real deployment, nodes leave/join different clusters. Thus, future work can consider designing a scheme with node mobility.
- The proposed cross-validation mechanism was designed for clustered WSN. How to extend this mechanism to a flat WSN is another important direction.

Author Contributions: Conceptualization, C.L. and X.L.; methodology, X.L.; software, X.L; validation, C.L. and X.L.; formal analysis, X.L.; investigation, X.L.; resources, X.L.; data curation, X.L.; writing–original draft preparation, C.L. and X.L.; writing–review and editing, C.L. and X.L.; visualization, C.L. and X.L.; supervision, X.L.; project administration, X.L.; funding acquisition, X.L. All authors have read and agreed to the published version of the manuscript.

Funding: This work was supported by the National Nature Science Foundation of China (61872110, 61672111), the Joint Fund of NSFC-General Technology Fundamental Research (U1836215), and Capital Science and Technology Leading Talent Training Project, China (Z191100006119030).

Acknowledgments: The authors would like to convey their heartfelt gratefulness to the reviewers and the editor for the valuable suggestions and important comments which greatly helped them to improve the presentation of this paper.

Conflicts of Interest: The authors declare no conflict of interest.

References

1. Iyengar, S.S.; Brooks, R.R. (Eds.) *Distributed Sensor Networks: Sensor Networking and Applications*; CRC Press: Boca Raton, FL, USA, 2016.
2. Kobo, H.I.; Abu-Mahfouz, A.M.; Hancke, G.P. Fragmentation-Based Distributed Control System for Software-Defined Wireless Sensor Networks. *IEEE Trans. Ind. Inform.* **2019**, *15*, 901–910. [CrossRef]
3. Wu, M.; Tan, L.; Xiong, N. Data prediction, compression, and recovery in clustered wireless sensor networks for environmental monitoring applications. *Inf. Sci.* **2016**, *329*, 800–818. [CrossRef]
4. Tian, Z.; Gao, X.; Su, S.; Qiu, J. Vcash: A Novel Reputation Framework for Identifying Denial of Traffic Service in Internet of Connected Vehicles. *IEEE Internet Things J.* **2020**. [CrossRef]
5. Wu, L.; Du, X.; Wang, W.; Lin, B. An Out-of-band Authentication Scheme for Internet of Things Using Blockchain Technology. In Proceedings of the IEEE ICNC 2018, Maui, HI, USA, 5–8 March 2018.
6. Kumar, D.; Aseri, T.C.; Patel, R.B. EEHC: Energy efficient heterogeneous clustered scheme for wireless sensor networks. *Comput. Commun.* **2009**, *32*, 662–667. [CrossRef]
7. Jin, Y.; Vural, S.; Moessner, K.; Tafazolli, R. An Energy-Efficient Clustering Solution for Wireless Sensor Networks. *IEEE Trans. Wirel. Commun.* **2011**, *10*, 3973–3983.
8. Younis, O.; Fahmy, S. HEED: A Hybrid, Energy-Efficient, Distributed Clustering Approach for Ad-Hoc Sensor Networks. *IEEE Trans. Mob. Comput.* **2004**, *3*, 366–379. [CrossRef]
9. Mali, G.; Misra, S. TRAST: Trust-based distributed topology management for wireless multimedia sensor networks. *IEEE Trans. Comput.* **2016**, *65*, 1978–1991. [CrossRef]

10. Li, X.; Zhou, F.; Du, J. LDTS: A Lightweight and Dependable Trust System for Clustered Wireless Sensor Networks. *IEEE Trans. Inf. Forensics Secur.* **2013**, *8*, 924–935. [CrossRef]

11. Lu, H.; Li, J.; Guizani, M. Secure and efficient data transmission for cluster-based wireless sensor networks. *IEEE Trans. Parallel Distrib. Syst.* **2014**, *25*, 750–761.

12. Desai, S.S.; Nene, M.J. Node-Level Trust Evaluation in Wireless Sensor Networks. *IEEE Trans. Inf. Forensics Secur.* **2019**, *14*, 2139–2152. [CrossRef]

13. Qiu, J.; Du, L.; Zhang, D.; Su, S.; Tian, Z. Nei-TTE: Intelligent Traffic Time Estimation Based on Fine-grained Time Derivation of Road Segments for Smart City. *IEEE Trans. Ind. Inform.* **2019**, *16*, 2659–2666. [CrossRef]

14. Tian, Z.; Luo, C.; Qiu, J.; Du, X.; Guizani, M. A Distributed Deep Learning System for Web Attack Detection on Edge Devices. *IEEE Trans. Ind. Inform.* **2019**, *16*, 1963–1971. [CrossRef]

15. Xiao, L.; Wan, X.; Dai, C.; Du, X.; Chen, X.; Guizani, M. Security in mobile edge caching with reinforcement learning. *IEEE Wirel. Commun.* **2018**, *25*, 116–122. [CrossRef]

16. Sun, Y.; Han, Z.; Liu, K.J.R. Defense of Trust Management Vulnerabilities in Distributed Networks. *IEEE Commun. Mag.* **2009**, *46*, 112–119. [CrossRef]

17. Yu, H.; Shen, Z.; Miao, C.; Leung, C.; Niyato, D. A Survey of Trust and Reputation Management Systems in Wireless Communications. *Proc. IEEE* **2010**, *98*, 1752–1754. [CrossRef]

18. Shaikh, R.A.; Jameel, H.; d'Auriol, B.J.; Lee, H.; Lee, S. Group-Based Trust Management Scheme for Clustered Wireless Sensor Networks. *IEEE Trans. Parallel Distrib. Syst.* **2009**, *20*, 1698–1712. [CrossRef]

19. Reddy, V.B.; Negi, A.; Venkataraman, S. Communication and Data Trust for Wireless Sensor Networks using DS Theory. *IEEE Sens. J.* **2017**, *7*, 3921–3929. [CrossRef]

20. Ren, Y.; Zadorozhny, V.I.; Oleshchuk, V.A.; Li, F.Y. A novel approach to trust management in unattended wireless sensor networks. *IEEE Trans. Mob. Comput.* **2014**, *13*, 1409–1423. [CrossRef]

21. Yin, L.; Luo, X.; Zhu, C.; Wang, L.; Xu, Z.; Lu, H. ConnSpoiler: Disrupting C&C Communication of IoT-Based Botnet through Fast Detection of Anomalous Domain Queries. *IEEE Trans. Ind. Inform.* **2019**. [CrossRef]

22. Shen, M.; Ma, B.; Zhu, L.; Mijumbi, R.; Du, X.; Hu, J. Cloud-Based Approximate Constrained Shortest Distance Queries over Encrypted Graphs with Privacy Protection. *IEEE Trans. Inf. Forensics Secur.* **2018**, *13*, 940–953. [CrossRef]

23. Tian, Z.; Shi, W.; Wang, Y.; Zhu, C.; Du, X.; Su, S.; Sun, Y.; Guizani, N. Real Time Lateral Movement Detection based on Evidence Reasoning Network for Edge Computing Environment. *IEEE Trans. Ind. Inform.* **2019**, *15*, 4285–4294. [CrossRef]

24. Xiao, L.; Li, Y.; Huang, X.; Du, X. Cloud-based Malware Detection Game for Mobile Devices with Offloading. *IEEE Trans. Mob. Comput.* **2017**, *16*, 2742–2750. [CrossRef]

25. Dong, P.; Du, X.; Zhang, H.; Xu, T. A Detection Method for a Novel DDoS Attack against SDN Controllers by Vast New Low-Traffic Flows. In Proceedings of the IEEE ICC 2016, Kuala Lumpur, Malaysia, 22–27 May 2016.

26. Fernández-Gago, M.C.; Roman, R.; Lopez, J. A survey on the applicability of trust management systems for wireless sensor networks. In Proceedings of the Third International Workshop on Security, Privacy and Trust in Pervasive and Ubiquitous Computing, Istanbul, Turkey, 19 July 2007; pp. 25–30.

27. Li, X.; Zhou, F.; Yang, X. Scalable Feedback Aggregating (SFA) Overlay for Large-Scale P2P Trust Management. *IEEE Trans. Parallel Distrib. Syst.* **2012**, *23*, 1944–1957. [CrossRef]

28. Li, X.; Ma, H.; Zhou, F.; Gui, X. Data-driven and Feedback-Enhanced Trust Computing Pattern for Large-scale Multi-Cloud Collaborative Services. *IEEE Trans. Serv. Comput.* **2018**, *11*, 671–684. [CrossRef]

29. Li, X.; Ma, H.; Zhou, F.; Gui, X. Service Operator-aware Trust Scheme for Resource Matchmaking across Multiple Clouds. *IEEE Trans. Parallel Distrib. Syst.* **2015**, *26*, 1419–429. [CrossRef]

30. Li, X.; Ma, H.; Zhou, F.; Yao, W. T-broker: A Trust-aware Service Brokering Scheme for Multiple Cloud Collaborative Services. *IEEE Trans. Inf. Forensics Secur.* **2015**, *10*, 1402–1415. [CrossRef]

31. Yuan, J.; Li, X. A Reliable and Lightweight Trust Computing Mechanism for IoT Edge Devices Based on Multi-Source Feedback Information Fusion. *IEEE Access* **2018**, *6*, 23626–23638. [CrossRef]

32. Yu, H.; Shen, Z.; Miao, C.; An, B. Challenges and opportunities for trust management in crowdsourcing. In Proceedings of the 2012 IEEE/WIC/ACM International Joint Conferences on Web Intelligence and Intelligent Agent Technology, Macau, China, 4–7 December 2012; pp. 486–493.

33. Dragoni, N. A Survey on Trust-Based Web Service Provision Approaches. In Proceedings of the 2010 Third International Conference on Dependability, Venice, Italy, 18–25 July 2010; pp. 83–99.

34. Anwar, R.W.; Zainal, A.; Outay, F.; Yasar, A.; Iqbal, S. BTEM: Belief based trust evaluvation mechanism for Wireless Sensor Networks. *Future Gener. Comput. Syst.* **2019**, *96*, 605–616. [CrossRef]

35. Zhan, G.; Shi, W.; Deng, J. Design and Implementation of TARF: A Trust-Aware Routing Framework for WSNs. *IEEE Trans. Depend. Secure Comput.* **2012**, *9*, 184–197. [CrossRef]

36. Boukerche, A.; Li, X.; L-Khatib, K.E. Trust-Based Security for Wireless Ad Hoc and Sensor Networks. *Comput. Commun.* **2007**, *30*, 2413–2427. [CrossRef]

37. Crosby, G.V.; Pissinou, N.; Gadze, J. A framework for trust-based cluster head election in wireless sensor networks. In Proceedings of the Second IEEE Workshop on Dependability and Security in Sensor Networks and Systems, Columbia, MD, USA, 24–28 April 2006; pp. 10–22.

38. Li, X.; Du, J. An Adaptive and Attribute-based Trust Model for SLA Guarantee in Cloud Computing. *IET Inf. Secur.* **2013**, *7*, 39–50. [CrossRef]

39. Liu, C.; Li, X.; Sun, M.; Gao, Y.; Duan, S. Bi-TCCS: Trustworthy Cloud Collaboration Service Scheme Based on Bilateral Social Feedback. *IEEE Trans. Cloud Comput.* **2020**. [CrossRef]

40. Deviation Analysis. Available online: https://en.wikipedia.org/wiki/ (accessed on 10 March 2017).

41. Yu, Y.; Li, K.; Zhou, W.; Li, P. Trust mechanisms in wireless sensor networks: Attack analysis and countermeasures. *J. Netw. Comput. Appl.* **2012**, *35*, 867–880. [CrossRef]

42. Whitby, A.; Jang, A.; Indulska, J. Filtering out unfair ratings in bayesian reputation systems. In Proceedings of the 7th International Workshop on Trust in Agent Societies, Autonomous Agents and Multi Agent Systems, New York, NY, USA, 19–23 August 2004; pp. 106–117.

43. Guerrero, D.A.; Jimenez, R.M.; Rodriguez-Colina, E. WSN simulation model with a complex systems approach. In Proceedings of the 2013 Summer Computer Simulation Conference, Society for Modeling & Simulation International, Toronto, ON, Canada, 7–10 July 2013; p. 41.

44. Abo-Zahhad, M.; Amin, O.; Farrag, M.; Ali, A. A survey on protocols, platforms and simulation tools for wireless sensor networks. *Int. J. Energy Inf. Commun.* **2014**, *5*, 17–34. [CrossRef]

45. Batool, K.; Niazi, M.A.; Sadik, S.; Shakil, A.R. Towards modeling complex wireless sensor networks using agents and networks: A systematic approach. In Proceedings of the TENCON 2014 IEEE Region 10 Conference, Bangkok, Thailand, 22–25 October 2014; pp. 1–6.

 *sensors*

Article

A Blockchain Framework for Securing Connected and Autonomous Vehicles

Geetanjali Rathee [1], Ashutosh Sharma [2,*], Razi Iqbal [3], Moayad Aloqaily [4], Naveen Jaglan [2] and Rajiv Kumar [2]

[1] Department of Computer Science and Engineering, Jaypee University of Information Technology, Waknaghat, Solan 173234, India
[2] Department of Electronics and Communication, Jaypee University of Information Technology, Waknaghat, Solan 173234, India
[3] College of Computer Information Technology, American University in the Emirates, Dubai 503000, UAE
[4] Gnowit Inc., 308 Legget Drive, Suite 206, Ottawa, ON K2K 1Y6, Canada
* Correspondence: sharmaashutosh1326@gmail.com

Received: 29 May 2019; Accepted: 3 July 2019; Published: 18 July 2019

Abstract: Recently, connected vehicles (CV) are becoming a promising research area leading to the concept of CV as a Service (CVaaS). With the increase of connected vehicles and an exponential growth in the field of online cab booking services, new requirements such as secure, seamless and robust information exchange among vehicles of vehicular networks are emerging. In this context, the original concept of vehicular networks is being transformed into a new concept known as connected and autonomous vehicles. Autonomous vehicular use yields a better experience and helps in reducing congestion by allowing current information to be obtained by the vehicles instantly. However, malicious users in the internet of vehicles may mislead the whole communication where intruders may compromise smart devices with the purpose of executing a malicious ploy. In order to prevent these issues, a blockchain technique is considered the best technique that provides secrecy and protection to the control system in real time conditions. In this paper, the issue of security in smart sensors of connected vehicles that can be compromised by expert intruders is addressed by proposing a blockchain framework. This study has further identified and validated the proposed mechanism based on various security criteria, such as fake requests of the user, compromise of smart devices, probabilistic authentication scenarios and alteration in stored user's ratings. The results have been analyzed against some existing approach and validated with improved simulated results that offer 79% success rate over the above-mentioned issues.

Keywords: connected vehicles, internet of vehicles; security; IoT; blockchain; vehicular ad-hoc network

1. Introduction

In order to increase their own comforts, human daily life routines have been partially or completed replaced by embedded or automated machines in all aspects of business. Embedded systems monitor their environmental surroundings and subsequently respond or control the situation without any human intervention [1]. Recently, with the development of wireless communications and the advancement in vehicular industry, vehicular ad-hoc networks (VANETs) have become a mature research area. A VANET consists of a group of stationary and moving vehicles connected via a wireless network. Until recent times, the main use of VANETs was to provide comfort and safety to drivers in a vehicular environment [2]. However, this view is changing the infrastructure towards intelligent transportation systems where vehicles are connected and communicate using smart devices. Connected and Autonomous Electric Vehicles (CAEVs) is the most emerging vehicular technology among them all [3].The optimum value of this destructive technology has been seen a big successful business model for auto makers [4] [Vehicles that may connect to the internet and provide improved data sharing in the form of risk data, sensory and localization data and environmental perception is known as the internet of vehicles (IoV) or connected autonomous vehicles (CAV) [5]. In addition, with the continuous increase in urban population and rapid expansion of cities, vehicle ownership has been increasing at an exponential rate. CAV has been considered as one of the essential applications of VANETs where vehicles are becoming smarter by having sensors, adapters and control units for monitoring and communicating with their surroundings. A potential area for the application of CAV that has witnessed an unprecedented growth is in online cab booking services. The use of CAV yields a better result in vehicle entertainment experiences and helps in reducing the congestion by allowing the current information to be obtained by the vehicles instantly. Recently, online cab facilities or ride sharing services have drastically changed the public transportation industry and have been widely accepted by the users to avail the services at any time [6]. Further, it reduces the overhead of money negotiation between driver and customer and allows the customers to book their ride on a phone by tracking driver availability through global positioning systems (GPS). In spite of several advantages of using these services, there are various issues that need to be tackled, such as any person may register its vehicle number and provide online cab sharing services with the means of benefitting his/her personnel concerns by compromising the smart/IoT (internet of things) sensors/devices. However, until now, there exists no tracking systems or recording mechanisms that keep a check on compromised sensors or misbehavior with the customers during the rides, especially at night. A group of technical experts may forge the network system and increase or decrease their service ratings in order to continue their misconduct with the customers. Further, in case of any mishap or misbehaving, the registered taxi driver is punished with a low rating depending upon the behavior [7]. Further, the online cab services which provide cab booking facilities and customized pick up taxi convenience can be further automated and secured by connecting and analyzing using sensors. Every vehicle may connect to the IoT devices so that any malicious activity can be analyzed and averted. An intruder can try to alter the stored information or compromise the vehicle or IoT device for their selfish interests. However, using a secure mechanism, almost all activities can be traced in real time, such as traffic jams, weather conditions that hamper the drive, vehicular damage and repair etc. In addition, intruders may further compromise some IoT sensors in order to increase their credit points or to disable their location. Further, any alteration or change in stored data or information may not be transparently reflected to other drivers or users in the network that further encourages these people to continue their misbehavior [8]. As soon as the driver uses this application, his/her necessary information is registered somewhere and the vehicle is tracked through sensors during the nights for customer safety. Therefore, a blockchain technique has engrossed the attention of organization associates across a broad spectrum of industries, such as healthcare, real estates, transportation, government sectors and finance [9–11].The demand for transparency is rising at an astounding pace. In addition, it is able to ensure the security and transparency among the users in spite of IoT devices being vulnerable to intruders. Further, a blockchain technique is able to track, organize and bear out communications by storing the data from a large number of devices and facilitating the formation of parties without any federal cloud. The blockchain can confine the devices or CAV activities and trace the location from IoT objects when the cab moves from one place to another. Supply chain usage is the most relevant part of a blockchain technique for resolving the tangible harms to businesses due to the requirement of analysis of IoT devices or a vehicle's legal or illegal activity information. Furthermore, the need of the blockchain in CAV is that it would capture the vehicle's location, trace the vehicle's position, record information or cab riding ratings phenomenon from IoT objects committed to the components or vehicle. Supply chain usage is the most general application of a blockchain for resolving real business issues due to the dearth of traceability of vehicle locations or in relation to the users or vehicles moving through the supply chain [12].

Figure 1 depicts a typical vehicular blockchain network where IoT objects (I1, I2 … I6) are connected among several peers. Further, the vehicles are considered as various peer nodes in the network which are further divided into miner nodes depending upon their service criteria. Miner nodes are responsible for validating the trust of remaining nodes or IoT devices while peer nodes are part of the entire blockchain.

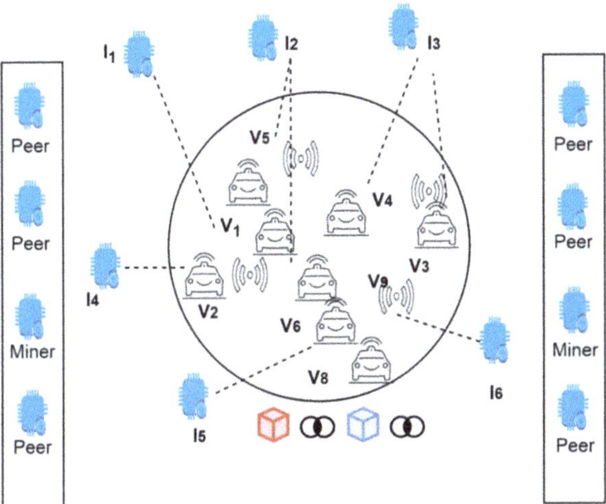

Figure 1. The blockchain framework.

1.1. Motivation of the Paper

The motivation of this manuscript is to provide security in the CAV network through IoT devices, safety and transparency among the users/customers during cab riding through a blockchain technique. During the movement of a user from one place to another, the vehicle number, current and previous vehicle ratings, captured through IoT devices are stored at the blockchain network. Therefore, even if intruders hack one or more IoT objects in order to gain their benefits, the users or vehicles present in the network are aware of the information registered under that compromised IoT device. Although, most of the researchers and scientists have proposed various smart frameworks in VANETs, however, most of the work is still at an early stage of CAV development. Further, many surveys have not affirmed the security concerns in the current progress of CAV applications.

1.2. Research Significance

Therefore, the scope of this paper is to ensure the transparency and security among drivers and customers using the blockchain. Further, the IoT sensors that automatically track the cab location and their identity using various IoT systems are traced by the blockchain network so that any compromise in any IoT device are recorded. The proposed blockchain framework has been validated against various security concerns such as user's fake requests, the compromise of IoT devices, probabilistic authentication scenarios and alteration in stored users' ratings. Further, the impact of using blockchain technology in IoV systems benefitted the society in variety of ways, such as ensuring the security and traceability of IoT devices or a vehicle's legal or illegal activity information. The experimental analysis of the proposed framework has been measured upon the illegal activities or communications done by malevolent IoT objects. The percentage of vehicles data security upon compromising of IoT devices has been discussed.

The rest of the paper is organized as follows. The survey of literature related to VANETs, vehicles and the blockchain along with various applications is presented in Section 2. The proposed blockchain framework for online cab services is described in Section 3. Further, Section 4 analyzes the performance metrics of the proposed framework against certain networking scenarios. Finally, Section 5 clinches the work and highlights the future scope of the paper.

2. Related Work

Various authors and researchers have formerly presented reviews on several use cases of VANETs and the blockchain in different areas. However, very few researchers have introduced a blockchain technique in CAV. In this section, a detailed description of the CAV along with various security techniques and role of the blockchain in one of the VANETs known as IoV has been elaborated.

Lin et al. [13] have proposed an intelligent transportation system in order to assist the drivers for efficient traffic schemes, optimal routes and dynamic guidance of routes during their travel. In this paper, during real time scenarios for reducing the fuel consumption and travel time with less road congestion, the authors have proposed a dynamic en-route decision real-time route guidance scheme. Further, the author's approach considers the real time traffic generation and transmission processes of the vehicular networks and predicts the traffic conditions based on multiple metrics by computing their trust probability. Furthermore, the authors have validated the proposed approach and shown the improved traffic efficiency by simulating the results in terms of efficient fuel, time and traffic parameters. Wang [14] introduced the basic methods and their major issues with their current applications for controlling and managing the traffic in parallel transportation management systems. The authors claim that the parallel transportation and management system is very effective for analyzing complex traffic networks. In this paper, the authors have described the parallel transportation and management system architecture, components, processes including Dyna, CAS, Itop, Adapts and Trans World. Finally, the author's proposed framework has been experimented and validated by analyzing real-world applications.

Hamid et al. [15] illustrated the overview of IoV by explaining its emergence, history and current applications of IoV in autonomous vehicles. Further, the authors pointed out certain issues occurred during IoV connectivity with the environment and its security concerns. In order to verify the CAV importance, a case study with computational simulation is done. In addition, various research ideas and future work directions are listed in smart city highways. The authors omitted a detailed technical specification of CAV. Wu et al. [16] introduced the IoV with its various applications by describing the background, notion, the IoV network architecture and their characteristics analysis along with its new research and challenges. Further, the authors described some enabling technologies by illustrating MAC standards and routing protocols. In addition, the core of this paper is to present a complete taxonomy with several categories, such as efficiency services, driving safety, informative services and intelligent traffic management system. Finally, Wu discussed the future directions in IoV research. With the continuous increase in urban population and rapid expansion in cities, vehicle ownership has been increasing at an exponential rate. In order to keep this view, traffic management has become a great issue in day to day life of human beings. The motivation of Dandala et al [17] in this paper is to provide a traffic management solution using CAV for overcoming the issues prevailing in daily life. Further, Liang et al. [18] provided an overview of VANETs from a research viewpoint. The paper begins with describing the basic network architecture by discussing three popular research issues and general methods. At last, the authors end with the analysis on research challenges and future drifts. Rawat et al. [19] presented data falsification threat detection using hashes for improving the network performance and security by acclimatizing the contention window size to broadcast accurate information to neighboring vehicles in timely manner. Further, the authors have proposed a clustering scheme to overcome travel time during traffic jams. The proposed mechanism is validated through numerical results attained from virtual simulations. Qian et al. [20] added cognitive engines in traditional CAV by restricting security strategies and transmission delays. Further, the study specified the switches of path selection as 0-1 programming issues and non-convex optimization problems. In addition, the 0-1 programming problem is converted into non-convex optimization via a log-det heuristics algorithm. The proposed mechanism is validated through experimental results. Sharma [21] proposed an efficient model proficient of handling energy demands of the blockchain enabled IoV by optimally controlling the number of transactions through distributed clustering. The simulated numerical results suggest that the proposed approach is 40.16% better in terms of energy conservation and 82.06% better in terms of transactions required to share the entire blockchain data compared with the traditional blockchain. Castillo et al. [22] discussed the IoV benefits along with topical industry standards expanded to endorse its implementations. They further present proposed communication protocols to facilitate operation and seamless integration of CAV. At last, IoV future research work was presented by requiring further deliberation from the vehicular research community.

Furthermore, Pustisek et al. [23] briefly explained the blockchain technique by outline architecture in the automatic selection of an electric vehicle charging station. Several blockchain use cases for prototypic implementation were presented. Various security concerns exist due to high exposure of information and data flow between vehicles to intersection and vehicle to vehicle. Buzachis et al. [24] proposed a blockchain framework for verifying, negotiating and facilitating among the consent entities. In this paper, the authors proposed a multi agent vehicle to intersection and vice versa communication to secure the vehicles through intersections. Further, Kuzmin et al. [25] introduced the concept of blockchain in unnamed aerial vehicles where each vehicle is considered as a node in which the functionality for reading and creating the transactions or communication exchange in done through the blockchain network. Yang et al. [26] used the concept of blockchain during the sharing of traffic flow among vehicles by ensuring the tamper resistant and data correctness in the agreement mechanism. The proof-of-event agreement is used to collect the traffic data bypassing of roadside vehicles. A two-phase transaction is introduced to access the warnings through the blockchain. The simulated result validates the proposed mechanism against tracing the events with trust verification. Further, in order to ensure the vehicles security, the authors proposed various authentication mechanisms by restricting the several attacks. By identifying various attacks such as replay, location spoofing, guessing and authentication time requirement, Chen et al. [27] proposed an improved security mechanism by forming a formal proof. Furthermore, the proposed mechanism is validated by comparing with existing results in terms of performance and security. Moreover few researchers have focused on secure information exchange between various vehicles where any intruder has the capability to disrupt integrity, authenticity, confidentiality. In a smart city for ensuring a secure message exchange, Dua et al. [28] proposed a novel elliptic curve cryptographic mechanism for providing two level authentications. For validating the proposed mechanism, the analysis is done using burrows logic along with formal and informal analysis using internet security protocols. Further, the proposed mechanism is compared with existing security schemes against high reliability, latency and overheads.

Malicious users in CAV may mislead the whole communication and create chaos on the road. Further, data falsification attack is one of the main security issues in CAV where vehicles rely on information received from other vehicles or peers. Until now, the numbers of secure CAV mechanisms have been proposed by different researchers and scientists, however, very few works have presented CAV with a blockchain technique. This paper has proposed the issue of IoT sensors which are compromised by expert intruders by proposing a blockchain framework.

3. Proposed Blockchain Framework for CAV Services Delivery

This section describes the blockchain framework of CAV that ensures the security and transparency of users and vehicles. In order to trace each and every activity of malevolent resources, a security mechanism is proposed that keeps track of each activity done by IoT sensors. Therefore, for providing and ensuring the security during ride sharing in CAV, each transmission among entities through smart devices is tracked. Although it would be easy to trace or record each and every activity of vehicles, however it may further enhance the complexity of computational communication during tractability in real time scenarios where upon the mobility of vehicles, intruders may attack through denial of service or man-in-middle threats. Whereas, in a case where IoT devices keep record of each and every vehicle, any attempt by intruders to compromise the IoT devices can be easily traced and identified. Also, since IoT devices form an upper layer in the network, the probability of attack is significantly reduced as compared to edge level comprising vehicles. For ensuring the security of smart devices, registered providers are verified, so that nobody changes, alters or tracks the information or IoT devices after they are casted. Also, individuality in money bank vehicles is completed, so that nobody can steel any tracking information. These issues can be easily resolved by a blockchain technique where the required smart contracts are defined which is the same as writing the rules, models, objects and code among the parties. Smart contracts are considered as a consensus or an agreement between the two parties. Once the smart contracts are designed, they cannot be further deleted or altered from the blockchain network. In this mechanism, there is no need of a central authority to provide validation of the work. All the nodes or vehicles may compute their results of contracts without any outside interference.

In the proposed blockchain framework, every automated vehicle or IoT device is registered or logged into the network before providing or accessing the vehicular services. Further, the necessary information of both vehicles and IoT devices are entered into an ordinary database initially and then stored in the blockchain permanently in order to track each and every activity of both the entities. Figure 2 depicts the architectural framework of CAV using the blockchain technique where all the vehicles are connected to IoT sensors or smart devices in order to control, monitor and guide the drivers on the road. In the proposed framework, the number of vehicles connected to the IoT devices or sensors depend upon their communication and transmission ranges. The vehicle number, ratings given by customers or users along with their IoT device are stored in ordinary tables as well as in the blockchain network to keep track and record each legal and illegal activity of the vehicle or IoT devices. In case of any IoT device being compromised by the intruders, the respective authorities which are part of the blockchain may be able to identify and take immediate actions against that compromised IoT device. Instead of recording IoT devices, each vehicle can be traced, analyzed and recorded over the blockchain. However, the keeping of records of such huge vehicular data during their mobility in real time scenarios may increase the possibility of computational power and time. Therefore, in order to limit the storage and computational power, it can be easy to record, analyze and store the activities of only IoT devices in the blockchain. The devices which trace a certain number of vehicles and provide services according to the user's request can be easily traceable and recordable in the blockchain. Each IoT device containing its vehicle record and providing services to different vehicles can be used as mining information to store over the block. Any change or alteration in information communication in vehicles or devices by intruders can alter the history or previous interactions that may further punish the devices or vehicles by blocking or reducing the ratings of vehicles. This paper details the security of both vehicles and IoT device through the blockchain in two different cases.

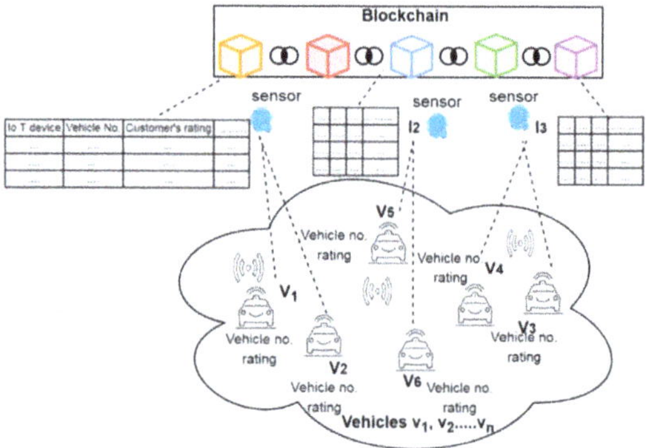

Figure 2. The architectural framework of a connected vehicle blockchain.

3.1. Vehicle Security: Registration of Every IoT Device on the Blockchain Network

In order to ensure security and transparency during a ride, every IoT device that provides the information about the vehicles registers itself on the blockchain network before providing the services to the vehicles. Further, every vehicle number or rating given by the customers is stored on the blockchain network. In CAV, smart objects continuously monitor and control the cab services and each IoT device authenticates to a peer in the blockchain network as depicted in Figure 3.

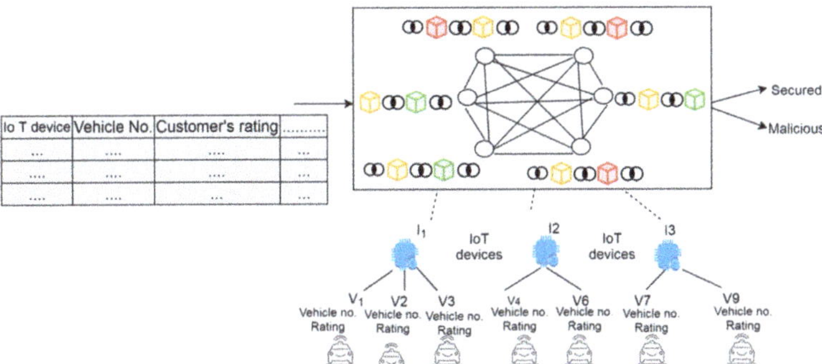

Figure 3. The blockchain network.

The blockchain network is a combination of peer and miner nodes that are responsible for generating the cryptographic keys and verifying the authenticity of a new vehicle or IoT device joining the network in order to avoid network failures. More than one manager is elected to ensure the security in the network for a particular period of time. Once the manager is selected, a secondary blockchain manager is chosen in order to recover from the failure of the primary manager. All the IoT devices or vehicles (providers) are registered in the blockchain network, by sending a subscription request to the peer manager. Further, the authenticity or legitimacy of each IoT device or provider is verified by the miner nodes with the help of device information, such as the international device identity, device identifier, innovation technology crop etc. Once the authentication is successful, miners generate a shared key that will help in further validation. At last, all IoT devices connect with their assigned peer nodes using their shared key between peer nodes and IoT devices/objects.

Each vehicle acts as a node connected to its subscribed or nearby peer nodes. The flowchart of the proposed framework is depicted in Figure 4. As depicted in Figure 4, whenever a user X needs to book a ride, he makes a ride request by sharing the time, pickup and drop-off points of the ride. In this paper, the users or customers are considered as legitimate and need not submit their identity in the blockchain network. This ride request is visible to all the registered providers in the network who are the part of the blockchain. A rider may get positive or negative reviews from other users based on their behavior. Various parameters are used in order to compute the ranking of providers such as trust factor or rating. The provider with a high trust factor (TF) or rating is considered to be most trustworthy. The user may choose its provider depending upon the rating or TF. In the cab ride service, various communications are performed between the service provider and the requester. If provider Y wants to respond to this request, it can share its intent to X.

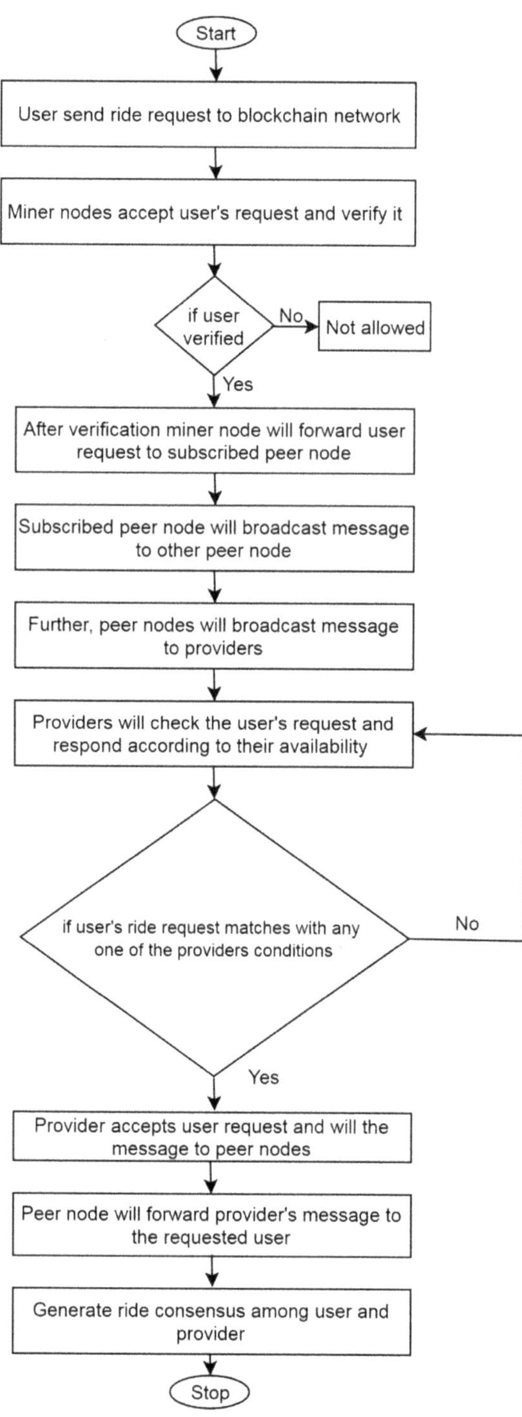

Figure 4. Flow work of the proposed framework.

The provider Y chooses to respond to a ride request based on certain criteria such as the user's route, where if the travel route of Y matches with the route of X, then Y accesses the ride request. Whenever a user X or provider Y' agree upon a ride request, a blockchain can be maintained along with a hash so that any misbehavior or alteration in the location pick or drop point can subsequently be identified in the network as depicted in Figure 5. Each block contains the information about the IoT devices attached with a previous block through a hash as depicted in Figure 6 so that any alteration or deletion of any information from the intruder can come to the notice of other devices.

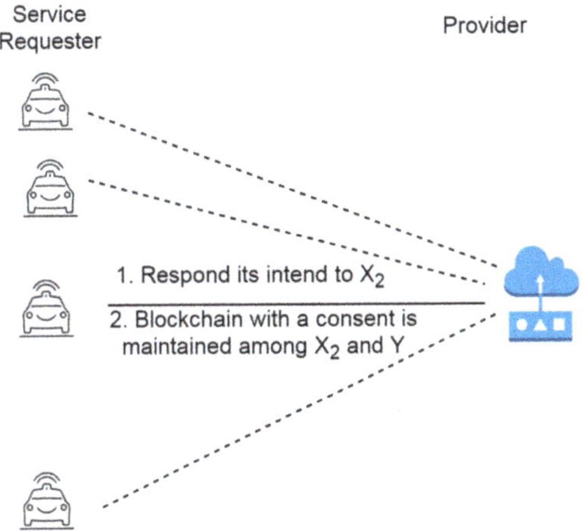

Figure 5. Consent through the blockchain among provider and ride requester.

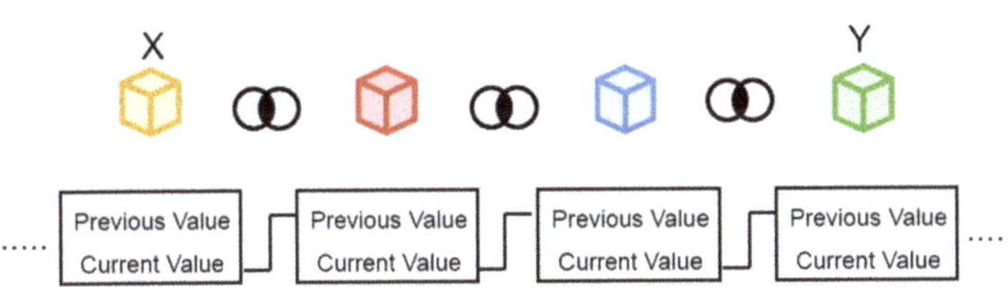

Figure 6. The blockchain among sender and receiver.

3.2. Attacking Scenarios

Whenever an intruder wants to perform some malicious activities in the network, it may adopt a number of attacking strategies. The compromise of IoT devices or sensors, modification of ratings given by riders, data falsification and traffic jams are some issues that can be easily generated by the intruders in order to fulfill their own interests. Attacking scenarios that can be possible during a ride service between user and vehicle are detailed as follows:

- Addition of compromised IoT by intruders: When the intruder registers its compromised IoT for executing its active or passive attacks, the blockchain peer nodes immediately identify by checking its illegal actions, like stealing or compromising of legitimate IoT devices.
- Misbehaving with the user: For example, a user, Alice, asks for a ride and a cab driver (provider) agrees to give the ride. However, during the ride the provider starts misbehaving with the user either by changing the route as chosen by the user, Alice or by stopping unnecessarily. Then, the IoT sensors which continuously monitor or trace the location of that cab takes action in order to prevent the user from any mishap. Simultaneously, the cab driver should be at the receiving end of the punishment with degradation in its rank or other necessary actions.

- Modification of ratings: Once the ratings have been submitted corresponding to any cab driver, it cannot be altered even after successfully compromising the IoT devices.
- Data falsification attack: It is one of the main security issues in CAV where vehicles rely on information received from other vehicles or peers.
- Traffic jam: In this, the intruders may try to divert the path suggestions on the roads for their own benefits.

However, in order to prevent these attacking strategies, this study has proposed a secure cab riding and sharing mechanism through the blockchain. Further, in order to validate the proposed mechanism, a numerical simulation is done on various parameters that show the outperformance of the proposed framework.

4. Performance Analysis

For validating the proposed framework, the simulation of the CAV blockchain framework has been ensured using the blockchain technique through NS2 simulator. In this paper, the possibility of attacks encountered at IoT devices or vehicles of the proposed framework has been analyzed. Initially, a network area of 700×700m was created having network sizes of 50 numbers of nodes where the vehicles are dynamic in nature and can abscond and join any other device's range as depicted in Table 1. For the deployment of network establishment, an initial random rating or TF (such as 70% and 5) was also assigned on the network to each device/vehicles and 5 nodes were created that act as blockchain nodes. In order to measure the validity of the proposed phenomenon, the performance was measured against several security metrics, such as the user's fake message alteration, the attack possibility on IoT devices and the modification of users' record information, such as the ratings. In order to measure validity or verify the proposed blockchain's CAV framework, NS2 simulator was used where numbers of attacking scenarios at IoT devices were considered. An attacking scenario or adversary model of the proposed framework is depicted in Figure 7 where intruders compromise the IoT devices either by forging the identity of legitimate devices or hacking the existing legitimate IoT sensors in the network. In order to validate or measure the authenticity of the proposed phenomenon, the attacking nodes were added with the rate of 10% for legitimate nodes in the network.

Table 1. Parameters.

Number of Nodes in a CRN	25, 500
Grid facet	700×700 m
Transmission Range	140 m (approx.)
Data Size or users request	256 Bytes
Simulation time	80 s
Physical Layer	PHY 802.11

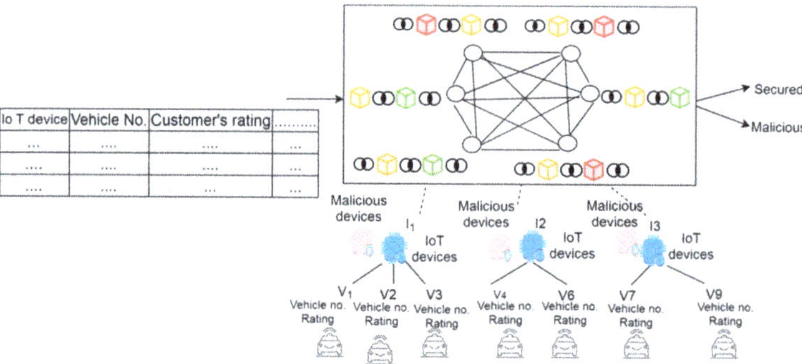

Figure 7. An adversary network model of the proposed phenomenon.

The intruders that try to compromise the IoT devices either by hacking the legitimate devices' identity (ID) in order to perform man-in-middle attack or behave like legitimate devices are considered for analyzing the proposed framework. Further, the proposed phenomenon has been verified against network congestion and compromising ability where intruders consume network resources by broadcasting fake requests. In addition, the proposed framework was measured against authentication probabilistic scenarios where attackers try to compromise IoT devices and showed how the possibility of attacks can still be analyzed and measured where intruders compromise the IoT devices. The proposed phenomenon showed false, no and correct authentication scenarios values that depicted how the proposed phenomenon efficiently measured the attacks where intruders compromised the devices. Further, in order to verify the framework against threats, malicious devices or nodes were added into the network based on normal distribution during the communication process. Further, the established blockchain environment is a combination of miners and peer nodes for validating and adding the new nodes (devices) in the network. Amongst them, some miner nodes were also converted to malevolent nodes to see the security recovery process. In addition, IoT devices were considered which are under the threat of intruders. The invasion of IoT signified that in a single unit of time, 2 out of 5, 10 out of 20 and 20 out of 50 devices were compromised as depicted in Table 2. Further, the user's fake request was considered another threat where insertion of a fake request by the intruders caused network congestion in the network. Taking all these assumptions, performance analysis was done for 60 s. The proposed framework was validated and compared against conventional approach as discussed in the below subsection.

Table 2. The configuration of NS2 for a different network environment.

S. No.	Transmitting Nodes	IoT Nodes	Compromised Miners	Attack Probability
1	25	5, 10, 20	2, 10, 20	5%
2	100	25, 50, 75	15, 25, 50	25%

Existing Method

Rawat et al. [17] presented data falsification threat detection using hashes for improving the network performance and security by acclimatizing contention window size to broadcast accurate information to neighboring vehicles in a timely manner. Further, the authors proposed a clustering scheme to overcome travel time during traffic jams. The existing mechanism was validated through numerical results attained from virtual simulation. The proposed paper analyzes the blockchain mechanism of IoT framework for CAVs over various networking parameters, such as the users' fake requests, compromise of IoT devices and alteration in the stored user's ratings. The proposed framework was measured against Rawat et al. [17] where the authors ensured the security by generating the hashes of information transmitted among the entities. However, the encrypted messages can be easily hacked and altered by the intruders. Further, a single change or compromise of IoT devices in the CAV network may be unaware of the entire network. However, in our proposed mechanism a single change in any information or device may immediately alert the remaining networks.

The proposed phenomenon results are measured against two attacking scenarios, i.e. network congestion and compromising ability that is further compared against existing approaches explained in subsection A of Section 4. The proposed method tries to improve the security of vehicles through the blockchain where every IoT device is recorded and analyzed for detecting the threat possibility. Rawat et al. on the other hand proposed a phenomenon where a block of hash chains of each vehicle was recorded over the network that may further have enhanced the possibility of an attack due to network congestion and computational power at the lower level of vehicles. The experimental evaluation of the proposed and conventional approaches was accomplished successfully and multiple results regarding various parameters have been recorded. The performance and system state parameters results are presented in previous subsections of the performance analysis. The system behaved as expected and all performance parameters for any CAV data were positive for the proposed framework. The movement and recording of activity is done by IoT devices that are static and can analyze and detect efficiently. The movement of vehicles allows new connection to the IoT device of their range where devices may collaborate among each other to further analyze their interactions.

Further, the accuracy of the proposed approach was close to 86% which may improve with the time because of removal of detected malicious nodes (MNs) from the system. The detection of MNs is based on trust where removal of detected MNs does not hinder the performance of other nodes. The proposed mechanism computes the trust of other nodes after every specific interval of time where nodes that are compromised and behave maliciously can have lower rating and trust because of high product loss ratios, black holes, and falsification attacks and may never be considered in the future. The depicted results showed the outperformance of the proposed mechanism against existing approaches with a success rate of 86%. This can be further improved if the experiment runs for a longer period. The measuring parameters in the proposed framework performed better in comparison to existing systems. Further, the obtained framework accuracy can be further improved with time because of the removal of detected MNs from the network. The detection of MNs followed by their removal does not alter the trust or hinder the performance of other nodes. The proposed mechanism computes the trust factor of their nodes after a specific interval of time. The nodes that are compromised and behave maliciously can have low ratings and trust, and may never be considered for the path formation. In all the depicted graphs from Figures 8–10, the proposed security framework outperforms better results against existing mechanisms. In case of the user's fake request graph as shown in Figure 8 corresponding to network congestion, the existing scheme performs less efficiently as the number of fake requests increase with the network size. The congestion of fake requests overloads the network and communication between the sender and the receiver and can become very difficult to maintain. Further, the increase of network congestion may consume the necessary resources that further leads to drastic degradation in network performance. Furthermore, corresponding to compromised devices, the data monitoring and controlling mechanism was affected highly as shown in Figure 9. In case of compromised IoT devices, the intruders not only affect the network performance, but also gain access to restricted areas or may further steal the confidential information for their own benefits. However, in the case of Figure 10, the intruders may alter the stored ratings of users and continue their misbehavior with their customers.

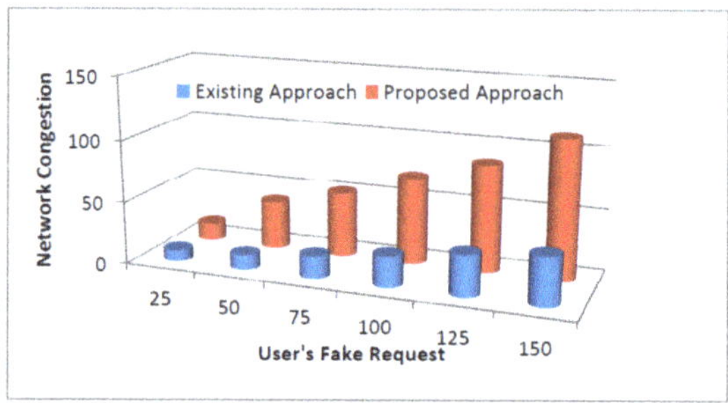

Figure 8. The users' fake requests corresponding to network congestion.

Along with the blockchain technology, all the necessary documenting or monitoring or controlling records are stored at the blockchain so that a single alteration, modification, deletion or compromise of any IoT device may quickly get under the surveillance and be known to other devices in order to secure or prevent from future possible harms. In our proposed phenomenon, a vehicular security framework was projected with the blockchain technique that enhanced the network performance and secured the online cab services. The performance analysis of proposed framework was further explained in detail along with verification time depending upon its various probabilistic attacking strategies. Figure 11 depicts the probabilistic strategies in terms of false, no and correct authentication and illustrates the validation of the proposed phenomenon with fake requests and compromising ability. The approach can be applied efficiently in real time scenarios by measuring its attacking possibilities.

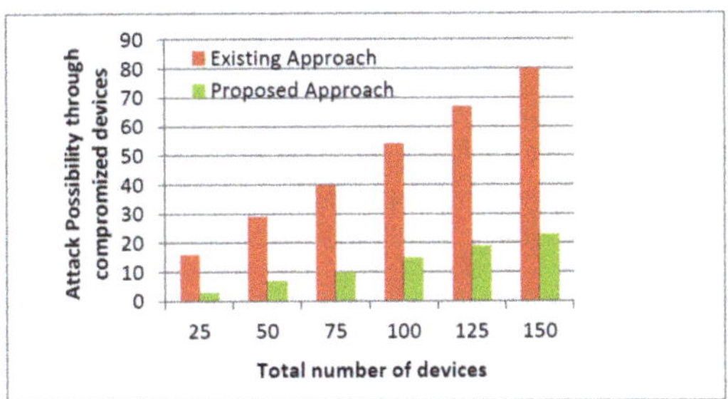

Figure 9. The attack possibility against compromised devices.

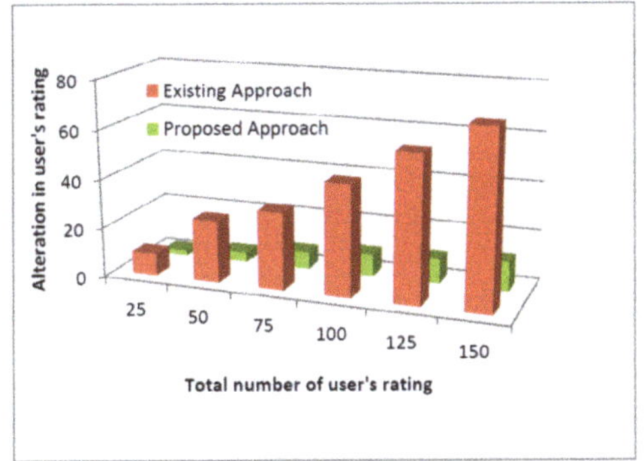

Figure 10. In user's stored ratings by intruders.

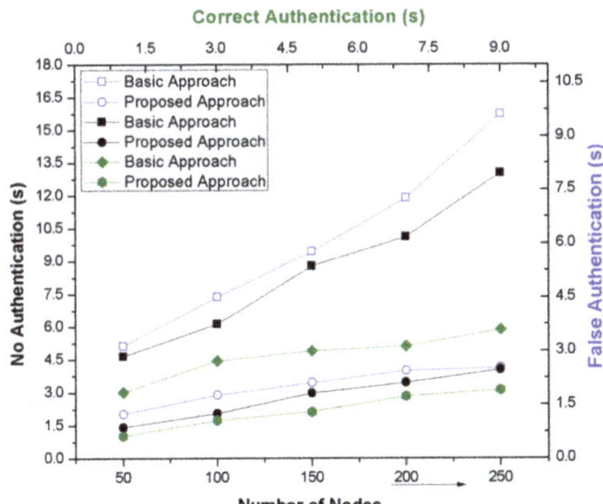

Figure 11. Probabilistic scenarios of attack possibility.

5. Conclusions

This paper has considered the IoV application and proposed a security mechanism for connected autonomous vehicles services framework using the blockchain technique. In order to provide the secrecy and transparency among the customers and cab drivers, each activity of the entities regarding vehicles or IoT devices is traced and recorded inside the blockchain. The blockchain mechanism is used to extract the information from IoT devices and store the extracted records in order to ensure the customer's safety and the devices' security by providing transparency among various authorities. The proposed framework significantly reduced the users' fake requests, the compromise of IoT devices and the alteration in the stored user's ratings. The simulated results against various parameters showed a 79% success rate in the proposed framework as compared to the existing approach against mentioned parameters. The proposed phenomenon against a larger number of nodes and the transaction alteration already stored at the blockchain network will be reported in future communications. Further, technology such as deep and reinforcement learning will be adopted to increase the system intelligent [29].

Author Contributions: The need of blockchain framework in vehicular technology along with literature survey is done by R.K. and A.S. The issue of security in smart sensors of connected vehicles that can be compromised by expert intruders addressed by a blockchain framework is proposed by G.R. and N.J. The identification and validation of the proposed mechanism based on various security criteria, such as fake requests of the user, compromise of smart devices, probabilistic authentication scenarios and alteration in stored user's ratings is detailed by R.I. The results validation against existing mechanisms along with success rate has been analyzed by M.A.

Acknowledgments: The authors are sincerely thankful to the anonymous reviewers for critical comments and suggestions to improve the quality of the manuscript.

Conflicts of Interest: The authors declare no conflict of interest.

References

1. Tonguz, O.; Wisitpongphan, N.; Bait, F.; Mudaliget, P.; Sadekart, V. Broadcasting in VANET. In Proceedings of the 2007 Mobile Networking for Vehicular Environments, Anchorage, AK, USA, 11 May 2007; pp. 7–12.

2. Hartenstein, H.; Laberteaux, K. *VANET: Vehicular Applications and Inter-Networking Technologies*; Wiley: Chichester, UK, 2010; Volume 1.

3. Alkheir, A.A.; Aloqaily, M.; Hussein, T.M. *Connected and autonomous electric vehicles (caevs)*; IEEE: New York, NY, USA, 2018; Volume 20, pp. 54–61.

4. Toglaw, S.; Aloqaily, M.; Alkheir, A.A. Connected, Autonomous and Electric Vehicles: The Optimum Value for a Successful Business Model. In Proceedings of the 2018 Fifth International Conference on Internet of Things: Systems, Management and Security, Valencia, Spain, 15–18 October 2018; pp. 303–308.

5. Zhang, J.; Wang, F.-Y.; Wang, K.; Lin, W.-H.; Xu, X.; Chen, C. Data-Driven Intelligent Transportation Systems: A Survey. *IEEE Trans. Intell. Transp. Syst.* **2011**, *12*, 1624–1639. [CrossRef]

6. Rezazadeh, H.; Azar, S.M.; Kisomi, M.S.; Bagheri, R. Robust cooperative maximal covering location problem: A case study of the locating Tele-Taxi stations in Tabriz, Iran. *Int. J. Serv. Oper. Manag.* **2018**, *29*, 163–183. [CrossRef]

7. Chim, T.W.; Yiu, S.M.; Hui, L.C.; Li, V.O. VANET-based secure taxi service. *Ad Hoc Netw.* **2013**, *11*, 2381–2390. [CrossRef]

8. Ammar, M.; Russello, G.; Crispo, B. Internet of Things: A survey on the security of IoT frameworks. *J. Inf. Secur. Appl.* **2018**, *38*, 8–27. [CrossRef]

9. Xu, T.; Wendt, J.B.; Potkonjak, M. Security of IoT systems: Design challenges and opportunities. In Proceedings of the 2014 IEEE/ACM International Conference on Computer-Aided Design, San Jose, CA, USA, 2–6 November 2014; pp. 417–423.

10. Sharma, A.; Rathee, G.; Kumar, R.; Saini, H.; Vijaykumar, V.; Nam, Y.; Chilamkurti, N. A Secure, Energy- and SLA-Efficient (SESE) E-Healthcare Framework for Quickest Data Transmission Using Cyber-Physical System. *Sensors* **2019**, *19*, 2119. [CrossRef] [PubMed]

11. Rathee, G.; Sharma, A.; Saini, H.; Kumar, R.; Iqbal, R. A hybrid framework for multimedia data processing in IoT-healthcare using blockchain technology. *Multimed. Tools Appl.* **2019**, *78*, 1–23. [CrossRef]

12. Khan, M.A.; Salah, K. IoT security: Review, blockchain solutions, and open challenges. *Future Gener. Comput. Syst.* **2018**, *82*, 395–411. [CrossRef]

13. Lin, J.; Yu, W.; Yang, X.; Yang, Q.; Fu, X.; Zhao, W. A Real-Time En-Route Route Guidance Decision Scheme for Transportation-Based Cyberphysical Systems. *IEEE Trans. Veh. Technol.* **2017**, *66*, 2551–2566. [CrossRef]

14. Wang, F.Y. Parallel control and management for intelligent transportation systems: Concepts, architectures, and applications. *IEEE Trans. Intell. Transp. Syst.* **2010**, *11*, 630–638. [CrossRef]

15. Hamid, U.Z.A.; Zamzuri, H.; Limbu, D.K. Internet of vehicle (IoV) applications in expediting the implementation of smart highway of autonomous vehicle: A survey. In *Performability in Internet of Things*; Springer: Cham, Switzerland, 2019; pp. 137–157.

16. Wu, W.; Yang, Z.; Li, K. Internet of Vehicles and applications. In *Internet of Things*; Elsevier: Amsterdam, The Netherlands, 2016; pp. 299–317.

17. Dandala, T.T.; Krishnamurthy, V.; Alwan, R. Internet of Vehicles (IoV) for traffic management. In Proceedings of the 2017 International Conference on Computer, Communication and Signal Processing (ICCCSP), Chennai, India, 10–11 January 2017; pp. 1–4.

18. Liang, W.; Li, Z.; Zhang, H.; Wang, S.; Bie, R. Vehicular Ad Hoc Networks: Architectures, Research Issues, Methodologies, Challenges, and Trends. *Int. J. Distrib. Sens. Netw.* **2015**, *11*, 745303. [CrossRef]

19. Rawat, D.B.; Garuba, M.; Chen, L.; Yang, Q. On the security of information dissemination in the Internet-of-Vehicles. *Tsinghua Sci. Technol.* **2017**, *22*, 437–445. [CrossRef]

20. Qian, Y.; Chen, M.; Chen, J.; Hossain, M.S.; Alamri, A. Secure Enforcement in Cognitive Internet of Vehicles. *IEEE Internet Things J.* **2018**, *5*, 1242–1250. [CrossRef]

21. Sharma, V. An Energy-Efficient Transaction Model for the Blockchain-Enabled Internet of Vehicles (IoV). *IEEE Commun. Lett.* **2019**, *23*, 246–249. [CrossRef]

22. Contreras-Castillo, J.; Zeadally, S.; Guerrero-Ibanez, J.A.; Contreras, J. Internet of Vehicles: Architecture, Protocols, and Security. *IEEE Internet Things J.* **2018**, *5*, 3701–3709. [CrossRef]

23. Pustišek, M.; Kos, A.; Sedlar, U. Blockchain based autonomous selection of electric vehicle charging station. In Proceedings of the 2016 International Conference on Identification, Information and Knowledge in the Internet of Things (IIKI), Beijing, China, 20–21 October 2016; pp. 217–222.

24. Buzachis, A.; Celesti, A.; Galletta, A.; Fazio, M.; Villari, M. A secure and dependable multi-agent autonomous intersection management (MA-AIM) system leveraging blockchain facilities. In Proceedings of the 2018 IEEE/ACM International Conference on Utility and Cloud Computing Companion (UCC Companion), Zurich, Switzerland, 17–20 December 2018; pp. 226–231.

25. Kuzmin, A.; Znak, E. Blockchain-base structures for a secure and operate network of semi-autonomous Unmanned Aerial Vehicles. In Proceedings of the 2018 IEEE International Conference on Service Operations and Logistics, and Informatics (SOLI), Singapore, 31 July–2 August 2018; pp. 32–37.

26. Yang, H.-K.; Cha, H.-J.; Song, Y.-J. Secure Identifier Management Based on Blockchain Technology in NDN Environment. *IEEE Access* **2019**, *7*, 6262–6268. [CrossRef]

27. Chen, C.-M.; Xiang, B.; Liu, Y.; Wang, K.-H. A Secure Authentication Protocol for Internet of Vehicles. *IEEE Access* **2019**, *7*, 12047–12057. [CrossRef]

28. Dua, A.; Kumar, N.; Das, A.K.; Susilo, W. Secure Message Communication Protocol among Vehicles in Smart City. *IEEE Trans. Veh. Technol.* **2018**, *67*, 4359–4373. [CrossRef]

29. Otoum, S.; Kantarci, B.; Mouftah, H.T. On the feasibility of deep learning in sensor network intrusion detection. *IEEE Networking Lett.* **2019**, *1*, 68–71. [CrossRef]

Article

An Adaptive Wake-Up-Interval to Enhance Receiver-Based Ps-Mac Protocol for Wireless Sensor Networks

Mohammed Sani Adam [1,*]**, Lip Yee Por** [1,*]**, Mohammad Rashid Hussain** [2]**, Nawsher Khan** [2]**, Tan Fong Ang** [1]**, Mohammad Hossein Anisi** [3]**, Zhirui Huang** [1] **and Ihsan Ali** [1,*]

[1] Faculty of Computer Science & Information Technology, University of Malaya, Kuala Lumpur 50603, Malaysia
[2] College of Computer Science, King Khalid University Abha, Abha 61421, Saudi Arabia
[3] School of Computer Science and Electronic Engineering, University of Essex, Colchester CO4 3SQ, UK
* Correspondence: majorsani@gmail.com (M.S.A.); porlip@um.edu.my (L.Y.P.); ihsanalichd@siswa.um.edu.my (I.A.)

Received: 12 June 2019; Accepted: 2 August 2019; Published: 29 August 2019

Abstract: Many receiver-based Preamble Sampling Medium Access Control (PS-MAC) protocols have been proposed to provide better performance for variable traffic in a wireless sensor network (WSN). However, most of these protocols cannot prevent the occurrence of incorrect traffic convergence that causes the receiver node to wake-up more frequently than the transmitter node. In this research, a new protocol is proposed to prevent the problem mentioned above. The proposed mechanism has four components, and they are Initial control frame message, traffic estimation function, control frame message, and adaptive function. The initial control frame message is used to initiate the message transmission by the receiver node. The traffic estimation function is proposed to reduce the wake-up frequency of the receiver node by using the proposed traffic status register (TSR), idle listening times (ILTn, ILTk), and "number of wake-up without receiving beacon message" (NWwbm). The control frame message aims to supply the essential information to the receiver node to get the next wake-up-interval (WUI) time for the transmitter node using the proposed adaptive function. The proposed adaptive function is used by the receiver node to calculate the next WUI time of each of the transmitter nodes. Several simulations are conducted based on the benchmark protocols. The outcome of the simulation indicates that the proposed mechanism can prevent the incorrect traffic convergence problem that causes frequent wake-up of the receiver node compared to the transmitter node. Moreover, the simulation results also indicate that the proposed mechanism could reduce energy consumption, produce minor latency, improve the throughput, and produce higher packet delivery ratio compared to other related works.

Keywords: wireless sensor networks; wake-up radio; medium access control protocol; receiver-initiated MAC protocol; traffic adaptation

---

1. Introduction

Wireless sensor networks (WSN) have gained a prolific attention in both academy and industry because of their wide-ranging applications, for example, health and remote monitoring/sensing [1,2]. WSNs are mostly deployed randomly in hostile and inaccessible areas. The main components of sensor nodes are a sensing unit, a control unit, a memory unit, and a limited battery power supply. The sensing unit is responsible for detecting the environment regarding humidity, temperature, and vibration. The control unit processes the sensed data and stores it into the memory unit of the sensor nodes. The sensor nodes are mainly powered by a battery with limited energy supply [3–6]. Research studies reveal that, in traditional WSN networks, the most energy consumption is done by the radio receiver. Furthermore, the radio receiver is monitored by the Medium Access Control (MAC) [7–10]. To utilize the wireless medium, MAC protocol was used to attain higher energy efficiency and low packet delivery latency. However, the leading bases of energy consumption in this protocol are idle listening and overhearing [11–13]. To reduced energy consumption, duty cycling has been widely used to design an energy-efficient MAC protocol [7,14]. Historically, the duty cycle mechanism has been proposed in MAC protocol, which falls into synchronous or asynchronous [11,15–18].

The synchronous mechanism coordinates neighboring sensor nodes to minimize energy consumption. Nonetheless, multi-hop time synchronization leads to huge overhead [19,20]. In disparity, the asynchronous duty cycle mechanism does not depend on any previous synchronization. Under a less congested traffic load, a huge number of the available asynchronous MAC protocols minimize energy consumption. However, the energy depletion in the synchronous duty cycling declines momentously with one or multiple communications and parallel traffic flows, and as a result, the synchronous does not scale well in huge and condensed networks [21,22].

Furthermore, in the duty cycling approach, sensor nodes periodically wake-up and check for an incoming message from the available channel. If there is no message received or sent, the sensor device changes from "active" to "sleep" mode to minimize energy consumption. Therefore, energy efficiency is the major concern in WSNs, and the MAC protocol handles most of the energy-related issues. The concern is to use the MAC protocol to improve the energy utilization of the WSN. As a result, designing a new protocol that reduces energy consumption is essential.

An Adaptive Wake-up-interval to enhance Receiver-based Preamble Sampling MAC protocol (AWR-PS-MAC) is proposed in this paper. Our proposed mechanism is designed for preventing the incorrect Traffic Status Register (TSR) convergence problem that causes the receiver node to wake-up more frequently than the transmitter node. The remainder of the paper is structured as follows. Relevant work of the receiver-initiated and the adaptive MAC protocols is described in Section 2. Then, background on the AWR-PS-MAC and the receiver-initiated MAC protocol discourse is given in Section 3. The methodology used and the simulation are explained in Section 4. The simulation results and evaluation through network simulations of the proposed mechanism compared with other related works are illustrated in Section 5. The final section of this paper presents the conclusion.

2. Related Works

Over the years, a large number of receiver-based PS-MAC protocols have been proposed by [23–30] to overcome the problem of a receiver node, which might wake-up twice or more compared to the transmitter node due to wrong traffic estimation. The following are the selected related works that have been analyzed based on strengths and weaknesses.

In [23], the authors proposed a protocol named Receiver-Initiated MAC (RI-MAC). In this protocol, when the receiver node is in an "active" mode, it broadcasts a preamble beacon message to the transmitter nodes. Then, the transmitter node starts sending the data message. Once the receiver node receives the data message, the receiver node sends another preamble beacon message. The preamble beacon message has two roles: (i) to acknowledge the data message was successfully received, and (ii) to notify the transmitter node that it is prepared to receive another data message. According to authors in [23], RI-MAC could minimize the energy depletion of the transmitter nodes because the transmitter nodes might only take part in transmitting a packet once they received the data message. However, the receiver node in this protocol uses a fixed wake-up interval and wakes-up twice or more compared to the transmitter node due to the incorrect TSR convergence because of the traffic pattern. Traffic patterns that are varied and unpredictable (variable traffic) might not be suitable to use with the fixed wake-up method, because this method might produce high latency for the transmitter nodes.

In [24], the authors proposed a protocol named Predictive Wake-Up MAC (PW-MAC). In PW-MAC protocol, if the transmitter node has a data message to transmit, the transmitter node switches into an "active" mode and waits for a preamble beacon message from the receiver node. After receiving the preamble beacon message, the transmitter node starts sending the data message. Once the receiver node receives the data message, the receiver node sends another preamble beacon message. The preamble beacon message has three roles: (i) to acknowledgement the data message was successfully received, (ii) to inform the transmitter node that it is prepared to receive another data message, and (iii) to provide the current wake-up time to the transmitter node so that it can predict the future wake-up interval of the receiver node. According to the authors, PW-MAC could limit the Idle Listening Time of the transmitter nodes, because this method could forecast the wake-up interval of the receiver node. However, the prediction mechanism used in this method considered only the current wake-up time of the receiver node. Using only the wake-up interval of the receiver node causes more frequent wake-up than the transmitter node, because the prediction does not consider traffic estimation. To predict a traffic pattern that is varied and unpredictable (variable traffic), other parameters such as traffic estimation might need to be taken into consideration.

In [31], the authors proposed a protocol named Traffic-Aware Dynamic MAC (TAD-MAC). In this protocol, when the receiver node is in an "active" mode, it periodically broadcasts a preamble beacon message to the transmitter nodes. Then, the transmitter nodes only start sending the data message. Once the receiver node has received the data message, it aligns its wake-up interval based on the traffic rate of the transmitter node. Such information is kept at the specific register called the TSR. After that, the receiver node sends another preamble beacon message. This preamble beacon message has two roles: (i) to acknowledge the data message was successfully received, and (ii) to inform the transmitter node that it is prepared to receive another data message. According to authors, TAD-MAC could reduce overhearing and Idle Listening Time of the transmitter nodes, because the receiver node could use TSR to determine the wake-up interval for each of the transmitter node. However, the receiver node in this protocol might wake-up more frequently than the transmitter node. Moreover, the traffic pattern that is varied and unpredictable (variable traffic) might increase the overhead of the receiver node, because it might need to align its wake-up interval, often due to the fluctuation of the traffic data rate produced by the transmitter nodes.

In [25], the authors proposed a protocol named Receiver-Initiated X-MAC with Tree Topology (TRIX-MAC). In this protocol, when the receiver node is in an "active" mode, it appends a new field into the preamble beacon message and periodically broadcasts it to the transmitter nodes. This field consists of the information about the number of time-slots that are required for each transmitter node to transmit their data message. The purpose of this field is to let the transmitter nodes forecast the next wake-up interval of the receiver node. According to the authors, TRIX-MAC could reduce energy consumption by enabling the transmitter nodes to forecast the receiver nodes' wake-up times. Moreover, this protocol could decrease the number of data message exchanges between the receiver node and the transmitter nodes. However, the receiver node in this protocol might still wake-up more frequently than the transmitter node, because it is difficult and challenging to approximate the data transmission rate of each of the transmitter nodes.

In [32], a protocol named Heuristic Self-Adaptive MAC (HSA-MAC) was proposed. In this protocol, the receiver node uses open-loop and closed-loop to adapt its wake-up and sleep patterns. Open-loop is used to evaluate the behavior pattern of the static traffic rate, while closed-loop is used to bring up the next wake-up interval of the receiver node. According to the authors, HAS-MAC allows the receiver node to adapt its wake-up and sleep intervals in static and variable traffic rates. However, the feedback method used in the closed-loop needs more energy, especially in variable traffic networks, because the receiver node needs to wake-up more frequently than the transmitter node to predict and update its wake-up and sleep intervals.

In [26], a protocol named Adaptive Scheduling Predictive-Wakeup MAC (AS-PW-MAC) was proposed. AS-PW-MAC uses a similar predictive mechanism as in PW-MAC protocol, but there is an additional pseudo-random scheduling generator incorporated in the predictive wake-up mechanism. This pseudo-random scheduling generator is used to enable the transmitter node to forecast the wake-up time of the receiver node so that the transmitter node can wake-up before the receiver node.

According to the authors, AS-PW-MAC is able to achieve a better packet delivery ratio in the maximum range of traffic loads compared to PW-MAC. However, AS-PW-MAC requires additional computation energy to forecast the wake-up time of the receiver node. Moreover, the pseudo-random scheduling generator might not be suitable to be used in variable traffic in WSNs, because the receiver node in AS-PW-MAC has to wake-up more frequently due to incorrect TSR convergence, and this protocol cannot detect traffic rate changes.

In [27], a protocol named Prediction-Based Asynchronous MAC (PBA-MAC) was proposed. The receiver node in this protocol sporadically wakes-up to broadcast a preamble beacon message to inform the transmitter nodes to start the data message transmission. Upon receiving the preamble beacon message by the transmitter node, the transmitter node can send the data message to the receiver node. PBA-MAC used an advanced mechanism to enable the transmitter node to forecast the wake-up time of the receiver node. According to the authors, the use of an exponential advance mechanism in PBA-MAC could reduce the communication cost by allowing the transmitter node to forecast the wake-up time of the receiver node. Although the transmitter node can forecast the wake-up time of the receiver node using the exponential advance mechanism, the receiver node might not be able to coordinate its wake-up interval with the traffic rate of the transmitter nodes. Therefore, the receiver node might wake-up more frequently than the transmitter node, and this might lead to high energy consumption because of avoidable wake-ups of the receiver node.

In [30], a protocol named A Low Duty Cycle Efficient MAC Protocol Based on Self-Adaption and Predictive Strategy (AP-MAC) was proposed. In this protocol, an information table is used to store the receiver nodes and the neighboring transmitter nodes information. The transmitter nodes use the information table to control the wake-up time of the receiver node. The transmitter nodes wake-up to listen to a preamble beacon message from the receiver node. After receiving the preamble beacon message, the transmitter nodes set up a channel to transmit the data message. The authors claimed that this protocol could reduce the crosstalk problem. Crosstalk happens when there is communication interference between transmitter nodes. Nonetheless, AP-MAC protocol is not proper to be implemented in variable traffic, because the traffic changes are unpredictable. Therefore, this protocol might result in unnecessary wake-ups that incur extra energy consumption.

In [33], an adaptive contention window MAC protocol was proposed to provide better throughput under a heavy load. This protocol chooses from the history of the collision to reflect the communication status and the usage of the wireless channel. A huge collision reflects greater competition in the wireless channel, and a big contention window is needed based on this condition. The authors claimed to prolong the access time to get rid of the competition. However, this exponential increase in speed when the traffic load is dense could possibly lead to avoidable delays. Therefore, this protocol might lead to heavy traffic load that incurs latency.

In [34], a protocol named self-adaptive sleep/wake-up scheduling was proposed. In this scheduling protocol, a reinforcement learning technique is used to allow every sensor node to independently choose its own operation mode—sleep, listen, or transmission mode—and every time slot is in a distributed system. In this proposed protocol, the time axis is distributed into the time slots. However, in each of the time slots, each sensor node independently chooses to either sleep or wake-up. Furthermore, it is primarily focused on the abstract level with some assumptions that the problem addressed has not been solved. Therefore, the aforementioned problem still exists under these assumptions.

In [35], a novel receiver-initiated MAC protocol was proposed to improve the data delivery packet delay. In this protocol, all sensor nodes choose a different time slot for beacon message transmission in a distributed system in respect to their transmission range, and hop counts from the based station improve the data delivery delay. The proposed protocol extended RI-MAC to achieve the aforementioned property. In addition, the author extended the back-off mechanism to support control messaging. Differently from the RI-MAC, when a node attempts to send a control message, it sends the message instantly after the beacon reception. However, the receiver node in this protocol uses a time slot, and it may not be suitable for variable traffic patterns. A traffic pattern that is varied and unpredictable (variable traffic) might not be suitable to use with the time slot method, because this method might produce high delay.

In [36], the authors presented a novel MAC protocol for energy-harvesting based WSNs using the advantage of ultralow-power wake-up radios. To eliminate this issue of a small range of wake-up radios, more than one hop wake-up technique based on a two radio system is proposed for aiding the communications between destination and any sensor nodes while upholding low latency and minimal energy consumption. To minimize the energy depletion, wake-up calls and data packets are sent using two different data rates with the address of the destination node. Moreover, by spending high data rate for data transmission, it reduces the risk of data transmission collisions. However, only three sensor nodes are used for evaluating the performance. As a result, conclusions could not be made, because the investigation was not carried out in larger and denser networks.

In [14], a Bird-MAC protocol for the Internet of Thing (IoT) applications was proposed that extremely minimizes the energy consumption of IoT applications in which sensor nodes report monitored status in a quasi-periodic manner, as in organized health-related and static environmental monitoring. The Bird-MAC protocol boasts extremely minimal energy consumption because it periodically sends the monitored information via a very low data rate. The high energy-saving impact of Bird-MAC (regardless of either a transmitter or a receiver) is because sensor nodes only wake-up for the duration of the actual clock drift among the transmitter–receiver pair, while the time duration of the wake-up is relative to the extreme clock drift. However, Bird-MAC requires additional computation energy to determine the maximum clock drift, and this protocol cannot be applied to variable traffic.

3. Background on Receiver-Initiated MAC and the Proposed Mechanism

The core idea of the receiver-based PS-MAC protocol is that the receiver node initiates the communication, assuming there is a single receiver node. In this protocol, the receiver node occasionally wakes-up and broadcasts a preamble beacon message to indicate that the receiver node is ready to accept a data message from any of the transmitter nodes. If no transmitter node has a data message to send, the receiver node switches back to the sleep "mode". This protocol addresses the high delay issue that occurred at the other transmitter nodes that have data messages to send, because this protocol prevents the communication channel from occupying the preamble message sent by the transmitter nodes [25]. However, a receiver node that uses this protocol to transmit a packet might wake-up twice or more compared to the transmitter node due to wrong traffic estimation [32]. This situation could cause energy wastage and produce high latency. Therefore, this research work was carried out to overcome the aforementioned problem. Figure 1 elucidates the receiver-initiated PS-MAC protocols. Note that the main problem in receiver-initiated PS-MAC protocols is how the transmitter and the receiver node pair define there mutually agreed upon wake-up time.

The figure is divided into two phases—before and after the convergence. In part (a), the receiver node periodically sends a Wake-Up Beacon (WUB) message to notify the transmitter nodes of its wake-up. In part (b), the transmitter node periodically wakes-up during its Wake-Up Interval (WUI) time, and the WUI is stored by the transmitter node. Before sending the control message, the transmitter node waits for the WUB from the receiver. This period is called Idle Listening Time (ILT), which is the activity that consumes the most energy in the receiver-initiated MAC protocols. After receiving the WB, the transmitter node sends the control message after sensing the medium. The communication ends with an acknowledgement (ACK) message from the receiver node to the transmitter node after it has successfully received the control message [37].

In this paper, we propose the AWR-PS-MAC designed to prevent the incorrect TSR convergence problem, which causes the receiver node to wake-up more frequently than the transmitter node.

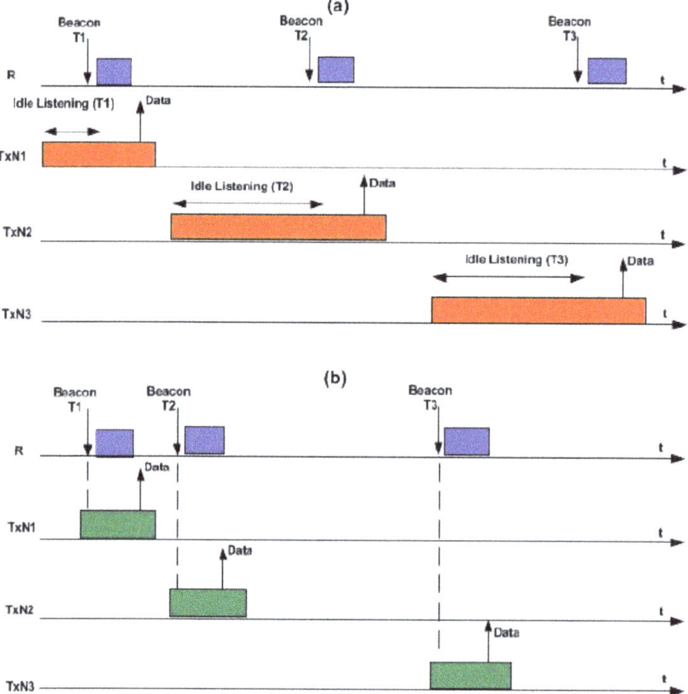

Figure 1. Receiver-initiated Preamble Sampling Medium Access Control (PS-MAC) general procedure [37]. (**a**) before convergence and (**b**) after convergence), and shows the communication of the three transmit nodes "TxN1, TxN2, and TxN3" trying to send data packet to a coordinating node (R). During the first phase (which we termed as an 'evolution phase'), before reaching a steady state (Figure 1a), each TxNi will waits for the WUB message from the receiver node before sending its data packet. The wake-up beacon packet is implored to an explicit transmit node containing its unique node ID (identifier). Whereas, other intending transmit nodes continue to wait for their respective wake-up beacons time. After several wake-ups, the receiver node adapts its WUI time based on the traffic it receives from each of the transmit node. In the second phase (i.e., after reaching the convergence as shown in Figure 1b), the receiver node has adapted its WUI time in such a way that ILT is minimized. To accommodate for the clock drift and hardware latencies, the receive node sends the WUB message slightly after its scheduled time to guarantee that the anticipating transmit node is already awake.

Figure 2 exemplifies the main components of the proposed mechanism. The proposed mechanism adopts and amends the WUI time and the TSR proposed by [31]. The improvised WUI time and TSR are used to minimize the energy wastage and the high latency problems caused by the wake-up time of a receiver node due to wrong traffic estimation. To simplify the explanation, a receiver node and a transmitter node are used to illustrate how our proposed mechanism works. Our proposed mechanism has four components: Initial Control Frame Message, Traffic Estimation Function, Control Frame Message, and Adaptive Function.

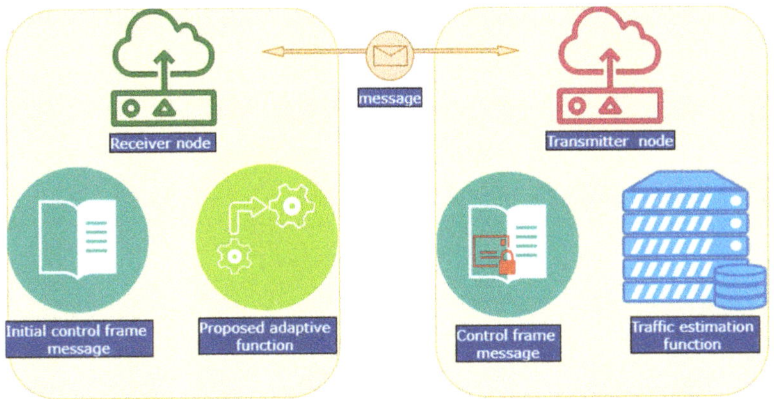

Figure 2. The main components of the proposed Adaptive Wake-Up Preamble Sampling MAC Protocol (AWR-PS-MAC).

3.1. Initial Control Frame Message

Initially, a receiver node initiates communication by sending an initial control frame message to a transmitter node. The initial control frame message consists of a WUB message with the size of two bytes. The WUB is a series of messages periodically sent by the receiver node to notify the transmitter node that there is a future data frame message. After every transmission, the receiver node waits for the transmitter node to respond.

Proposed Adaptive Function

Under the proposed adaptive function, the receiver node receives the Control Frame Message. The receiver node then replays an ACK message to the transmitter node. Then, the receiver node checks whether the WUB message is in an "active" mode. If the WUB message is in the "active" mode, the receiver node uses the proposed Adaptive Function stated in Equation (1) to calculate the WUI time. Otherwise, Equation (2) is used.

$$WUI_t(n+1) = \sum_{i=j}^{n} \frac{WUI_t(j) + ILT_n - ILT_k}{NW_{wbm} + 1} \tag{1}$$

$$WUI_t(n+1) = \sum_{i=j}^{n} \frac{WUI_t(j) + N_0(i) * t_{ref}}{NW_{wbm} + 1} \tag{2}$$

ILT_n and ILT_k, represent the Idle Listening Time, and n and k are auto-increment variables, which are used to keep track of the first two successful data packet transmissions. ILT_n and ILT_k are the ILTs for the first two successful data packet transmissions. NW_{wbm} represents the "Number of Wake-up without Receiving Beacon Message" and is used to store the statistical information of the received WUB Message. N_0 is the number of occurrence of zeros in the TSR for a transmitter node, tref is the simulation time, and j and i are two successful data packets received.

In the current system, the Adaptive Function cannot detect the changes in the traffic rate if the traffic rate is high [24,31]. In the proposed AWR-PS-MAC protocol, the receiver node adapts its WUI time based on the traffic information received using the proposed Adaptive Function. The proposed Adaptive Function uses the traffic information collected to reduce the wake-up frequency of a receiver node to minimize the energy depletion as well as the message overhead.

After going through the proposed Adaptive Function, the receiver node learns about the traffic information of each transmitter node. The receiver node updates the TSR corresponding to all the transmitter nodes. After that, the receiver node stores the transmitter node identification number, the WUI time, and the ILT. For the next data packet transmission, the receiver node uses the latest TSR information to determine the next WUI time for a transmitter node. With the proposed mechanism, the receiver node wakes-up close to the transmitter's WUI time. Thus, the proposed mechanism can reduce energy consumption as well as the message overhead.

3.2. Transmitter Node

The transmitter node consists of two functions—control frame message and traffic estimation function (see Figure 2). Below are the explanations of each of the functions.

3.2.1. Control Frame Message

After going through the proposed traffic estimation function, the transmitter node then transmits the Control Frame Message to the receiver node. The Control Frame Message consists of Frame Control, Address Information, Idle Listening Time, "Number of Wake-up without Beacon Message," Data Payload Field, and Check Sum fields (see Table 1). The Frame Control field belongs to the MAC header section. It has 1 byte. It also contains the type of message used by the proposed mechanism, such as WUB, Data Packet, and ACK. The Address Information field has 4 bytes. It stores the address information of the destination and source nodes. ILT has 1 byte. It is used to store the listening time when a sensor node is in "active" mode. The "Number of Wake-up without Beacon Message" field has 1 byte. It stores the information of the number of times the transmitter node wakes-up before it receives the WUB message. The Data Payload field has a variable data size. It holds the actual data to be sent by the transmitter node. The Check Sum has 2 bytes. It is used to determine whether an error has occurred during the data packet transmission. The receiver node then calculates the next wake-up interval time using the Adaptive Function.

Table 1. Control Frame Message (send by transmitter node).

Description	Size (Bytes)
Frame Control	1
Address Information	4
Idle Listening Time	1
Number of Wake-up without Beacon Message	1
Data Payload	Variable
Check Sum	2

3.2.2. Traffic Estimation Function

In the current approach, the process of storing traffic information does not consider the data rate of the transmitter node. An incorrect TSR convergence problem might occur when the receiver node wrongly predicts the WUI time based on the TSR record that stores the traffic information of the sensor nodes. Therefore, it makes the receiver node wake-up more frequently than the transmitter node. To overcome this problem, a traffic estimation function is proposed.

After the transmitter node has received the initial control frame message, the transmitter node calculates ILTn, ILTk, and NWwbm. n and k are auto-increment variables, which are used to keep track of the first two successful data packet transmissions. ILTn and ILTk are the ILTs for the first two successful data packet transmissions. NWwbm is the "Number of Wake-up without Receiving Beacon Message" from the receiver node. NWwbm is used to store the statistical information of the received WUB message. This information is used to prevent the incorrect TSR convergence problem that can cause the receiver node to wake-up more frequently than the transmitter node.

4. Simulation Experiments

Sensor nodes are randomly deployed to evaluate the proposed techniques using the parameters shown in Table 2. We used Objective Modular Network Testbed in C++ (OMNeT++) and mixed simulator (MiXiM) to simulate the transmission of the packets. Table 2 shows the simulation parameters and the values used for simulating the proposed mechanism under variable traffic. To have a fair comparison with the benchmark work, the configuration settings that are commonly used for simulating variable traffic were adopted from [32]. The simulation was run for a period of 2000 s, where 5 to 50 nodes were randomly deployed in an area of 500 m². The WUI time was set from 0.5 s to 2 s, and each of the simulation results were calculated after 100 stochastic simulations. The traffic rate used was based on the increase or the decrease of the ILT of the sender nodes, and the traffic rate range was from 1 to 10 frames/s. The TSR length used was set to 1 byte for storing the values of messages received by the sender node. One receiver node was used, and the constant bit rate (CBR) traffic model was used to generate traffic.

Table 2. The simulation settings.

Field	Values
Simulation Time	2000 s
Number of Nodes	Data
Node Distribution	Randomly
WUI time	0.5 s–2 s
Number of Ransom Simulation	100
Traffic Rates	1–10 frames/s
Bitrate	250 kbit/s
Rx (receive) Current	18.8 mA
Tx (transmit) Current	17.4 mA
Sleep Current	0.03A
Number of Receiver	1
Traffic Model	Constant Bit Rate

5. Result and Discussion

5.1. The Performance Analysis for Energy Consumption Proposed a Method and Other Related Works

The energy consumption is calculated using Equation (3). This equation is adopted from [38]. It is the standard formula used for calculating the energy consumed.

$$E = E_{tx}^{WUB} + E_{rx}^{WUB} + E_{tx}^{d} + E_{rx}^{d} + E_{sam} + E_{oh} + E_{s} + E_{i} \tag{3}$$

where E_{tx}^{WUB} is the amount of energy used in transmitting WUB. E_{rx}^{WUB} is the amount of energy used in receiving WUB. E_{tx}^{d} is the amount of energy that is essential to communicate a data packet. E_{rx}^{d} is the amount of the energy required to receive a data packet. E_{sam} is the amount of energy necessary to sample the channel for an ongoing transmission. E_{oh} is the energy of overhearing when a sensor node overhears a message that is destined to another sensor node. E_{s} is the amount of energy used when the mode of the sensor is in "sleep mode". E_{i} is the energy of ILT.

Figure 3 illustrates the energy consumption analysis results from the comparison among the proposed method, AP-MAC, PBA-MAC, AS-PW-MAC, HSA-MAC, TRIX-MAC, TAD-MAC, PW-MAC, and RI-MAC. The figure illustrates that the proposed mechanism had a very significant decrease in energy consumption compared to other related works. On average, the proposed method produced approximately 14%, 25%, 34%, 43%, 56%, 60%, 67%, and 77% less energy consumption compared to AP-MAC, PBA-MAC, AS-PW-MAC, HSA-MAC, TRIX-MAC, TAD-MAC, PW-MAC, and RI-MAC, respectively.

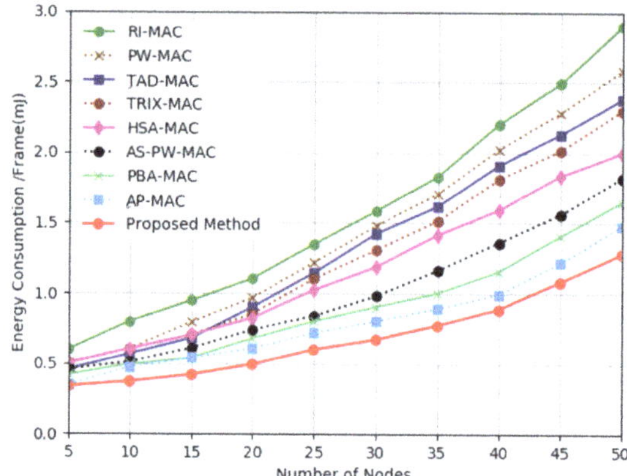

Figure 3. Energy consumption analysis.

On average, AP-MAC protocol produced the second least energy consumption compared to other related works. AP-MAC protocol managed to consume less energy because it utilized the information table to know the wake-up time of the receiver node. However, this protocol was still less effective than the proposed mechanism except for the number nodes at five (which tied with the proposed method).

On average, PBA-MAC protocol produced the third least energy consumption compared to other related works. However, it provided the same energy consumption as AP-MAC for the number nodes at 10 and 15. The reason that PAB-MAC managed to deliver less energy depletion compared to other protocols was that it used an advance mechanism to enable the transmitter node to forecast the wake-up time of the receiver node. However, this protocol still produced higher energy consumption compared to the proposed mechanism.

Generally, AS-PW-MAC protocol produced the fourth least energy consumption compared to other related works. However, it produced the same energy consumption as PBA-MAC protocol for the number nodes at 5 and 10. The reason that AS-PW-MAC managed to deliver less energy depletion compared to other protocols was that it used a pseudo-random scheduling generator to enable the transmitter node to forecast the wake-up time of the receiver node so that the transmitter node could wake-up before the receiver node. However, this protocol still consumed high energy depletion compared to the proposed mechanism.

TAD-MAC and TRIX-MAC protocol produced the fifth least energy consumption compared to other related works for the number nodes at 5 until 15. TRIX-MAC protocol dropped to sixth place when the number of sensor nodes increased to 20 until 50. The reason that TRIX-MAC managed to produce less energy depletion compared to other protocols was that TRIX-MAC enabled the transmitter nodes to forecast the receiver nodes' wake-up times. On the other hand, TAD-MAC protocol dropped to seventh place when the number of sensor nodes increased to 20 until 50. The reason that TAD-MAC protocol could not perform well compared to other protocols was that the receiver node in this protocol woke-up more frequently than the transmitter node.

HSA-MAC protocol and PW-MAC protocol produced the sixth least energy consumption compared to other related works for the number nodes at 5 until 10. PW-MAC protocol started to drop to seventh place for the number nodes at 15 and then to second to last place for the number nodes at 20 until 50. The reason that PW-MAC protocol produced higher energy depletion when the number of nodes increased was due to the prediction mechanism used in this protocol that made the receiver node wake-up more frequently than the transmitter node. Ironically, HSA-MAC protocol not only managed to retain sixth place for the number nodes at 15, but it also managed to improve its place, moving to fifth place for the number nodes at 20 until 50. The reason that HAS-MAC could reduce the energy depletion when the number of nodes increased was that this protocol allowed the receiver node to adapt its wake-up and sleep intervals in static and variable traffic rates.

In general, RI-MAC protocol produced the highest energy consumption compared to all the related works. The reason that RI-MAC protocol produced the highest energy consumption was that this protocol used a fixed wake-up interval, which made the receiver node wake-up more frequently than the transmitter node.

To recap, the results from the simulation show that the proposed mechanism can outperform other related works. This means that the proposed mechanism that uses ILT and NWwbm can prevent the incorrect TSR convergence problem that causes the receiver node to wake-up more frequently than the transmitter node.

5.2. The Performance Analysis for Energy Consumption Proposed a Method and Other Related Works

Equation (4) is used to determine the latency of all the protocols. This equation is the standard formula, and it is adapted from [38].

$$L = \frac{\sum_{i=1}^{P_{rx}} \left(T_{rxi} - T_{pgi} \right)}{P_{rx}} \tag{4}$$

where L is the latency. P_{rx} is the packet received. T_{rxi} is the time when the packet is received. T_{pgi} is the time when a generated packet is received.

Figure 4 shows the latency analysis results from the comparison among the proposed mechanism, AP-MAC, PBA-MAC, AS-PW-MAC, HSA-MAC, TRIX-MAC, TAD-MAC, PW-MAC, and RI-MAC. In general, the latency increased when the number of nodes increased for all the protocols. However, the proposed mechanism had the least latency compared to AP-MAC, PBA-MAC, AS-PW-MAC, HSA-MAC, TRIX-MAC, TAD-MAC, PW-MAC, and RI-MAC. The proposed mechanism reported 6%, 11%, 19%, 26%, 31%, 35%, 40%, and 45% lesser latency compared to AP-MAC, PBA-MAC, AS-PW-MAC, HSA-MAC, TRIX-MAC, TAD-MAC, PW-MAC, and RI-MAC, respectively. This analysis shows that the proposed mechanism can reduce the latency for the variable traffic by using the proposed adaptive function. The proposed adaptive function uses the latest TSR information to determine the next WUI time for the transmitter's WUI time. Therefore, it can help the receiver node to adapt its WUI time based on the traffic information received.

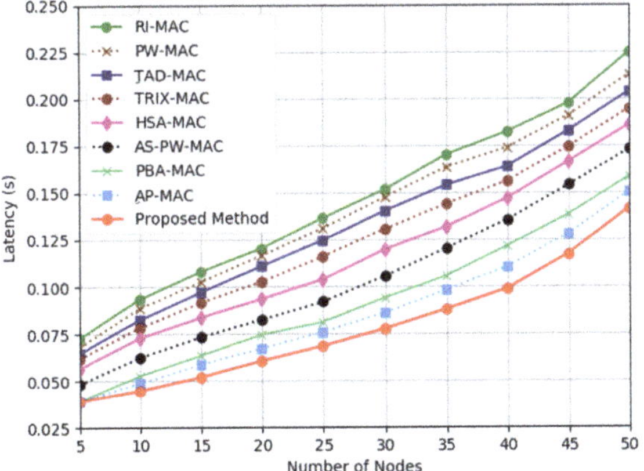

Figure 4. Latency analysis.

5.3. The Performance Analysis for Energy Consumption Proposed a Method and Other Related Works

A standard formula for calculating the throughput is adopted from [39] to determine the throughput of all the protocols. Hence, Equation (5) is used.

$$T = \frac{N_{rp}S_p}{T_s} \tag{5}$$

where T is the throughput. N_{rp} is the number of packets received. S_p is the size of the packet. T_s is the simulation time.

Figure 5 shows the throughput results from the analysis for the proposed mechanism, AP-MAC, PBA-MAC, AS-PW-MAC, HSA-MAC, TRIX-MAC, TAD-MAC, PW-MAC, and RI-MAC. The throughputs of the proposed mechanism, AP-MAC, PBA-MAC, AS-PW-MAC, HSA-MAC, TRIX-MAC, TAD-MAC, PW-MAC, and RI-MAC showed a linear increment when the number of notes increased. From the graph, the throughput produced by the proposed mechanism was about 11%, 21%, 27%, 36%, 46%, 54%, 59%, and 74% higher than AP-MAC, PBA-MAC, AS-PW-MAC, HSA-MAC, TRIX-MAC, TAD-MAC, PW-MAC, and RI-MAC respectively. The reason that the proposed mechanism can outperform the other methods is that the proposed adaptive function uses WUB message to notify the transmitter node so that the receiver node is ready to receive any incoming data packet. With this modification, it enables the proposed mechanism to produce higher throughput.

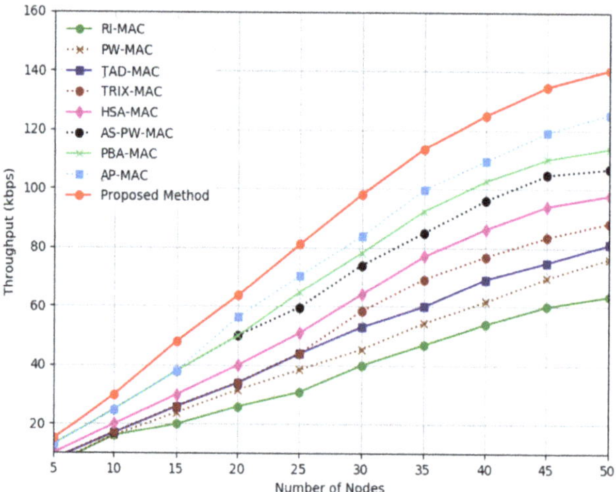

Figure 5. Throughput analysis.

5.4. The Performance Analysis for Energy Consumption Proposed a Method and Other Related Works

The packet delivery ratio is calculated using Equation (6). This equation is the standard formula used by [39] to compute the packet delivery ratio.

$$PDR = \frac{P_{rx} * 100}{\sum_{i=1}^{n} P_g} \tag{6}$$

where PDR is the packet delivery ratio. P_{rx} is the number of packets received. n is the number of sensor nodes. P_g is the total number of the packet generated.

Figure 6 presents the results with respect to the Packet Delivery Ratio for the proposed mechanism, AP-MAC, PBA-MAC, AS-PW-MAC, HSA-MAC, TRIX-MAC, TAD-MAC, PW-MAC, and RI-MAC. In general, the results obtained from the simulation demonstrated that the proposed mechanism had a higher Packet Delivery Ratio compared to AP-MAC, PBA-MAC, AS-PW-MAC, HSA-MAC,

TRIX-MAC, TAD-MAC, PW-MAC, and RI-MAC. The Packet Delivery Ratio of the proposed mechanism was about 7%, 10%, 13%, 18%, 22%, 25%, 44%, and 54% higher compared to AP-MAC, PBA-MAC, AS-PW-MAC, HSA-MAC, TRIX-MAC, TAD-MAC, PW-MAC, and RI-MAC, respectively. The reason that the proposed mechanism can provide better Packet Delivery Ratio is that the proposed adaptive function uses ILT_n, ILT_k, and NW_{wbm} to schedule the receiver node so that the transmitter node can receive an incoming data packet in a shorter waiting time.

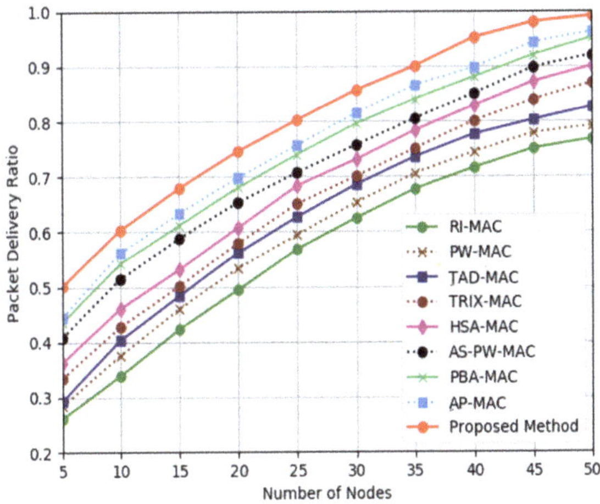

Figure 6. Packet Delivery Ratio analysis.

The simulation results show that, on average, the proposed mechanism can outperform other related works in terms of energy consumption, latency, throughput, and Packet Delivery Ratio by 14%, 6%, 11%, and 7%, respectively. The reason that the proposed mechanism can achieve better results is because the proposed Adaptive Function uses the latest TSR information to determine the next WUI time for the transmitter's WUI time. Therefore, it can help the receiver node to adapt its WUI time based on the traffic information received and prevent the incorrect TSR convergence problem that causes the receiver node to wake-up more frequently than the transmitter node.

6. Conclusions

The AW-RB-PS-MAC protocol was proposed. The proposed AW-RB-PS-MAC protocol consists of four components: Initial Control Frame Message, Traffic Estimation Function, Control Frame Message, and Adaptive Function. Initial Control Frame Message is the initial message sent by a receiver node. It consists of WUB message and has two bytes. Traffic Estimation Function is used by a transmitter node to estimate the traffic rate by using the TSR and the proposed variables. Control Frame Message consists of Frame Control, Address Information, Idle Listening Time, "Number of Wake-up without Beacon Message," Data Payload Field, and Check Sum fields. The Frame Control field belongs to the MAC header section. It has 1 byte. The Address Information field has 4 bytes. Idle Listening Time (ILT) has 1 byte. The "Number of Wake-up without Beacon Message" field has 1 byte. The Data Payload field has a variable data size. The Check Sum has 2 bytes. Finally, the receiver node to calculate the next WUI time of each of the transmitter nodes uses the Adaptive Function. The proposed Adaptive Function uses the proposed function in two ways. When the WUB is in "active" mode, Equation (1) is used; otherwise, Equation (2) is used. Furthermore, the proposed Adaptive Function uses new features to reduce energy consumption and high latency. The implementation of the proposed AW-RB-PS-MAC protocol was carried out in the OMNeT++ network simulator and the MiXiM framework. The results showed that the proposed mechanism could prevent the incorrect TSR convergence problem that causes the receiver node to wake-up more frequently than the transmitter node.

In the future, we will conduct research on preventing other issues, such as wake-up collision [32] and data packet collision [26] in variable traffic. The wake-up collision occurs when one or more transmitter nodes share the same wake-up time while they are within transmission range of each other. This means that these transmitter nodes send a wake-up beacon at the same time to the receiver node and collide in the transmission channel. This problem causes more energy consumption at the transmitter nodes. Data packet collision happens when more than one transmitter node sends a data packet concurrently to the receiver node using the same channel. This problem causes more packet retransmission and high energy consumption at the transmitter nodes. We believe that, by looking into these issues, the adaptation of the variable traffic method could be improved.

Author Contributions: M.S.A. wrote the manuscript under the supervision of L.Y.P., M.H.A. and I.A., T.F.A. and Z.H. contributed in organizing and streamlining the flow of the manuscript. M.R.H. and N.K. refined the overall manuscript and responses to the reviews.

Funding: The authors extend their appreciation to the Deanship of Scientific Research at King Khalid University for funding this work through Research Group Project under grant number R.G.P. 1/166/40, University of Malaya Postgraduate Research Grant (PG035-2016A) and Fundamental Research Grant Scheme (FRGS) (FP114-2018A) from the Ministry of Higher Education, Malaysia.

Conflicts of Interest: The authors declare that there are no conflict of interest.

References

1. Stojkoska, B.L.R.; Trivodaliev, K.V. A review of Internet of Things for smart home: Challenges and solutions. *J. Clean. Prod.* **2017**, *140*, 1454–1464. [CrossRef]
2. Scuotto, V.; Ferraris, A.; Bresciani, S. Internet of Things: Applications and challenges in smart cities: A case study of IBM smart city projects. *Bus. Process Manag. J.* **2016**, *22*, 357–367. [CrossRef]
3. Dinh, T.; Kim, Y.; Gu, T.; Vasilakos, A.V. L-MAC: A wake-up time self-learning MAC protocol for wireless sensor networks. *Comput. Netw.* **2016**, *105*, 33–46. [CrossRef]
4. Memon, K.A.; Memon, M.A.; Shaikh, M.M.; Das, B.; Zuhaib, K.M.; Koondhar, I.A.; Memon, N.U.A. Optimal Transmit Power for Channel Access Based WSN MAC Protocols. *Int. J. Comput. Sci. Netw. Secur.* **2018**, *18*, 51–60.
5. Malik, M.; Sharma, M. A Novel Approach for Comparative Analysis on Energy Effectiveness of H-MAC and S-MAC Protocols for Wireless Sensor Networks. *Adv. Sci. Eng. Med.* **2019**, *11*, 29–35. [CrossRef]
6. Yazdi, F.R.; Hosseinzadeh, M.; Jabbehdari, S. A Priority-Based MAC Protocol for Energy Consumption and Delay Guaranteed in Wireless Body Area Networks. *Wirel. Pers. Commun.* **2019**, 1–20. [CrossRef]
7. Carrano, R.C.; Passos, D.; Magalhaes, L.C.; Albuquerque, C.V. Survey and taxonomy of duty cycling mechanisms in Wireless Sensor Networks. *IEEE Commun. Surv. Tutor.* **2014**, *16*, 181–194. [CrossRef]
8. Rasheed, M.B.; Javaid, N.; Imran, M.; Khan, Z.A.; Qasim, U.; Vasilakos, A. Delay and energy consumption analysis of priority guaranteed MAC protocol for wireless body area networks. *Wirel. Netw.* **2017**, *23*, 1249–1266. [CrossRef]
9. Subramanian, A.K.; Paramasivam, I. PRIN: A priority-based energy efficient MAC protocol for wireless sensor networks varying the sample inter-arrival time. *Wirel. Personal Commun.* **2017**, *92*, 863–881. [CrossRef]
10. Kumar, A.; Zhao, M.; Wong, K.J.; Guan, Y.L.; Chong, P.H.J. A Comprehensive Study of IoT and WSN MAC Protocols: Research Issues, Challenges and Opportunities. *IEEE Access* **2018**, *6*, 76228–76262. [CrossRef]
11. Asudeh, A.; Záruba, G.V.; Das, S.K. A general model for MAC protocol selection in wireless sensor networks. *Ad Hoc Netw.* **2016**, *36*, 189–202. [CrossRef]
12. Tolins, J.; Zeamer, C.; Fox Tree, J.E. Overhearing dialogues and monologues: How does entrainment lead to more comprehensible referring expressions. *Discourse Process* **2018**, *55*, 545–565. [CrossRef]
13. Ramadan, K.F.; Dessouky, M.I.; Abd-Elnaby, M.; El-Samie, F.E.A. Node-power-based MAC protocol with adaptive listening period for wireless sensor networks. *AEU Int. J. Electron. Commun.* **2018**, *84*, 46–56. [CrossRef]

14. Kim, D.; Jung, J.; Koo, Y.; Yi, Y. Bird-MAC: Energy-Efficient MAC for Quasi-Periodic IoT Applications by Avoiding Early Wake-up. *IEEE Trans. Mobile Comput.* **2019**. [CrossRef]
15. Fernandes, R.F.; Brandão, D. Proposal of Receiver Initiated MAC Protocol for WSN in urban environment using IoT. *IFAC-PapersOnLine* **2016**, *49*, 102–107. [CrossRef]
16. Pan, G.; Feng, Q. Performance Analysis of a Multichannel Dynamically Aggregative MAC Protocol for WSNs. *J. Internet Technol.* **2014**, *15*, 441–450.
17. Mickus, T.; Mitchell, P.; Clarke, T. The Emergence MAC (E-MAC) protocol for wireless sensor networks. *Eng. Appl. Artif. Intell.* **2017**, *62*, 17–25. [CrossRef]
18. Duan, R.; Zhao, Q.; Zhang, H.; Zhang, Y.; Li, Z. Modeling and performance analysis of RI-MAC under a star topology. *Comput. Commun.* **2017**, *104*, 134–144. [CrossRef]
19. Medani, K.; Aliouat, M.; Aliouat, Z. Fault tolerant time synchronization using offsets table robust broadcasting protocol for vehicular ad hoc networks. *AEU Int. J. Electron. Commun.* **2017**, *81*, 192–204. [CrossRef]
20. Panigrahi, N.; Khilar, P.M. Multi-hop consensus time synchronization algorithm for sparse wireless sensor network: A distributed constraint-based dynamic programming approach. *Ad Hoc Netw.* **2017**, *61*, 124–138. [CrossRef]
21. Sahoo, P.K.; Sheu, J.P. Design and analysis of collision free MAC for wireless sensor networks with or without data retransmission. *J. Netw. Comput. Appl.* **2017**, *80*, 10–21. [CrossRef]
22. Le, D.T.; Le Duc, T.; Zalyubovskiy, V.V.; Kim, D.S.; Choo, H. Collision-tolerant broadcast scheduling in duty-cycled wireless sensor networks. *J. Parallel Distrib. Comput.* **2017**, *100*, 42–56. [CrossRef]
23. Sun, Y.; Gurewitz, O.; Johnson, D.B. RI-MAC: A receiver-initiated asynchronous duty cycle MAC protocol for dynamic traffic loads in wireless sensor networks. In Proceedings of the 6th ACM Conference on Embedded Network Sensor Systems, Raleigh, NC, USA, 5–7 November 2008.
24. Tang, L.; Sun, Y.; Gurewitz, O.; Johnson, D.B. PW-MAC: An energy-efficient predictive-wakeup MAC protocol for wireless sensor networks. In Proceedings of the INFOCOM IEEE, Shanghai, China, 10–15 April 2011.
25. Park, I.; Yi, J.; Lee, H. A receiver-initiated MAC protocol for wireless sensor networks based on tree topology. *Int. J. Distrib. Sens. Netw.* **2015**, *11*, 950656. [CrossRef]
26. Hamada, L.; Henna, S. AS-PW-MAC: An adaptive scheduling predictive wake-up MAC protocol for wireless sensor networks. In Proceedings of the 2016 IEEE Sixth International Conference on Innovative Computing Technology (INTECH), Dublin, Ireland, 24–26 August 2016.
27. Dong, C.; Yu, F. A prediction-based asynchronous MAC protocol for heavy traffic load in wireless sensor networks. *AEU Int. J. Electron. Commun.* **2017**, *82*, 241–250. [CrossRef]
28. Dinh, T.; Kim, Y.; Gu, T.; Vasilakos, A.V. An Adaptive Low-Power Listening Protocol for Wireless Sensor Networks in Noisy Environments. *IEEE Syst. J.* **2018**, *12*, 2162–2173. [CrossRef]
29. Siddiqui, S.; Ghani, S.; Khan, A.A. ADP-MAC: An Adaptive and Dynamic Polling-Based MAC Protocol for Wireless Sensor Networks. *IEEE Sens. J.* **2018**, *18*, 860–874. [CrossRef]
30. Zhang, D.G.; Zhou, S.; Tang, Y.M. A low duty cycle efficient MAC protocol based on self-adaption and predictive strategy. *Mobile Netw. Appl.* **2018**, *23*, 828–839. [CrossRef]
31. Nguyen, V.T.; Gautier, M.; Berder, O. Implementation of an adaptive energy-efficient MAC protocol in OMNeT++/MiXiM. *arXiv* **2014**, arXiv:1409.0991.
32. Alam, M.M.; Hamida, E.B.; Berder, O.; Menard, D.; Sentieys, O. A heuristic self-adaptive medium access control for resource-constrained WBAN Systems. *IEEE Access* **2016**, *4*, 1287–1300. [CrossRef]
33. Sun, P.; Li, G.; Wang, F. An Adaptive Back-Off Mechanism for Wireless Sensor Networks. *Future Internet* **2017**, *9*, 19. [CrossRef]
34. Ye, D. A self-adaptive sleep/wake-up scheduling approach for wireless sensor networks. *IEEE Trans. Cybern.* **2018**, *48*, 979–992. [CrossRef] [PubMed]
35. Fujimoto, A.; Masui, Y.; Yoshihiro, T. Scheduling Beacon Transmission to Improve Delay for Receiver-initiated-MAC Based Wireless Sensor Networks. *J. Inf. Process.* **2018**, *26*, 140–147. [CrossRef]
36. Pegatoquet, A.; Le, T.N.; Magno, M. A Wake-Up Radio-Based MAC Protocol for Autonomous Wireless Sensor Networks. *IEEE ACM Trans. Netw.* **2019**, *27*, 56–70. [CrossRef]
37. Alam, M.M.; Berder, O.; Menard, D.; Sentieys, O. TAD-MAC: Traffic-aware dynamic MAC protocol for wireless body area sensor networks. *IEEE J. Emerg. Sel. Top. Circuits Syst.* **2012**, *2*, 109–119. [CrossRef]

38. Buettner, M.; Yee, G.V.; Anderson, E.; Han, R. X-MAC: A short preamble MAC protocol for duty-cycled wireless sensor networks. In Proceedings of the 4th International Conference on Embedded Networked Sensor Systems, Boulder, CO, USA, 31 October–3 November 2006.
39. Jang, B.; Lim, J.B.; Sichitiu, M.L. An asynchronous scheduled MAC protocol for wireless sensor networks. *Comput. Netw.* **2013**, *57*, 85–98. [CrossRef]

Article

Barrier Access Control Using Sensors Platform and Vehicle License Plate Characters Recognition

Farman Ullah [1], Hafeez Anwar [1], Iram Shahzadi [1], Ata Ur Rehman [1], Shizra Mehmood [1], Sania Niaz [1], Khalid Mahmood Awan [2], Ajmal Khan [1] and Daehan Kwak [3,*]

[1] Department of Electrical & Computer Engineering, COMSATS University Islamabad-Attock Campus, Attock 43600, Pakistan
[2] Department of Computer Sciences, COMSATS University Islamabad-Attock Campus, Attock 43600, Pakistan
[3] Department of Computer Science, Kean University, Union, NJ 07083, USA
* Correspondence: dkwak@kean.edu

Received: 16 April 2019; Accepted: 18 June 2019; Published: 9 July 2019

Abstract: The paper proposes a sensors platform to control a barrier that is installed for vehicles entrance. This platform is automatized by image-based license plate recognition of the vehicle. However, in situations where standardized license plates are not used, such image-based recognition becomes non-trivial and challenging due to the variations in license plate background, fonts and deformations. The proposed method first detects the approaching vehicle via ultrasonic sensors and, at the same time, captures its image via a camera installed along with the barrier. From this image, the license plate is automatically extracted and further processed to segment the license plate characters. Finally, these characters are recognized with the help of a standard optical character recognition (OCR) pipeline. The evaluation of the proposed system shows an accuracy of 98% for license plates extraction, 96% for character segmentation and 93% for character recognition.

Keywords: barrier control; sensors platform; vehicle detection; license plate recognition; raspberry-pi; features extraction; machine learning algorithms

1. Introduction

The security sensitive areas of a country, such as classified defense areas, government buildings and military installations, are under constant surveillance to avoid potential threats. Such surveillance also extends to the vehicles that constantly access these areas. A vast majority of the currently installed systems use barrier gates that are either manually operated [1] or use vehicle identification based on radio frequency identification (RFID) technology [2]. In RFID-based systems, every vehicle has an RFID tag and RFID reader installed at a gate to identify authorized vehicles. Such systems automatize the access control process; however, the installation of RFID tag in each vehicle makes such systems costly. Alternatively, we propose using a combination of a sensors platform and camera system for automatic barrier access control. The approaching vehicle is automatically detected via the ultrasonic sensors while a camera captures the image of the front side of the vehicle. This image is then further processed to extract and recognize the license plate (LP) of vehicle for authorization. Consequently, the barrier is opened only for authorized vehicles.

An automatic license plate recognition (ALPR) system is instrumental in identifying a vehicle from the image of its LP. As a common rule in various parts of the world, the government issues LPs with fixed aspect ratios, fonts and backgrounds. However, arbitrarily designed LPs is an ever growing problem in countries like Pakistan, where the ALPR becomes a challenging task for a few reasons. First, the position of an LP is not fixed on the front side of the vehicle. Second, there exists a huge variation in the aspect ratios of the LPs. Third, the backgrounds of the LPs vary from on to the other. Finally, variations are also found due to non-uniform font styles and font sizes. Some of these variations are depicted in Figure 1.

A typical ALPR system mainly consists of the following steps [3,4]:

1. Image capturing device acquires an image or extract an image from a video
2. Localization and extraction of license plate in the acquired image
3. Character segmentation and recognition OCR in the extracted LP

Figure 1. Overview of Pakistani license plates (LPs) with various background, foreground, characters fonts and font-sizes.

The ALPR process begins with LP localization and extraction from the vehicle image. LP localization techniques extract the rectangular bounding box or the text regions directly from the image [5]. Without any prior knowledge about the LP size and its location on vehicle, the entire image must be examined to extract the required LP region. We use a Canny edge detector-based method [6] followed by morphological operations and connected components detection to find the rectangular bounding box around the LP in vehicle image. The next step is the segmentation of the desired LP to extract individual characters for recognition. We propose a segmentation approach for characters that have variations in font size, style, and color. Finally, optical character recognition OCR is used to recognize the letters and digits of the extracted LP. To this end, we adapt the feature-based approach that extracts the features of each individual character. These feature include character contour, zoning of solid binary image character, and a skeleton of thin characters [7,8]. These features are used to train the model of a machine learning algorithm or classifier for character recognition. We evaluate a number of classifiers such as the support vector machine (SVM) [9,10], K-nearest neighbors (KNN) [11,12], artificial neural network (ANN) [13] and Decision Trees [14]. The main contributions of this paper are as follows:

1. Development of a prototype for barrier access control and then deploying it in a real world scenario.
2. An image dataset of challenging number plates commonly used in Pakistan with variations in background, position on vehicle, fonts and font styles.
3. Development of an algorithm for character extraction and segmentation of LPs having different background, position on vehicle, fonts and font styles.
4. An extensive performance evaluation of classifiers for optical character recognition.
5. A performance evaluation of the proposed system on two different hardware environments to select the one which is favorable for real-time application.

The rest of this paper is structured as follows: Section 2 outlines related work; Section 3 explains the proposed methodology; the dataset description, results and performance evaluation are discussed in Section 4; finally, Section 5 concludes the paper and outlines the future directions of the current research.

2. Related Work

In this section, we briefly introduce the related work about LP Localization in vehicle images, characters segmentation and characters recognition.

2.1. LP Detection and Localization

In an ALPR system, the starting step is LP detection and extraction. If the LP is not properly extracted, then the LP segmentation will be severely affected [15]. As a common practice, an LP has a rectangular shape. However, in the captured vehicle image there may be other rectangular objects such as the headlights. Therefore, for an effective segmentation, the properties and features of an LP such as its area and aspect-ratio, should be known beforehand. Tarabek et al. [16] proposed a connectivity based rectangular bounding-box extraction with fixed properties. The combination of edge detection and morphological operations is used for LP detection and localization [17–20]. Wang et al. [19] converted the RGB image to HSV color space and proposed a two-stage process for LP localization using color and edge information. Dun et al. [21] proposed an ALPR system for specifically yellow and blue Chinese LPs. A special threshold function was proposed to convert the RGB image to gray to highlight the yellow and blue colors. The transition between the LP background and characters are then used to remove the fake plates and reserve the real plate. In the final step, the accurate location is determined using character size and stroke width. Safaei et al. [22] proposed LP localization based on hierarchical saliency. The proposed algorithm has two steps: in the first step, the algorithm finds the saliency map and then using the connected component analysis detects the LP region. After finding connected components, a Sobel filter and a closing morphological operation is applied. It eliminates many non-number plate regions and then finds the most populated region using L_1-norm. Its result is then binarized using Otsu's method. The largest connected component covering the plate number is then cropped from the vehicle image.

2.2. Characters Segmentation and Extraction form LP

Character segmentation divides the LP into individual characters and digits. Character segmentation becomes challenging due to multi-color background and foreground of an LP. Tabrizi et al. [23] proposed LP segmentation using morphological operations such as dilation, hole filling, erosion, and characters width and height. Gazcón et al. [24] proposed a bounding box technique and its properties to extract characters from the cropped LP. A Convolutional Neural Network (CNN) based two-stage process is proposed [25] to segment and recognize characters (0–9, A–Z). Tarigan et al. [26] proposed an LP segmentation technique consisting of horizontal character segmentation, connected component labeling, verification and scaling. Horizontal and vertical projections of characters are used to segment the cropped LP [27]. Zheng et al. [28] proposed an improved blob detection algorithm to segment LP characters. The segmentation process consists of three steps: first, character height is estimated using the lower and upper boundaries; character width is estimated; and finally, the character is labeled using the block extraction algorithm.

2.3. LP Extracted Characters Recognition

One of the main components of ALPR is the automatic recognition of characters. Chen et al. [29] proposed SIFT based features extraction and matching these features in order to recognize the Chinese characters. A template matching based LP characters' recognition [30] has been proposed for Arabic characters, to recognize 27 alphanumeric characters (17 alphabets and 10 numeric) of fixed size 50×25. A tesseract OCR engine [31] with modification is used in Reference [28] for LP characters' recognition. Tabrizi et al. [23] proposed a hybrid approach of k-nearest neighbor (KNN) and multi-class support vector machine (SVM) for Iranian LP recognition. First, the KNN classifies the characters using the structural, horizontal and vertical features. Then the SVM classifier is applied to the zoning features. Gazcón et al. [24] compared the proposed intelligent template matching (ITM) with the artificial neural network (ANN). Compared to the traditional template matching technique the ITM constructs trees of the character's skeleton. These trees are used to compare with the tree obtained from the testing character skeleton. ITM showed higher accuracy and also minimized the recognition time. Wang et al. [32] proposed LP detection and recognition simultaneously in a single forward pass by using a deep neural network algorithm. In the first step of this algorithm, a number of convolutional layers are used to extract and discriminate the features of LP. After this, the proposed network detects the objects on a LP. This technique takes the low level convolutional features and generates a set of bounding boxes. In the last step, a bidirectional recurrent neural network (BRNN) with Connectionist Temporal Classification recognizes the LP characters. Björklund [33] proposed an ALPR system trained on synthetic data that has varying pose conditions and illumination levels and showed precision and recall of 93%. Table 1 presents the overall literature review of LP detection, LP region of interest extraction, characters' segmentation, and character recognition.

Table 1. Literature summary of LP detection, extraction, characters segmentation, and characters recognition.

Ref.	Methodology	Character Language (0–9, A–Z)	Dataset Size	Platform	Plate Localization	Character Segmentation	Character Recognition	Processing Time
[28]	Plate Detection:Viola and jones algorithm, Character Segmentation: Blob detection, Characters Recognition: OCR	(0–9) (A–Z)	Testing dataset (160) Training dataset (300)	PC with Pentium 2.8 GHz CPU	96.4%	98.2%	99%	0.1 s
[26]	Recognition: Genetic algorithms	(0–9) (A–Z)	220 Images	PC	N/A	N/A	85.97%	0.2 s
[23]	Characters Extraction & Segmentation: Morphological operations, Recognition: KNN and SVM	Persian characters	257 Images	PC	96.01%	95.24%	97.03%	N/A
[34]	Segmentation: Otsu's method, Character Extraction:Vertical projection, Recognition: Backward propagation Neural Network (BP-ANN)	(0–9) (A–Z)	Training: 2700 & Testing: 354 characters	CPU i3 2.9G and 2G memory	N/A	N/A	93.5%	N/A
[35]	Detection, Extraction, and Recognition: Convolutional Neural Network (CNN)	(0–9) (A–Z) Excluding 'O'	Testing USA (328) Europe (550)	PC	99.0% USA 93.64% Europe	N/A	93.445% USA 94.54% Europe	N/A
[36]	Plate Detection:Modified visual attention model, Segmentation: Vertical and horizontal projection, Recognition:CNN and SVM	Chinese (0–9) (A–Z)	Testing: Chinese (620), 0–9 and A–Z (680): Training Chinese (930), 0-9 and A-Z (1020)	PC	98.9%	N/A	98.3% Chinese characters 99.1% (numerals and alphabets)	0.14 s
[24]	Through searching rectangles (plate extraction) Bounding Box technique (character extraction) ANN and ITM (character recognition)	(0–9) (A–Z)	Testing data (73)	Pentium Dual core, 1.73 GHz, 2 GB RAM	N/A	N/A	99.09% (ANN) 91.1% (ITM)	0.5 ms (ANN) 0.75 ms (ITM)
[27]	YOLO detector (vehicle and LP detection) CNN (character segmentation and recognition)	(0–9) (A–Z)	Data set (4500)	NVIDIA Titan XP GPU	100%	99.75%	97.83%	28.3 ms

3. Proposed System

This section explains the proposed architecture including the main functions from vehicle detection to the barrier control mechanism. Figure 2 illustrates the block diagram of the proposed system while Figure 3 depicts the algorithm flowchart of the proposed system. Following are the main steps.

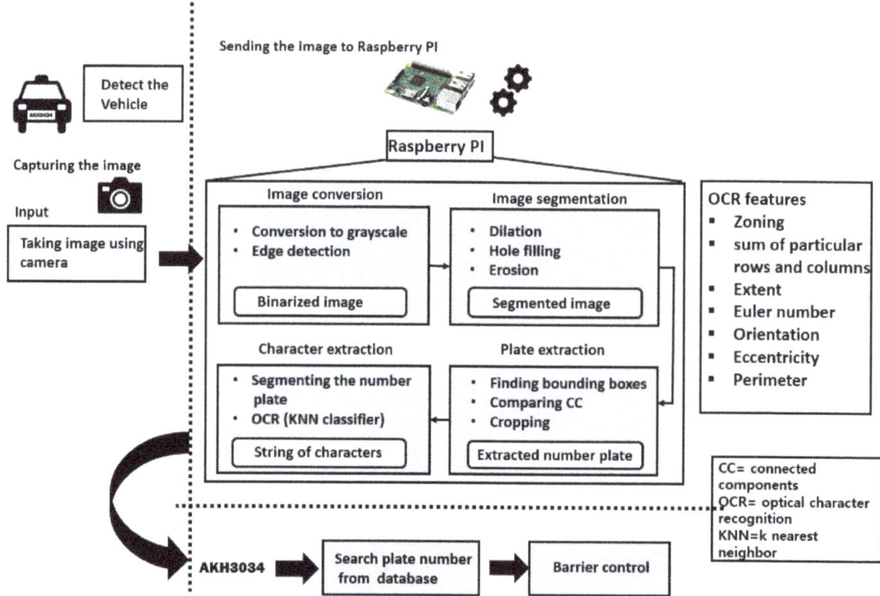

Figure 2. The proposed architecture of barrier access control for vehicle entrance using sensors platform and an image-based LP recognition.

1. Vehicle arrival detection and image acquisition
2. Image pre-processing and edge image generation
3. Image segmentation based on detected edges
4. LP extraction via the count of connected components
5. Character segmentation and features calculation
6. Optical character recognition
7. Vehicle authorization and barrier control system

These steps are further explained in the following subsections.

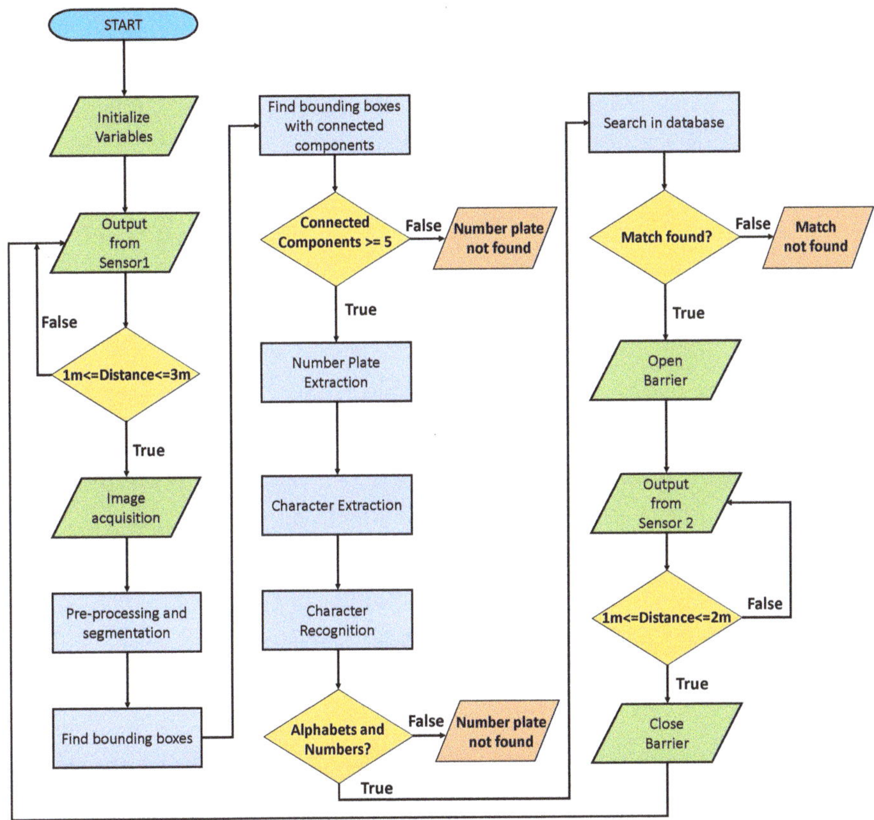

Figure 3. Algorithm flow chart of the proposed smart access control for vehicle entrance using sensors platform and an image-based LP recognition.

3.1. Vehicle Arrival Detection and Image Acquisition

Figure 4 shows the proposed hardware architecture for barrier access entrance control. Ultrasonic sensors installed at the barrier detect the approaching vehicle. The sensor emits 8-pulses of 40 KHz for 10 μs and listens to the echo signal for 100 μs to 36 ms. Using $S = \frac{Vt}{2}$, we find the distance between barrier and the vehicle where S is the distance, V is the speed of sound: .034 m/μs and t is the time in μs for transmission and its echo signal. The camera is only activated for image acquisition when the ultrasonic sensors detect the vehicle in a specific range of distance which is set from 1 to 3 m. As a common practice on gate entrances, a lane is built for the entering vehicle so that they are almost straight when the image is taken by the camera. Due to this reason, the image of the entering vehicle is taken with negligible rotations.

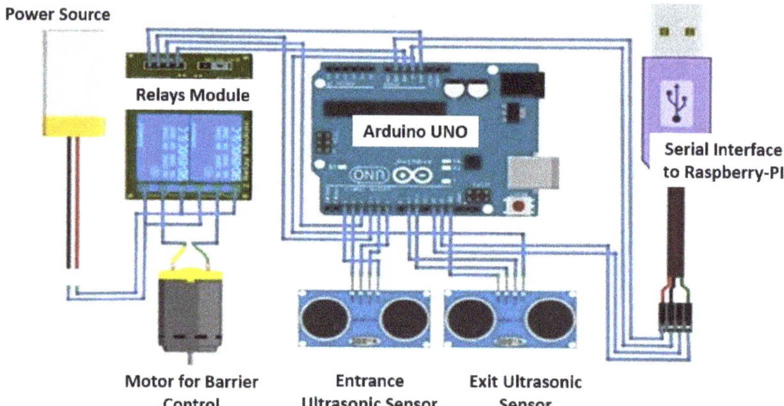

Figure 4. Hardware architecture of an ultrasonic sensors-based vehicle entrance and exit detection.

3.2. Image Pre-Processing and Edge Image Generation

In the proposed ALPR method, we convert the captured image into grayscale. It reduces the processing complexity and processing time and is robust to color changes due to different lighting conditions. A canny edge detector is applied to this image to detect all the edges. The Canny edge detector is a combination of a Gaussian filter for smoothing and a Sobel filter for edge detection. Equation (1) shows the Gaussian filter that suppresses the noise in an image with σ as the standard deviation of the Gaussian filter.

$$\mathbf{G}(x,y) = \frac{1}{2\pi\sigma^2}e^{\left(-\frac{x^2+y^2}{2\sigma^2}\right)} \tag{1}$$

$$\mathbf{S_x} = \begin{bmatrix} +1 & 0 & -1 \\ +2 & 0 & -2 \\ +1 & 0 & -1 \end{bmatrix} \quad \mathbf{S_y} = \begin{bmatrix} +1 & +2 & +1 \\ 0 & 0 & 0 \\ -1 & -2 & -1 \end{bmatrix} \tag{2}$$

$$|S| = \sqrt{\mathbf{S_x}^2 + \mathbf{S_y}^2} \tag{3}$$

$$\angle S = \tan^{-1}\left(\frac{|\mathbf{S_y}|}{|\mathbf{S_x}|}\right) \tag{4}$$

After the Gaussian filter, we apply the Sobel masks [37] to detect the horizontal and vertical edges as shown by Equation (2). Equations (3) and (4) show the magnitude and direction of the Sobel gradient respectively. Considering the pixel magnitude, direction, non-maximum suppression and thresholds, the pixel is marked as an edge if its magnitude is greater than the threshold in the gradient direction. Finally, at this stage, we get an edge segmented image.

3.3. Image Segmentation Based on Detected Edges via Morphological Operations

On the generated edge image, we perform various morphological operations such as dilation, horizontal erosion, vertical erosion and hole filling. Dilation adds the pixels to the boundary of edges to complete the boundary and increases the efficiency of LP extraction. Mathematically, Equation (5) shows the dilation.

$$\mathbf{I_{dilated}} = \mathbf{I} \oplus \mathbf{B} = \{z|(\hat{\mathbf{B}})_z \cap \mathbf{I} \neq \varnothing\} \tag{5}$$

where **I** is edge segmented image and **B** is structure element. After dilation, we filled the closed boundaries and remove unnecessary parts of the image without affecting the LP area. A hole filling technique is used for this purpose and its mathematical expression is given by Equation (6).

$$\mathbf{I}_{\text{holefilled}} = \mathbf{X_k} = (\mathbf{X}_{(k-1)} \oplus \mathbf{B}) \cap \mathbf{I}^c_{\text{dilated}} \tag{6}$$

We use vertical and horizontal erosion to remove those pixels, which makes it difficult to extract the LP. All the unnecessary lines and parts connected to the LP area create problems for the LP extraction. Equation (7) shows the mathematical expression used for erosion.

$$\mathbf{I}_{\text{eroded}} = \mathbf{I}_{\text{holefilled}} \ominus \mathbf{B} = \{z | (\mathbf{B})_z \in \mathbf{I}_{\text{eroded}}\} \tag{7}$$

3.4. LP Extraction via the Count of Connected Components

We find the 8-connectivity components based rectangular bounding box objects in the eroded image. In addition to the LP, there are other rectangular objects such as headlights, radiators, grille and bumper. Therefore, it is likely that these objects are also segmented along with the LP. Due to this reason, we use the count of connected components in each segment as a clue to differentiate between the LP and other rectangular objects. To this end, for a segment to be considered an LP, the number of objects inside that segment should be more than five. This is due to the fact that the Pakistani LP consists of at least five characters as shown in Figure 3. Once the mask of the LP is generated in this way, it is used to extract the LP from the RGB image.

3.5. Characters Segmentation from LP Segment and Features Calculation

Once the LP region is extracted, character segmentation is employed to extract the LP characters. For this purpose, as a first step, the LP region is binarized using the algorithm shown in Figure 5. First, we calculate the intensity histogram of the LP region image and then find the two highest peaks in this histogram. We considered the two highest peak because the LP mostly consists of two colors, that is, the LP background color and the characters' color. The threshold is the average value of these two peaks. An LP grayscale image is then binarized using this threshold. We extract characters from the binary image using 8-connectivity, considering a character height of 30 to 90, a width of 10 to 40 and an area of 700 to 800 pixels.

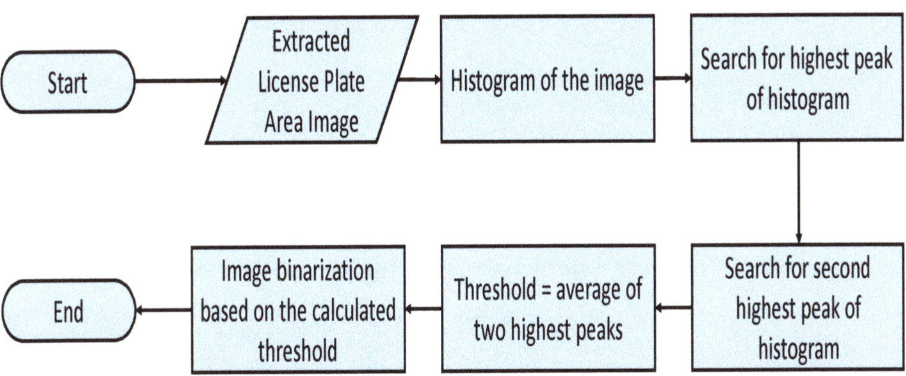

Figure 5. Thresholding algorithm to convert the LP grayscale image to binary.

In this paper, we focus on the features-based approach for character recognition. We extract the following features of a character.

- **Zoning:** It divides the character image into various sub-images. Figure 6 shows the overview of zoning a character image into 3×3 sub-images. The white pixels are summed in each sub-image and become a feature.

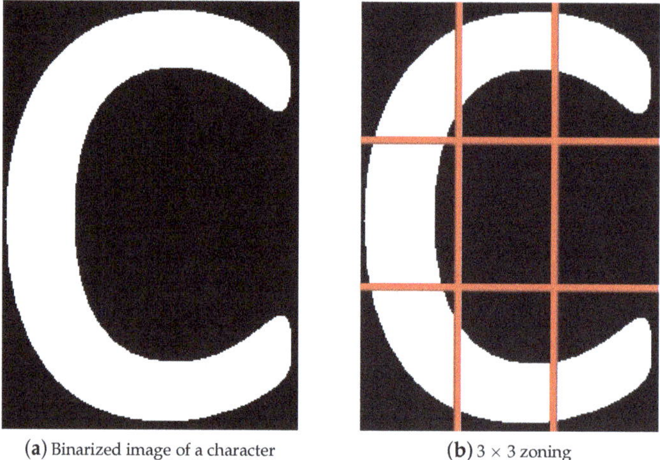

(a) Binarized image of a character   (b) 3 × 3 zoning

Figure 6. Overview of zoning of an image into sub-images.

Mathematical it can be calculated by Equation (8).

$$zoning_{feature} = \sum_{i=1}^{M} \sum_{j=1}^{N} \mathbf{subimage}(i, j) \tag{8}$$

where $M \times N$ is the size of sub-image. We considered a character 42×24 size of image and then divided it into nine sub-images of 14×8 each.

- **Perimeter:** The set of interior boundary pixels of a connected component (character image (**C**)) [33]. We considered 8-connectivity to find the perimeter. Equation (9) finds the perimeter of a character.

$$Permiter_8 = \{(a, b) \in \mathbf{C} | N_4(a, b) - \mathbf{C} \neq \varnothing\} \tag{9}$$

where (a, b) is the pixel location.

- **Extent:** It is the ratio of white pixels in an image to the total number of pixel in the binary image. Equation (10) finds the Extent value of a character image.

$$Extent = \frac{Number\_of\_White_{pixels}}{Total\_Pixels\_Image} \tag{10}$$

- **Euler Number:** Euler number is the topology measure of an image. It is the number of objects in an image minus the number of holes in the image. Equation (11) finds the Euler number:

$$Euler\_number = 1 - number\_of\_holes \tag{11}$$

- **Particular Rows and Columns Pixels Summation:** In the paper, we consider some particular rows and columns to add their pixels. That particular row or column pixels summation is considered as a feature. Equation (12) finds the sum of a particular row i.

$$sum\_row\_i = \sum_{col=1}^{TotalCol} \mathbf{C}(i, col) \tag{12}$$

where *TotalCol* shows the number of columns in the character image **C**. We considered the summation of rows third, fifteen, twenty-seven and thirty-seven as features. The summation of the column is given by Equation (13).

$$sum\_column\_j = \sum_{r=1}^{R} \mathbf{C}(r,j) \tag{13}$$

where R is rows in character image **C** and we find the summation of second, twelve and seventeen columns.

- **Eccentricity:** Finds how close an object is to being a circle. It is ratio of the linear eccentricity to the semi-major axis.
- **Orientation:** The major axis of an ellipse around the object and then finding the angle which the major axis made with the x-axis.

3.6. Optical Character Recognition of LP Characters

In this paper, we evaluated various supervised learning algorithms (classifiers) to recognize characters on the LP. Figure 7 shows the process of LP character recognition. As a first step, these characters are manually extracted from the images. The aforementioned features of each extracted character are calculated in order to represent each of them in a single feature vector of length 21. A given classifier is then trained on these features. For testing, the proposed extraction algorithm first extracts the LP characters automatically while the trained classifier recognizes the characters by predicting their labels. We used KNN, Decision Trees, Random Forest, SVM, and ANN for LP character recognition.

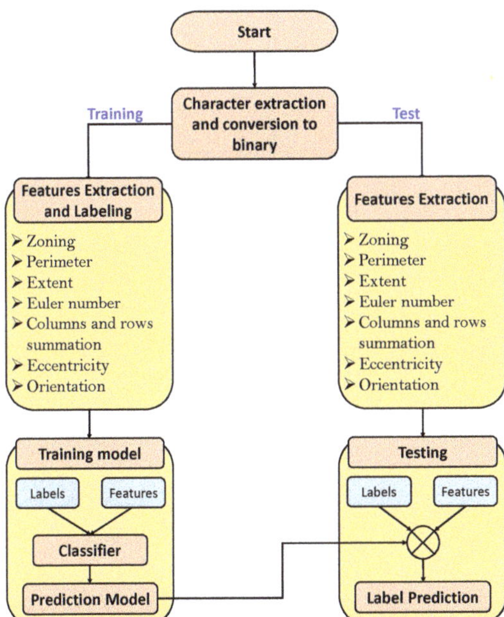

Figure 7. An overview of LP characters recognition process.

3.7. Vehicle Authorization and Barrier Control System

The real-time system for vehicle detection and authorization is implemented on a Raspberry Pi. The ultrasonic sensors interfaced to the Raspberry Pi detect the approaching vehicle on entrance. The

LP of this vehicle is then verified using its image. If the vehicle is permitted then the Raspberry Pi sends a command to open the barrier. Figure 8 shows the circuit, schematic, hardware setup and access mechanism of the barrier control system. Figure 8c shows the real-time hardware setup used to detect the vehicle, recognize the LP and control the barrier position. A camera and two ultrasonic sensors installed on the barrier are also shown. The front ultrasonic sensor detects the vehicle at the entrance and the rear ultrasonic sensor detects the exited vehicle. The barrier control circuitry is interfaced using the RS-232 serial port to the LP processing system. We used a DC motor [38] to control the barrier access that rotates between $0°$ and $360°$. We used $90°$ and $+180°$ for closed and open barrier systems respectively, as shown in Figure 8d. The motor rotates the barrier bar from open to closing when the first relay is active and second is de-active and vice-versa. In real implementation, we used the DC motor rated 24 V of high torque which can easily move a barrier bar that weighs upto 8 kg. However, in a PC based simulation, we used 9 V to simulate the controlling of the motor.

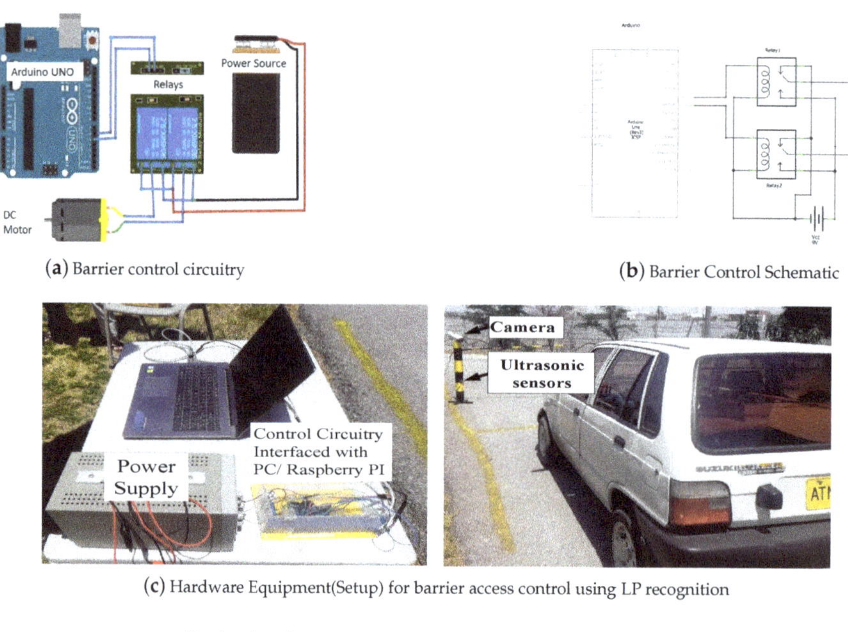

Figure 8. Hardware Setup and mechanism of barrier system control using DC motor.

4. Results & Discussion

The proposed system is implemented on the following frameworks.

1. PC(Intel(R), Core(TM) i3-4010U CPU 1.70GHz,RAM: 4.00GB) running Matlab(*R2013a*, 64-bits) and interfaced the Arduino using an RS232 serial port for the barrier control system. Matlab programming and Arduino C-based code are used to implement the system
2. Raspberry Pi- system on chip single board computer with 1.4 GHz 64-bits quad-core processor and interfaced the Arduino using RS232 serial port for the barrier control system. Python 3, OpenCV 3.4.0, and Arduino C-based code is used to implement the system.

Table 2 shows the details of the acquired dataset used as training and test images. Images were taken with a camera in daylight conditions.

Table 2. Dataset description for LP extraction, characters segmentation and recognition.

Sequence No.	Description	No. of Images
1	Images for LP extraction	500
2	Images for characters segmentation	500
3	Characters (0–9) & (A–Z)	3643
4	Features vector dimension	21

4.1. Results of Pre-Processing, Edge Detection, and LP Area of Interest Extraction

Figure 9 shows the qualitative results of pre-processing before LP extraction. Figure 9a shows the original RGB captured image and the resized image when detected by the ultrasonic sensor in the specified range. Figure 9b shows the RGB image converted to grayscale and Figure 9c shows the detected edges in the image via Canny edge detector. Figure 9d–f shows various morphological operations applied to the edge image. A 5×1 structure element of Dilation enlarges the edges. Hole filling fills the connected objects and erosion removes the pixels on object boundaries and the single pixel objects (lines). Connected components based segments are extracted using a constraint of connected objects on a segment as shown in Figure 9g. Finally, Figure 9h shows the extracted LP area of interest.

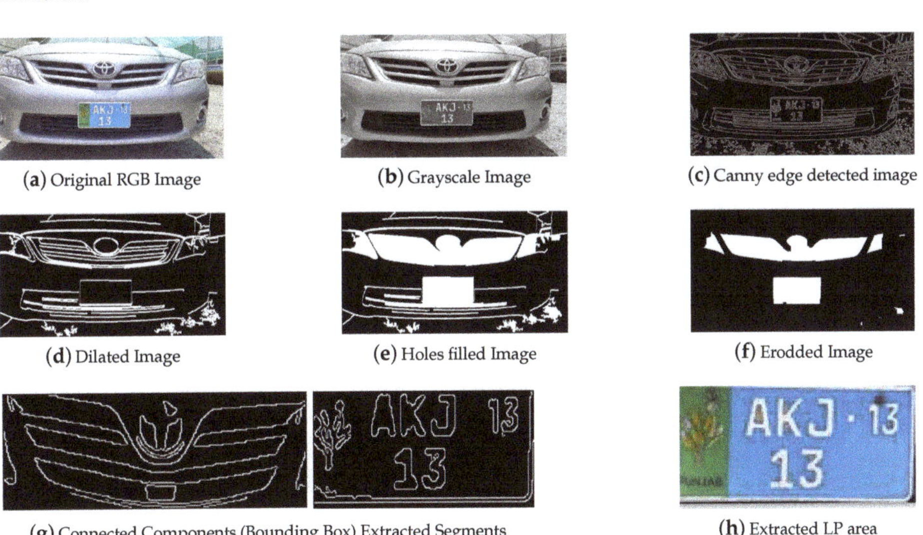

(**a**) Original RGB Image   (**b**) Grayscale Image   (**c**) Canny edge detected image

(**d**) Dilated Image   (**e**) Holes filled Image   (**f**) Erodded Image

(**g**) Connected Components (Bounding Box) Extracted Segments   (**h**) Extracted LP area

Figure 9. Results of pre-processing, edge detection, morphological operations, connected components extraction and final LP area extraction.

For 500 images, the LP extraction accuracy of the proposed method is 98%. Figure 10 shows images where the LP is not correctly extracted due to various reasons such as character occlusion due to dirt, non-rectangular LP and broken LP.

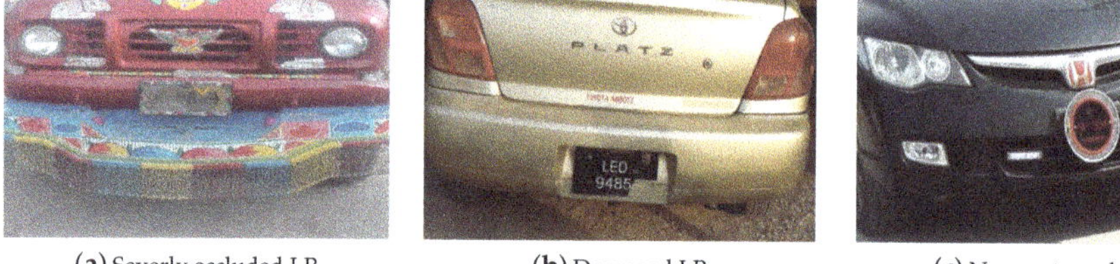

(**a**) Severly occluded LP   (**b**) Damaged LP   (**c**) Non-rectangular LP

Figure 10. Vehicle images where LPs are not correctly segmented due to various reasons.

4.2. Results of LP Characters Segmentation

The step-by-step result of the LP segmentation and character recognition are visually shown in Figure 11. The variations in the LP background, font sizes and styles of the characters' positions can be observed in the different types of LPs. There are also additional numbers and characters in the LPs. The proposed method clearly shows its robustness to such challenges and extracts the bounding boxes that enclose only those characters that belong to the license plate's number. The LP character segmentation of the proposed method for an image dataset of 500 images is 96%.

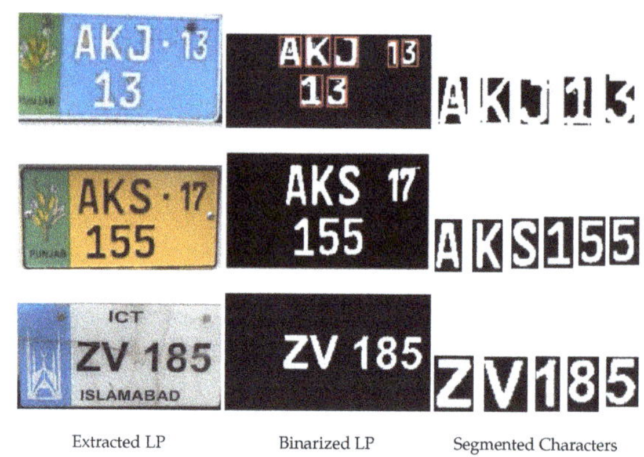

Extracted LP   Binarized LP   Segmented Characters

Figure 11. LP extraction, binarization and character segmentation.

4.3. Results of LP Characters Recognition

We assigned class labels 0–9 to the digits and 10–35 for alphabets A–Z to recognize the LP characters. A total of 3643 characters were extracted from images of computerized and handwritten LPs. The aforementioned features were calculated for each of the extracted characters. In the current setting, we evaluated various classifiers such as KNN, Naive Bayes, Bayes Network, SVM using linear kernel, MLP, Decision Tree and RF for LP character recognition. The classifiers training and testing was done using 10-fold cross validation where a given dataset is split into 90% training set and 10% test set. The performances of classifiers were compared with respect to their classification accuracy, true positive rates, false positive rates, precision, recall, F-measure and ROC area. Figure 12 shows the classification accuracies of all the classifiers on the given dataset. Table 3 shows the detailed comparison of classifiers with respect to other metrics.

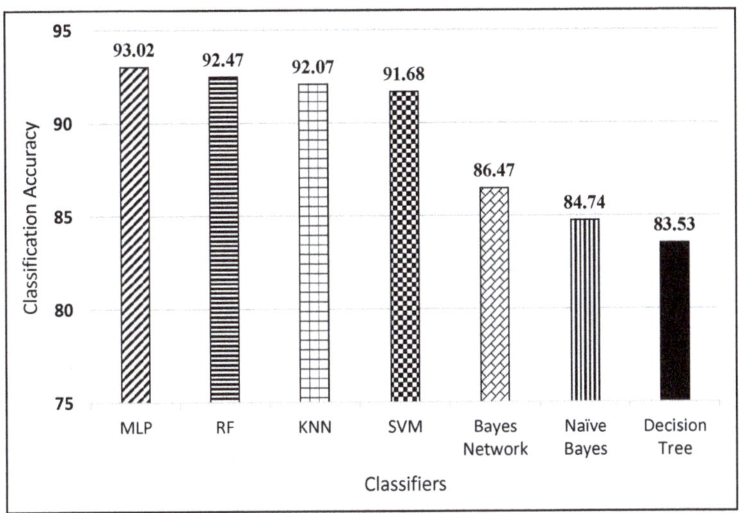

Figure 12. Classifiers accuracy performance comparison for OCR.

Table 3. Comparison of performance metrics (*TPR, FPR, Precision, Recall, F-measure*, and *ROC Area*) of classification algorithms for OCR.

Classifiers Parameters	TPR	FPR	Precision	Recall	F-measure	ROC Area
KNN	0.921	0.003	0.921	0.921	0.92	0.959
Bayes Network	0.865	0.005	0.868	0.865	0.865	0.994
Naive Bayes	0.847	0.006	0.859	0.847	0.85	0.99
SVM	0.917	0.004	0.913	0.929	0.916	0.956
MLP	**0.93**	**0.002**	**0.931**	**0.93**	**0.93**	**0.995**
Decision Tree	0.835	0.007	0.832	0.835	0.833	0.93
RF	0.93	0.004	0.93	0.925	0.926	0.965

Tables 4 and 5 show the confusion matrix of OCR using the KNN and MLP algorithms, respectively. The characters on the LPs are handwritten and computerized. Mostly, the handwritten characters such as 0 have higher similarity with *O* and also *Q* with 0. 5 and *S*, and *M* and *N* have higher similarity. If they are computerized based *O*, *Qw* and 0 are used it will increase the recognition accuracy.

The accuracy of MLP, KNN, SVM and RF are close to each other. Therefore, we implemented the KNN algorithm for real-time ALPR both on Matlab and Raspberry Pi based proposed systems. Table 6 shows the time analysis of the proposed system, both implemented using a PC with Matlab and Raspberry Pi with Python and OpenCV library. It depicts the time from vehicle detection to LP number recognition and barrier access control opening. The time consuming part of the proposed system is the pre-processing-LP localization and recognition of LP characters. We compared the timing for 100-LPs that had 5-characters, 6-characters, and 7-characters, respectively. The Raspberry Pi based system had the lowest computation time, a small size and low power requirements. It can be easily installed in the constraint area for vehicle detection and control access to a restricted area.

Table 4. Confusion matrix of OCR using KNN classifier algorithm.

Predicted Class \ Actual Class	0	1	2	3	4	5	6	7	8	9	A	B	C	D	E	F	G	H	I	J	K	L	M	N	O	P	Q	R	S	T	U	V	W	X	Y	Z
0	148	0	0	0	0	0	0	0	0	0	0	0	0	0	0	0	0	0	0	0	0	0	0	0	6	0	0	0	0	0	0	0	0	0	0	0
1	0	156	0	0	0	0	0	2	0	0	0	0	0	0	0	0	0	0	6	0	0	0	0	0	0	0	0	0	0	0	0	0	0	1	0	0
2	0	1	223	3	0	0	0	0	1	1	0	0	0	0	1	0	0	0	0	1	0	0	0	0	0	0	0	0	2	0	0	0	0	1	0	3
3	0	1	3	195	0	4	0	0	0	0	0	0	0	0	0	0	0	0	0	0	0	0	0	0	0	0	0	0	1	0	0	0	0	0	0	0
4	0	0	0	1	212	0	1	0	0	1	2	0	0	0	0	0	0	0	0	0	1	0	0	0	0	0	0	0	0	0	0	1	0	0	0	0
5	0	0	2	1	0	198	1	0	2	0	0	0	0	0	0	0	0	0	0	0	0	0	0	0	0	0	0	0	8	0	0	0	0	0	0	0
6	0	0	0	0	1	0	213	0	5	0	0	0	0	0	0	0	0	0	0	0	0	1	0	0	0	0	0	0	0	0	0	0	0	0	0	0
7	0	0	1	0	0	0	0	213	0	0	0	0	0	0	0	0	0	0	0	0	0	0	0	0	0	0	0	0	1	0	0	0	0	0	0	0
8	0	0	0	0	0	1	1	0	213	1	1	13	0	0	0	0	0	0	0	0	0	0	0	1	0	0	0	0	1	0	0	0	0	0	0	0
9	0	0	0	0	0	0	1	0	2	224	0	0	0	0	0	0	0	0	0	0	0	0	0	0	0	0	0	0	0	0	0	0	0	0	0	0
A	0	0	0	0	3	0	1	0	0	0	232	0	0	0	0	0	0	0	1	0	0	0	0	0	0	0	0	0	0	0	0	0	0	0	0	0
B	0	0	0	0	0	1	1	0	23	0	0	74	0	0	0	0	0	1	0	0	0	0	0	0	0	0	0	2	0	0	0	0	0	0	0	0
C	0	0	0	0	0	0	1	0	0	0	0	0	68	0	1	0	1	0	1	0	0	0	0	0	1	0	0	0	0	0	0	0	0	0	0	0
D	9	0	0	0	0	0	0	0	1	0	1	0	0	40	0	0	0	0	0	0	0	0	0	0	0	0	0	0	0	0	0	0	0	0	0	0
E	0	0	1	0	0	2	1	0	0	1	0	0	1	0	167	1	1	0	0	0	0	0	0	0	0	0	0	0	0	0	0	0	0	0	0	1
F	0	0	0	0	0	0	0	1	0	0	0	0	0	0	1	48	0	0	0	0	2	0	0	0	0	1	0	0	0	0	0	0	0	0	0	0
G	2	0	0	0	0	0	2	0	0	0	0	1	1	0	1	0	21	0	0	0	0	0	0	0	0	0	0	0	1	0	0	0	0	0	0	0
H	0	0	0	0	0	0	2	0	0	0	0	1	0	0	0	0	0	30	0	0	0	0	2	4	0	0	0	0	0	0	0	0	0	0	0	0
I	0	6	0	0	0	0	0	0	0	0	0	0	0	0	0	0	0	0	25	0	0	1	0	0	0	0	0	0	0	0	0	0	1	0	0	0
J	0	1	1	1	0	0	0	0	0	0	0	0	0	0	0	0	0	0	0	26	0	0	0	0	0	0	0	0	0	0	0	0	0	0	0	0
K	0	0	0	0	0	0	2	0	0	0	0	0	0	0	1	0	2	0	0	0	84	0	0	1	0	0	0	1	0	0	0	0	0	1	0	0
L	0	0	1	0	1	1	0	0	0	0	0	0	0	0	0	0	0	0	0	0	1	185	0	0	0	0	0	0	0	0	0	0	0	0	0	0
M	0	0	0	0	0	0	0	0	1	0	0	0	0	0	0	0	0	0	1	0	0	0	23	8	0	0	1	0	0	0	0	0	0	0	0	0
N	0	0	0	0	0	0	0	0	1	0	0	0	0	0	0	0	0	3	0	0	1	0	1	55	0	0	0	0	0	0	0	0	7	0	0	0
O	5	0	0	0	0	0	0	0	0	0	0	0	1	1	0	0	0	0	0	0	0	0	1	0	7	0	1	0	0	0	0	0	0	0	0	0
P	0	0	0	0	0	0	0	1	0	1	1	0	0	0	0	2	0	0	0	0	0	0	0	0	0	19	0	0	0	0	0	0	0	0	0	0
Q	5	0	0	0	0	0	0	0	1	1	0	0	0	1	0	0	1	0	0	0	0	0	0	0	1	0	3	0	0	0	0	0	0	0	0	0
R	0	0	0	0	0	0	2	0	3	1	1	1	0	0	1	0	0	0	0	0	0	2	0	0	0	0	0	47	0	0	0	0	0	0	0	0
S	0	0	3	1	0	6	1	0	0	0	0	0	0	0	0	0	0	0	0	0	0	0	0	0	0	0	0	0	37	0	0	0	0	0	0	0
T	0	0	0	0	0	0	0	1	0	0	0	0	0	0	0	0	0	0	0	0	0	0	0	0	0	0	0	0	0	32	0	0	0	0	0	0
U	0	0	0	0	0	0	0	0	0	0	0	0	1	0	0	1	0	0	0	0	0	0	0	0	0	0	0	0	0	0	27	0	0	0	0	0
V	0	0	0	0	0	0	0	0	0	0	0	0	0	0	0	0	0	0	0	0	0	0	2	0	0	0	0	0	0	0	0	24	0	0	1	0
W	0	0	0	0	0	1	0	0	0	0	0	0	0	0	0	0	0	0	0	0	0	0	12	0	0	0	0	1	0	0	0	0	23	0	0	0
X	0	1	2	1	0	0	0	0	0	0	0	0	0	0	0	0	0	0	0	0	0	0	0	0	0	0	0	0	0	0	0	0	0	39	0	0
Y	0	0	0	0	0	0	0	0	0	0	0	0	0	0	0	0	0	0	0	0	0	0	0	0	0	0	0	0	0	0	0	1	0	0	15	0
Z	0	0	5	0	0	0	0	0	0	0	0	0	0	0	0	0	0	0	0	0	0	0	0	0	0	0	0	0	0	1	0	0	0	0	0	42

Table 5. Confusion matrix of OCR using MLP classifier algorithm.

Actual Class → / ↓ Predicted Class	0	1	2	3	4	5	6	7	8	9	A	B	C	D	E	F	G	H	I	J	K	L	M	N	O	P	Q	R	S	T	U	V	W	X	Y	Z
0	145	0	0	0	0	0	1	0	1	0	0	0	0	2	0	0	0	0	0	1	0	0	0	0	3	0	1	0	0	0	0	0	0	0	0	0
1	0	154	1	0	0	0	0	3	0	0	0	0	0	0	1	0	0	0	4	1	0	1	0	0	0	0	0	0	0	0	0	0	0	0	0	0
2	0	1	209	1	0	0	0	4	0	2	0	2	0	0	1	0	1	0	0	2	0	0	2	1	0	0	0	0	4	0	0	0	0	2	0	5
3	0	0	0	200	0	2	0	0	0	0	0	0	0	0	0	0	0	0	0	0	0	0	0	0	0	0	0	0	1	0	0	0	0	1	0	0
4	0	0	0	0	211	0	0	0	0	1	1	0	0	0	0	0	0	1	0	0	1	1	0	1	0	0	0	0	0	0	0	1	0	0	0	0
5	0	1	0	1	0	199	0	0	1	0	0	2	0	1	0	0	0	0	2	0	0	0	0	2	0	0	0	0	3	0	0	0	0	0	0	0
6	0	0	0	0	1	0	210	0	1	0	0	1	0	0	0	0	0	0	0	0	2	2	0	0	0	0	0	2	1	0	0	0	0	0	0	0
7	0	2	0	0	0	0	0	213	0	0	0	0	0	0	0	0	0	0	0	0	0	0	0	0	0	0	0	0	0	0	0	0	0	0	0	0
8	0	0	0	0	0	2	0	0	218	1	0	7	0	0	0	0	0	0	0	0	0	0	0	1	1	0	0	2	0	0	0	0	0	0	0	0
9	0	0	0	1	0	0	0	0	1	221	0	0	0	0	0	0	0	0	1	0	0	0	0	0	0	1	0	0	1	0	0	1	0	0	0	0
A	0	0	0	0	5	0	2	0	2	0	225	0	0	0	0	0	0	0	1	0	0	1	0	1	0	0	0	0	0	0	0	0	0	0	0	0
B	0	0	0	0	0	1	0	0	13	0	0	81	1	1	0	0	0	0	0	0	0	0	0	0	0	0	0	5	0	0	0	0	0	0	0	0
C	0	0	0	0	0	0	0	0	0	0	0	0	69	0	0	2	0	0	0	0	0	1	0	0	0	0	0	0	0	0	0	0	0	0	0	0
D	2	0	0	0	0	0	0	0	0	0	0	2	0	47	0	0	0	0	0	0	0	0	0	0	0	0	0	0	0	0	0	0	0	0	0	0
E	0	0	0	0	0	0	0	0	0	0	0	0	0	0	172	2	0	0	2	0	0	0	0	0	0	0	0	0	0	0	0	0	0	0	0	0
F	0	0	0	0	0	0	0	0	0	0	0	1	0	0	0	52	0	0	0	0	0	0	0	0	0	0	0	0	0	0	0	0	0	0	0	0
G	0	0	0	0	0	0	0	0	0	0	0	0	1	0	0	0	28	0	0	0	0	0	0	0	0	0	0	0	0	0	0	0	0	0	0	0
H	0	0	0	0	0	1	0	0	0	0	0	0	0	0	0	0	0	34	0	0	0	0	2	1	0	0	0	0	0	0	0	0	1	0	0	0
I	0	4	0	0	0	1	0	0	0	0	0	0	0	0	0	0	0	0	24	0	0	1	0	0	0	0	0	0	0	0	0	0	1	0	0	2
J	0	1	1	1	0	0	0	0	0	0	0	0	0	0	0	0	0	0	0	26	0	0	0	0	0	0	0	0	0	0	0	0	0	0	0	0
K	0	0	0	0	0	0	0	0	1	0	0	0	0	0	0	0	0	1	0	0	84	0	0	0	0	0	0	4	0	0	0	0	0	1	0	1
L	0	0	0	0	0	0	0	0	0	0	0	0	0	0	1	0	0	0	0	0	0	188	0	0	0	0	0	0	0	0	0	0	0	0	0	0
M	0	0	0	0	0	0	0	0	0	0	0	0	0	0	0	0	0	0	0	0	0	0	28	5	0	0	0	0	0	0	0	0	0	1	0	0
N	0	0	0	0	0	0	0	0	1	0	0	0	0	0	0	1	1	1	0	0	0	0	1	60	0	0	0	0	0	0	1	0	2	0	0	0
O	8	0	0	0	0	0	0	0	0	0	0	0	1	0	0	0	0	0	0	0	0	0	0	0	6	0	0	0	0	0	0	0	0	0	0	0
P	0	0	0	0	0	0	0	0	3	0	0	0	0	0	0	1	0	0	0	0	0	0	0	0	01	9	0	1	0	0	0	0	0	0	0	0
Q	1	0	0	0	0	0	0	0	1	1	0	0	0	0	0	0	0	0	1	0	0	0	0	0	2	0	6	0	0	0	1	0	0	0	1	0
R	0	0	0	0	0	1	0	0	0	0	1	0	1	0	0	1	0	0	0	0	3	0	1	0	0	0	04	9	0	0	0	1	0	0	0	0
S	0	1	1	3	0	3	0	0	1	0	0	0	0	0	0	0	1	0	0	0	0	0	0	0	0	0	0	03	7	0	0	0	0	1	0	0
T	0	1	0	0	0	0	0	1	0	0	0	0	0	0	0	0	0	0	0	0	0	0	0	0	0	0	0	0	03	1	0	0	0	0	0	0
U	0	0	0	0	0	0	0	0	0	0	0	0	0	0	0	1	0	0	0	0	0	0	0	0	0	0	0	0	0	02	8	0	0	0	0	0
V	0	0	0	0	0	0	0	0	0	0	0	0	0	0	0	1	0	0	0	0	0	0	0	0	0	1	0	0	0	0	02	3	0	0	2	0
W	0	0	0	0	1	0	0	0	0	0	0	0	0	0	0	0	0	0	0	0	1	0	2	1	0	0	0	0	0	1	0	03	1	0	0	0
X	0	0	0	1	0	0	0	0	0	0	0	0	0	0	0	0	0	0	0	0	0	0	0	0	0	0	0	0	0	0	0	0	04	1	0	1
Y	0	0	0	1	0	0	0	0	0	0	0	0	0	0	0	0	0	0	0	0	0	0	0	0	0	0	0	0	0	0	0	0	0	01	5	0
Z	0	1	3	0	0	0	0	1	0	0	0	0	0	0	1	0	0	0	0	0	0	0	0	0	0	0	0	0	1	0	0	0	0	2	03	9

Table 6. Performance comparison of time (Seconds) taken by the proposed system implemented on a PC (running Matlab) and a Raspberry Pi (Python + OpenCV).

Characters on LP *Processing on LP*	PC[Matlab]			Raspberry Pi (Python)		
	5 Characters	**6 Characters**	**7 Characters**	**5 Characters**	**6 Characters**	**7 Characters**
Vehicle Detection and LP Extraction	3.2	3.19	3.21	0.13	0.11	0.14
LP Characters Segmentation	2.1	2.14	2.27	0.1	0.12	0.13
Characters Recognition and Barrier Control	3.22	3.26	3.37	0.16	0.17	0.21
Total Time Taken (Seconds)	8.52	8.59	8.85	0.39	0.4	0.48

5. Conclusions

We presented a robust, accurate, industrial barrier access control system using a sensor platform and vehicle license plate recognition. The proposed system automatically detects a vehicle at an entrance via ultrasonic sensors and then recognizes it by image-based recognition of its license plate, which can have various backgrounds, fonts and font styles. To this end, a performance evaluation of various classifiers was carried out to find out that which had the best recognition rate. Lastly, the proposed system was implemented both on a PC running Matlab and on a Raspberry Pi (system on chip) running Python with OpenCV. The Raspberry Pi-based system had low computational time, a smaller size, and low power consumption, due to which it was used in the real-time application. In future, we are working to increase the dataset of handwritten LP characters to improve accuracy and laser beam-based vehicle detection to increase the detection range.

Author Contributions: All authors contributed to the paper. F.U.: Conceptualization, Software, Hardware, Formal analysis, Writing; H.A.: Data curation, Software, Validation; I.S.: Hardware, Software, Writing; A.U.R.: Hardware, Writing; S.M.: Data curation, Software; S.N.: Data curation, Software; K.M.A.: Validation; A.K.: Data curation, Validation; D.K.: Conceptualization, Funding acquisition, Review and editing.

Funding: The research was funded by the Untenured Faculty Research Initiative (UFRI), Kean University and ICT Pakistan under the National Grassroot ICT Research Initiative (NGIRI).

Conflicts of Interest: The authors declare no conflict of interest.

Dataset Availabity: Dataset will be provided on request by the corresponding author.

References

1. Joshi, Y.; Gharate, P.; Ahire, C.; Alai, N.; Sonavane, S. Smart parking management system using RFID and OCR. In Proceedings of the 2015 International Conference on Energy Systems and Applications, Pune, India, 30 October–1 November 2015; pp. 729–734.

2. Zhou, H.; Li, Z. An intelligent parking management system based on RS485 and RFID. In Proceedings of the 2016 International Conference on Cyber-Enabled Distributed Computing and Knowledge Discovery (CyberC), Chengdu, China, 13–15 October 2016; pp. 355–359.

3. Rashedi, E.; Nezamabadi-Pour, H. A hierarchical algorithm for vehicle license plate localization. *Multimed. Tools Appl.* **2018**, *77*, 2771–2790. [CrossRef]

4. Jin, L.; Xian, H.; Bie, J.; Sun, Y.; Hou, H.; Niu, Q. License plate recognition algorithm for passenger cars in Chinese residential areas. *Sensors* **2012**, *12*, 8355–8370. [CrossRef] [PubMed]

5. Gou, C.; Wang, K.; Yao, Y.; Li, Z. Vehicle license plate recognition based on extremal regions and restricted Boltzmann machines. *IEEE Trans. Intell. Transp. Syst.* **2015**, *17*, 1096–1107. [CrossRef]

6. Canny, J. A computational approach to edge detection. *IEEE Trans. Pattern Anal. Mach. Intell.* **1986**, *6*, 679–698. [CrossRef]

7. Trier, Ø.D.; Jain, A.K.; Taxt, T. Feature extraction methods for character recognition-a survey. *Pattern Recognit.* **1996**, *29*, 641–662. [CrossRef]

8. Zhang, Z.; Shen, W.; Yao, C.; Bai, X. Symmetry-based text line detection in natural scenes. In Proceedings of the IEEE Conference on Computer Vision and Pattern Recognition, Boston, MA, USA, 7–12 June 2015; pp. 2558–2567.

9. Soora, N.R.; Deshpande, P.S. Review of Feature Extraction Techniques for Character Recognition. *IETE J. Res.* **2018**, *64*, 280–295. [CrossRef]

10. Liao, Y.; Zhang, J.; Wang, S.; Li, S.; Han, J. Study on Crash Injury Severity Prediction of Autonomous Vehicles for Different Emergency Decisions Based on Support Vector Machine Model. *Electronics* **2018**, *7*, 381. [CrossRef]

11. Liu, W.-C.; Lin, C. A hierarchical license plate recognition system using supervised K-means and Support Vector Machine. In Proceedings of the 2017 International Conference on Applied System Innovation (ICASI), Sapporo, Japan, 13–17 May 2017; pp. 1622–1625.

12. Ullah, F.; Sarwar, G.; Lee, S. N-Screen Aware Multicriteria Hybrid Recommender System Using Weight Based Subspace Clustering. *Sci. World J.* **2014**, *2014*, 679849. [CrossRef]

13. Anagnostopoulos, C.N.E.; Anagnostopoulos, I.E.; Loumos, V.; Kayafas, E. A license plate-recognition algorithm for intelligent transportation system applications. *IEEE Trans. Intell. Transp. Syst.* **2006**, *7*, 377–392. [CrossRef]

14. Pan, X.; Ye, X.; Zhang, S. A hybrid method for robust car plate character recognition. *Eng. Appl. Artif. Intell.* **2005**, *18*, 963–972. [CrossRef]

15. Liu, Y.; Huang, H.; Cao, J.; Huang, T. Convolutional neural networks-based intelligent recognition of Chinese license plates. *Soft Comput.* **2018**, *22*, 2403–2419. [CrossRef]

16. Tarabek, P. A real-time license plate localization method based on vertical edge analysis. In Proceedings of the 2012 Federated Conference on Computer Science and Information Systems (FedCSIS), Wroclaw, Poland, 9–12 September 2012; pp. 149–154.

17. Megalingam, R.K.; Krishna, P.; Pillai, V.A.; Hakkim, R.U. Extraction of license plate region in Automatic License Plate Recognition. In Proceedings of the 2010 2nd International Conference on Mechanical and Electrical Technology (ICMET), Singapore, 10–12 September 2010; pp. 496–501.

18. Bai, H.; Liu, C. A hybrid license plate extraction method based on edge statistics and morphology. In Proceedings of the 17th International Conference on Pattern Recognition, Cambridge, UK, 23–26 August 2004; Volume 2, pp. 831–834.

19. Wang, J.; Bacic, B.; Yan, W.Q. An effective method for plate number recognition. *Multimed. Tools Appl.* **2018**, *77*, 1679–1692. [CrossRef]

20. Zheng, D.; Zhao, Y.; Wang, J. An efficient method of license plate location. *Pattern Recognit. Lett.* **2005**, *26*, 2431–2438. [CrossRef]

21. Dun, J.; Zhang, S.; Ye, X.; Zhang, Y. Chinese license plate localization in multi-lane with complex background based on concomitant colors. *IEEE Intell. Transp. Syst. Mag.* **2015**, *7*, 51–61. [CrossRef]

22. Safaei, A.; Tang, H.L.; Sanei, S. Robust search-free car number plate localization incorporating hierarchical saliency. *J. Comput. Sci. Syst. Biol.* **2016**, *9*, 93–103.

23. Tabrizi, S.S.; Cavus, N. A hybrid KNN-SVM model for Iranian license plate recognition. *Procedia Comput. Sci.* **2016**, *102*, 588–594. [CrossRef]

24. Gazcón, N.F.; Chesñevar, C.I.; Castro, S.M. Automatic vehicle identification for Argentinean license plates using intelligent template matching. *Pattern Recognit. Lett.* **2012**, *33*, 1066–1074. [CrossRef]

25. Zhang, Y.; Wang, Y.; Zhou, G.; Jin, J.; Wang, B.; Wang, X.; Cichocki, A. Multi-kernel extreme learning machine for EEG classification in brain-computer interfaces. *Expert Syst. Appl.* **2018**, *96*, 302–310. [CrossRef]

26. Tarigan, J.; Diedan, R.; Suryana, Y. Plate Recognition Using Backpropagation Neural Network and Genetic Algorithm. *Procedia Comput. Sci.* **2017**, *116*, 365–372. [CrossRef]

27. Laroca, R.; Severo, E.; Zanlorensi, L.A.; Oliveira, L.S.; Gonçalves, G.R.; Schwartz, W.R.; Menotti, D. A Robust Real-Time Automatic License Plate Recognition based on the YOLO Detector. *arXiv* **2008**, arXiv:1802.09567.

28. Zheng, L.; He, X.; Samali, B.; Yang, L.T. An algorithm for accuracy enhancement of license plate recognition. *J. Comput. Syst. Sci.* **2013**, *79*, 245–255. [CrossRef]

29. Chen, H.; Hu, B.; Yang, X.; Yu, M.; Chen, J. Chinese character recognition for LPR application. *Opt. Int. J. Light Electron Opt.* **2014**, *125*, 5295–5302. [CrossRef]

30. Massoud, M.A.; Sabee, M.; Gergais, M.; Bakhit, R. Automated new license plate recognition in Egypt. *Alex. Eng. J.* **2013**, *52*, 319–326. [CrossRef]

31. Smith, R. An overview of the Tesseract OCR engine. In Proceedings of the Ninth International Conference on Document Analysis and Recognition, Parana, Brazil, 23–26 September 2007; Volume 2, pp. 629–633.

32. Li, H.; Wang, P.; Shen, C. Towards end-to-end car license plates detection and recognition with deep neural networks. *arXiv* **2017**, arXiv:1709.08828.

33. Björklund, T.; Fiandrotti, A.; Annarumma, M.; Francini, G.; Magli, E. Automatic license plate recognition with convolutional neural networks trained on synthetic data. In Proceedings of the 2017 IEEE 19th International Workshop on Multimedia Signal Processing (MMSP), Luton, UK, 16–18 October 2017; pp. 1–6.

34. Zhu, Y.; Huang, H.; Xu, Z.; He, Y.; Liu, S. Chinese-style plate recognition based on artificial neural network and statistics. *Procedia Eng.* **2011**, *15*, 3556–3561. [CrossRef]

35. Masood, S.Z.; Shu, G.; Dehghan, A.; Ortiz, E.G. License Plate Detection and Recognition Using Deeply Learned Convolutional Neural Networks. *arXiv* **2017**, arXiv:1703.07330.

36. Zang, D.; Chai, Z.; Zhang, J.; Zhang, D.; Cheng, J. Vehicle license plate recognition using visual attention model and deep learning. *J. Electron. Imaging* **2015**, *24*, 033001. [CrossRef]

37. Gonzalez, R.C.; Woods, R.E. *Digital Image Processing*; Publishing House of Electronics Industry: Beijing, China, 2002; Volume 141.

38. DC Motor Details. Available online: https://www.alibaba.com/product-detail/90-watt-12-volt-24-volt_60170246153.html?spm=a2700.7724857.normalList.46.170f239bl7CAAN (accessed on 15 May 2019).

MDPI

St. Alban-Anlage 66

4052 Basel

Switzerland

Tel. +41 61 683 77 34

Fax +41 61 302 89 18

www.mdpi.com

Sensors Editorial Office

E-mail: sensors@mdpi.com

www.mdpi.com/journal/sensors